Popular Astronomy

SIMON NEWCOMB

CAMBRIDGE UNIVERSITY PRESS

Cambridge, New York, Melbourne, Madrid, Cape Town,
Singapore, São Paolo, Delhi, Tokyo, Mexico City

Published in the United States of America by Cambridge University Press, New York

www.cambridge.org
Information on this title: www.cambridge.org/9781108037730

© in this compilation Cambridge University Press 2011

This edition first published 1878
This digitally printed version 2011

ISBN 978-1-108-03773-0 Paperback

This book reproduces the text of the original edition. The content and language reflect
the beliefs, practices and terminology of their time, and have not been updated.

Cambridge University Press wishes to make clear that the book, unless originally published
by Cambridge, is not being republished by, in association or collaboration with, or
with the endorsement or approval of, the original publisher or its successors in title.

CAMBRIDGE LIBRARY COLLECTION

Books of enduring scholarly value

Physical Sciences

From ancient times, humans have tried to understand the workings of the world around them. The roots of modern physical science go back to the very earliest mechanical devices such as levers and rollers, the mixing of paints and dyes, and the importance of the heavenly bodies in early religious observance and navigation. The physical sciences as we know them today began to emerge as independent academic subjects during the early modern period, in the work of Newton and other 'natural philosophers', and numerous sub-disciplines developed during the centuries that followed. This part of the Cambridge Library Collection is devoted to landmark publications in this area which will be of interest to historians of science concerned with individual scientists, particular discoveries, and advances in scientific method, or with the establishment and development of scientific institutions around the world.

Popular Astronomy

Furnished with more than a hundred figures, maps and tables, this book was first published in 1878 by Simon Newcomb (1835–1909), a noted mathematician and professor at the United States Naval Observatory. A meticulous work, originally intended to be of use to the general reader as well as the student, it provides a view of astronomy as it stood on the eve of General Relativity, and inevitably includes some theories which have since been disproved. Newcomb outlines a brief history of astronomy, from ancient Greece (when the planets were thought to be fixed in crystal spheres), to the application of the new laws of thermodynamics and the latest observations of the solar system. Included are a rejection of the then prevalent theory that the sun has a cool interior and its own inhabitants, details of the anomaly of Mercury's orbit according to Newtonian theory, and thorough observational guides.

Cambridge University Press has long been a pioneer in the reissuing of out-of-print titles from its own backlist, producing digital reprints of books that are still sought after by scholars and students but could not be reprinted economically using traditional technology. The Cambridge Library Collection extends this activity to a wider range of books which are still of importance to researchers and professionals, either for the source material they contain, or as landmarks in the history of their academic discipline.

Drawing from the world-renowned collections in the Cambridge University Library, and guided by the advice of experts in each subject area, Cambridge University Press is using state-of-the-art scanning machines in its own Printing House to capture the content of each book selected for inclusion. The files are processed to give a consistently clear, crisp image, and the books finished to the high quality standard for which the Press is recognised around the world. The latest print-on-demand technology ensures that the books will remain available indefinitely, and that orders for single or multiple copies can quickly be supplied.

The Cambridge Library Collection will bring back to life books of enduring scholarly value (including out-of-copyright works originally issued by other publishers) across a wide range of disciplines in the humanities and social sciences and in science and technology.

THE GREAT TELESCOPE OF THE UNITED STATES NAVAL OBSERVATORY, WASHINGTON.

CONSTRUCTED BY ALVAN CLARK AND SONS, 1873.

POPULAR ASTRONOMY.

BY

SIMON NEWCOMB, LL.D.,

PROFESSOR, U. S. NAVAL OBSERVATORY.

*WITH ONE HUNDRED AND TWELVE ENGRAVINGS,
AND FIVE MAPS OF THE STARS.*

London:
MACMILLAN AND CO.
1878.

LONDON:
PRINTED BY WILLIAM CLOWES AND SONS,
STAMFORD STREET AND CHARING CROSS.

PREFACE.

To prevent a possible misapprehension in scientific quarters, the author desires it understood that the present work is not designed either to instruct the professional investigator or to train the special student of astronomy. Its main object is to present the general reading public with a condensed view of the history, methods, and results of astronomical research, especially in those fields which are of most popular and philosophic interest at the present day, couched in such language as to be intelligible without mathematical study. He hopes that the earlier chapters will, for the most part, be readily understood by any one having clear geometrical ideas, and that the later ones will be intelligible to all. To diminish the difficulty which the reader may encounter from the unavoidable occasional use of technical terms, a Glossary has been added, including, it is believed, all that are used in the present work, as well as a number of others which may be met with elsewhere.

Respecting the general scope of the work, it may be said that the historic and philosophic sides of the subject have been treated with greater fulness than is usual in works of this character, while the purely technical side has been proportionately condensed. Of the four parts into which it is divided, the first two treat of the methods by which the motions and the mutual relations of the heavenly bodies have been investigated, and of the results of such investigation, while in the last two the individual peculiarities of those bodies are considered in greater detail. The subject of the general structure and probable development of the universe, which, in strictness, might be considered as belonging to the first part, is, of necessity, treated last of all, because it requires all the light that can be thrown upon it from every available source. Matter admitting of presentation in tabular form has, for the most part, been collected in the Appendix, where will be found a number of brief articles for the use of both the general reader and the amateur astronomer.

The author has to acknowledge the honor done him by several eminent astronomers in making his work more complete and interesting by their contributions. Owing to the great interest which now attaches to the question of the constitution of the sun, and the rapidity with which our knowledge in this direction is advancing, it was deemed desirable to present the latest views of the most distinguished investigators of this subject from their own pens. Four of these gentlemen—Rev. Father Secchi, of Rome; M. Faye, of Paris; Professor Young, of Dartmouth College; and Professor Langley, of Allegheny Observatory—have, at the author's request, presented brief expositions of their theories, which will be found in their own language in the chapter on the sun.

An Addendum gives the basis of the remarkable modification of the theory of the solar spectrum proposed by Dr. Henry Draper, which appeared while the sheets were passing through the press.

CONTENTS.

PART I.

THE SYSTEM OF THE WORLD HISTORICALLY DEVELOPED.

	PAGE
INTRODUCTION	1

CHAPTER I.

THE ANCIENT ASTRONOMY, OR THE APPARENT MOTIONS OF THE HEAVENLY BODIES.. 7
  § 1. The Celestial Sphere... 7
  § 2. The Diurnal Motion 9
  § 3. Motion of the Sun among the Stars 13
  § 4. Precession of the Equinoxes.—The Solar Year................... 19
  § 5. The Moon's Motion.. 21
  § 6. Eclipses of the Sun and Moon................................... 24
  § 7. The Ptolemaic System 32
  § 8. The Calendar 44

CHAPTER II.

THE COPERNICAN SYSTEM, OR THE TRUE MOTIONS OF THE HEAVENLY BODIES.. 51
  § 1. Copernicus.. 51
  § 2. Obliquity of the Ecliptic; Seasons, etc.; on the Copernican System... 61
  § 3. Tycho Brahe.. 66
  § 4. Kepler.—His Laws of Planetary Motion.......................... 68
  § 5. From Kepler to Newton.. 71

CHAPTER III.

	PAGE
UNIVERSAL GRAVITATION	74
§ 1. Newton.—Discovery of Gravitation	74
§ 2. Gravitation of Small Masses.—Density of the Earth	81
§ 3. Figure of the Earth.	86
§ 4. Precession of the Equinoxes	88
§ 5. The Tides	90
§ 6. Inequalities in the Motions of the Planets produced by their Mutual Attraction	93
§ 7. Relation of the Planets to the Stars	101

PART II.

PRACTICAL ASTRONOMY.

INTRODUCTORY REMARKS	103

CHAPTER I.

THE TELESCOPE	106
§ 1. The First Telescopes	106
§ 2. The Achromatic Telescope	114
§ 3. The Mounting of the Telescope	118
§ 4. The Reflecting Telescope	121
§ 5. The Principal Great Reflecting Telescopes of Modern Times	125
§ 6. Great Refracting Telescopes	135
§ 7. The Magnifying Powers of the Two Classes of Telescopes	139

CHAPTER II.

APPLICATION OF THE TELESCOPE TO CELESTIAL MEASUREMENTS	146
§ 1. Circles of the Celestial Sphere, and their Relations to Positions on the Earth	146
§ 2. The Meridian Circle, and its Use	152
§ 3. Determination of Terrestrial Longitudes	157
§ 4. Mean, or Clock, Time	162

CHAPTER III.

	PAGE
MEASURING DISTANCES IN THE HEAVENS	165
§ 1. Parallax in General	165
§ 2. Measures of the Distance of the Sun	171
§ 3. Solar Parallax from Transits of Venus	175
§ 4. Other Methods of Determining the Sun's Distance, and their Results	194
§ 5. Stellar Parallax	201

CHAPTER IV.

THE MOTION OF LIGHT 210

CHAPTER V.

THE SPECTROSCOPE 222

PART III.

THE SOLAR SYSTEM.

CHAPTER I.

GENERAL STRUCTURE OF THE SOLAR SYSTEM. 231

CHAPTER II.

THE SUN	237
§ 1. The Photosphere	237
§ 2. The Solar Spots and Rotation	242
§ 3. Periodicity of the Spots	248
§ 4. Law of Rotation of the Sun	249
§ 5. The Sun's Surroundings.—Phenomena of Total Eclipses	251
§ 6. Physical Constitution of the Sun	258
§ 7. Views of Distinguished Students of the Sun on the Subject of its Physical Constitution	265

CHAPTER III.

THE INNER GROUP OF PLANETS 283
   § 1. The Planet Mercury 283
   § 2. The Supposed Intra-Mercurial Planets 286
   § 3. The Planet Venus 289
   § 4. The Earth 298
   § 5. The Moon 306
   § 6. The Planet Mars 320
   § 7. The Small Planets 323

CHAPTER IV.

THE OUTER GROUP OF PLANETS 331
   § 1. The Planet Jupiter 331
   § 2. The Satellites of Jupiter 336
   § 3. Saturn and its System, Physical Aspect, Belts, Rotation 338
   § 4. The Rings of Saturn 341
   § 5. Constitution of the Ring 349
   § 6. The Satellites of Saturn 351
   § 7. Uranus and its Satellites 353
   § 8. Neptune and its Satellite 358

CHAPTER V.

COMETS AND METEORS 365
   § 1. Aspects and Forms of Comets 365
   § 2. Motions, Origin, and Number of Comets 369
   § 3. Remarkable Comets 374
   § 4. Encke's Comet, and the Resisting Medium 381
   § 5. Meteors and Shooting-stars 384
   § 6. Relations of Comets and Meteoroids 391
   § 7. The Physical Constitution of Comets 398
   § 8. The Zodiacal Light 405

PART IV.

THE STELLAR UNIVERSE.

INTRODUCTORY REMARKS 407

CHAPTER I.

THE STARS AS THEY ARE SEEN 410
    § 1. Number and Orders of Stars and Nebulæ 410
    § 2. Description of the Principal Constellations 417
    § 3. New and Variable Stars 426
    § 4. Double Stars 436
    § 5. Clusters of Stars 441
    § 6. Nebulæ 444
    § 7. Proper Motions of the Stars 452

CHAPTER II.

THE STRUCTURE OF THE UNIVERSE 460
    § 1. Views of Astronomers before Herschel 461
    § 2. Researches of Herschel and his Successors 465
    § 3. Probable Arrangement of the Visible Universe 478
    § 4. Do the Stars really form a System? 483

CHAPTER III.

THE COSMOGONY 491
    § 1. The Modern Nebular Hypothesis 493
    § 2. Progressive Changes in our System 499
    § 3. The Sources of the Sun's Heat 505
    § 4. Secular Cooling of the Earth 511
    § 5. General Conclusions respecting the Nebular Hypothesis 514
    § 6. The Plurality of Worlds 516

ADDENDUM TO PART III., CHAPTER II. 520

APPENDIX.

    I. LIST OF THE PRINCIPAL GREAT TELESCOPES OF THE WORLD 521
    II. LIST OF THE MORE REMARKABLE DOUBLE STARS 523
    III. LIST OF THE MORE INTERESTING AND REMARKABLE NEBULÆ AND
        STAR CLUSTERS 525
    IV. PERIODIC COMETS SEEN AT MORE THAN ONE RETURN 527

		PAGE
V.	ELEMENTS OF THE ORBITS OF THE EIGHT MAJOR PLANETS FOR 1850.	528
	ELEMENTS OF THE SATELLITES OF JUPITER	529
	ELEMENTS OF THE SATELLITES OF SATURN	529
	ELEMENTS OF THE SATELLITE OF NEPTUNE	529
	ELEMENTS OF THE SATELLITES OF URANUS	529
VI.	ELEMENTS OF THE SMALL PLANETS	530
VII.	DETERMINATIONS OF STELLAR PARALLAX	535
VIII.	SYNOPSIS OF PAPERS ON THE SOLAR PARALLAX, 1854–'77	538
IX.	LIST OF ASTRONOMICAL WORKS, MOST OF WHICH HAVE BEEN CONSULTED AS AUTHORITIES IN THE PREPARATION OF THE PRESENT WORK	542
X.	GLOSSARY OF TECHNICAL TERMS OF FREQUENT OCCURRENCE IN ASTRONOMICAL WORKS	549

INDEX 559

ADDENDUM II.—THE SATELLITES OF MARS 565

EXPLANATION OF THE STAR MAPS 568

LIST OF ILLUSTRATIONS.

FIG.   PAGE

    The Great Telescope of the United States Naval Observatory, Washington .... *Frontispiece*
1. Section of the Imaginary Celestial Sphere 8
2. Map illustrating the Diurnal Motion round the Pole 10
3. The Celestial Sphere and Diurnal Motion 12
4. Motion of the Sun past the Star Regulus 15
5. Showing the Sun to be farther than the Moon 22
6. Annular Eclipse of the Sun 26
7. Partial Eclipse of the Sun 26
8. Eclipse of the Sun, the Shadow of the Moon falling on the Earth 26
9. Eclipse of the Moon, in the Shadow of the Earth 27
10. Showing the Apparent Orbit of a Planet 38
11. Apparent Orbits of Jupiter and Saturn 39
12. Arrangement of the Seven Planets in the Ptolemaic System ... 41
13. The Eccentric 42
14. Showing the Astrological Division of the Seven Planets among the Days of the Week 46
15. Apparent Annual Motion of the Sun explained 55
16. Showing how the Apparent Epicyclic Motion of the Planets is accounted for 56
17. Relation of the Terrestrial and Celestial Poles and Equators 62
18. Causes of Changes of Seasons on the Copernican System 63
19. Enlarged View of the Earth, showing Winter in the Northern Hemisphere, and Summer in the Southern 65
20. Illustrating Kepler's First Two Laws of Planetary Motion ... 69
21. Illustrating the Fall of the Moon towards the Earth 78
22. Baily's Apparatus for determining the Density of the Earth 83
23. View of Baily's Apparatus 84
24. Diagram illustrating the Attraction of Mountains 85
25. Precession of the Equinoxes 88

LIST OF ILLUSTRATIONS.

FIG.		PAGE
26.	Attraction of the Moon tending to produce Tides	91
27.	Armillary Sphere as described by Ptolemy	105
28.	The Galilean Telescope	108
29.	Formation of an Image by a Lens	109
30.	Great Telescope of the Seventeenth Century	112
31.	Refraction through a Compound Prism	114
32.	Section of an Achromatic Objective	115
33.	Section of Eye-piece of a Telescope	118
34.	Mode of Mounting a Telescope	119
35.	Speculum Bringing Rays to a Single Focus by Reflection	122
36.	Herschelian Telescope	123
37.	Horizontal Section of a Newtonian Telescope	123
38.	Section of the Gregorian Telescope	124
39.	Herschel's Great Telescope	127
40.	Lord Rosse's Great Telescope	130
41.	Mr. Lassell's Great Four-foot Reflector	132
42.	The New Paris Reflector	134
43.	The Great Melbourne Reflector	136
44.	Circles of the Celestial Sphere	147
45.	The Washington Transit Circle	153
46.	Spider Lines in Field of View of a Meridian Circle	154
47.	Diagram illustrating Parallax	165
48.	Diagram illustrating Parallax	166
49.	Variation of Parallax with the Altitude	167
50.	Apparent Paths of Venus across the Sun	176
51.	Venus approaching Internal Contact on the Face of the Sun	178
52.	Internal Contact of Limb of Venus with that of the Sun	178
53.	The Black Drop, or Ligament	179
54.	Method of Photographing the Transit of Venus	186
55.	Artificial Transit of Venus	188
56.	Map of the Earth, showing the Areas of Visibility of the Transit of 1874	191
57.	Map of the World, showing the Regions in which the Transit of Venus will be visible on December 6th, 1882	195
58.	Effect of Stellar Parallax	202
59.	Aberration of Light	212
60.	Revolving Wheel for measuring the Velocity of Light	216
61.	Illustrating Foucault's Method of measuring the Velocity of Light	218
62.	Course of Rays through a Spectroscope	224

LIST OF ILLUSTRATIONS. xv

FIG.		PAGE
63.	RELATIVE SIZE OF SUN AND PLANETS	232
64.	ORBITS OF THE PLANETS FROM THE EARTH OUTWARD	236
65.	MAN HOLDING TELESCOPE, TO SHOW SUN ON SCREEN	243
66.	SOLAR SPOT, AFTER SECCHI	244
67.	CHANGES IN THE ASPECT OF A SOLAR SPOT AS IT CROSSES THE SUN'S DISK	246
68.	TOTAL ECLIPSE OF THE SUN, AS SEEN AT DES MOINES, IOWA, AUGUST 7TH, 1869	253
69.	SPECIMENS OF SOLAR PROTUBERANCES, AS DRAWN BY SECCHI	256
70.	THE SUN, WITH ITS CHROMOSPHERE AND RED FLAMES, ON JULY 23D, 1871	261
71.	ILLUSTRATING SECCHI'S THEORY OF SOLAR SPOTS	269
72.	SOLAR SPOT, AFTER LANGLEY	281
73.	ORBITS OF THE FOUR INNER PLANETS, ILLUSTRATING THE ECCENTRICITY OF THOSE OF MERCURY AND MARS	283
74.	PHASES OF VENUS	291
75.	SHOWING THE THICKNESS OF THE EARTH'S CRUST	299
76.	DISTRIBUTION OF AURORAS	302
77.	VIEW OF AURORA	303
78.	SPECTRUM OF TWO OF THE GREAT AURORAS OF 1871	305
79.	RELATIVE SIZE OF EARTH AND MOON	306
80.	VIEW OF MOON NEAR THE THIRD QUARTER	313
81.	LUNAR CRATER "COPERNICUS"	315
82.	THE PLANET MARS ON JUNE 23D, 1875	322
83.	MAP OF MARS	322
84.	NORTHERN HEMISPHERE OF MARS	323
85.	SOUTHERN HEMISPHERE OF MARS	323
86.	JUPITER, AS SEEN WITH THE GREAT WASHINGTON TELESCOPE, MARCH 21ST, 1876	331
87.	VIEW OF JUPITER, AS SEEN IN LORD ROSSE'S GREAT TELESCOPE, FEBRUARY 27TH, 1861	333
88.	VIEW OF SATURN AND HIS RINGS	339
89.	SPECIMENS OF DRAWINGS OF SATURN BY VARIOUS OBSERVERS	343
90.	VIEWS OF ENCKE'S COMET IN 1871	367
91.	HEAD OF DONATI'S GREAT COMET OF 1858	368
92.	PARABOLIC AND ELLIPTIC ORBIT OF A COMET	370
93.	ORBIT OF HALLEY'S COMET	377
94.	GREAT COMET OF 1858	380
95.	METEOR PATHS, ILLUSTRATING THE RADIANT POINT	390
96.	ORBIT OF NOVEMBER METEORS AND THE COMET OF 1861	394

FIG.		PAGE
97.	ORBIT OF THE THIRD COMET OF 1862	395
98.	MEASURE OF POSITION ANGLE OF DOUBLE STAR	438
99.	DISTANCE OF COMPONENTS OF DOUBLE STAR	438
100.	DIAGRAM TO ILLUSTRATE POSITION ANGLE	438
101.	TELESCOPIC VIEW OF THE PLEIADES	442
102.	CLUSTER OF 47 TOUCANI	444
103.	CLUSTER ω CENTAURI	444
104.	THE GREAT NEBULA OF ORION	446
105.	THE ANNULAR NEBULA IN LYRA	448
106.	THE OMEGA NEBULA	450
107.	NEBULA HERSCHEL 3722	451
108.	THE LOOPED NEBULA; HERSCHEL 2941	451
109.	HERSCHEL'S VIEW OF THE FORM OF THE UNIVERSE	469
110.	ILLUSTRATING HERSCHEL'S ORDERS OF DISTANCE OF THE STARS	471
111.	PROBABLE ARRANGEMENT OF THE STARS AND NEBULÆ VISIBLE WITH THE TELESCOPE	481
112.	DIAGRAM ILLUSTRATING ELLIPTIC ELEMENTS OF A PLANET	551

STAR MAPS.

MAP I.—THE NORTHERN CONSTELLATIONS WITHIN 50° OF THE POLE

" II.—SOUTHERN CONSTELLATIONS VISIBLE IN AUTUMN AND WINTER

" III.—SOUTHERN CONSTELLATIONS VISIBLE IN WINTER AND SPRING

" IV.—SOUTHERN CONSTELLATIONS VISIBLE IN SPRING AND SUMMER

" V.—SOUTHERN CONSTELLATIONS VISIBLE IN SUMMER AND AUTUMN

At End of Book.

POPULAR ASTRONOMY.

PART I.—THE SYSTEM OF THE WORLD HISTORICALLY DEVELOPED.

INTRODUCTION.

ASTRONOMY is the most ancient of the physical sciences, being distinguished among them by its slow and progressive development from the earliest ages until the present time. In no other science has each generation which advanced it been so much indebted to its predecessors for both the facts and the ideas necessary to make the advance. The conception of a globular and moving earth pursuing her course through the celestial spaces among her sister planets, which we see as stars, is one to the entire evolution of which no one mind and no one age can lay claim. It was the result of a gradual process of education, of which the subject was not an individual, but the human race. The great astronomers of all ages have built upon foundations laid by their predecessors; and when we attempt to search out the first founder, we find ourselves lost in the mists of antiquity. The theory of universal gravitation was founded by Newton upon the laws of Kepler, the observations and measurements of his French contemporaries, and the geometry of Apollonius. Kepler used as his material the observations of Tycho Brahe, and built upon the theory of Copernicus. When we seek the origin of the instruments used by Tycho, we soon find ourselves among

2 SYSTEM OF THE WORLD HISTORICALLY DEVELOPED.

the mediæval Arabs. The discovery of the true system of the world by Copernicus was only possible by a careful study of the laws of apparent motion of the planets as expressed in the epicycles of Ptolemy and Hipparchus. Indeed, the more carefully one studies the great work of Copernicus, the more surprised he will be to find how completely Ptolemy furnished him both ideas and material. If we seek the teachers and predecessors of Hipparchus, we find only the shadowy forms of Egyptian and Babylonian priests, whose names and writings are all entirely lost. In the earliest historic ages, men knew that the earth was round; that the sun appeared to make an annual revolution among the stars; and that eclipses were caused by the moon entering the shadow of the earth, or the earth that of the moon.

Indeed, each of the great civilizations of the ancient world seems to have had its own system of astronomy strongly marked by the peculiar character of the people among whom it was found. Several events recorded in the annals of China show that the movements of the sun and the laws of eclipses were studied in that country at a very early age. Some of these events must be entirely mythical; as, for instance, the despatch of astronomers to the four points of the compass for the purpose of determining the equinoxes and solstices. But there is another event which, even if we place it in the same category, must be regarded as indicating a considerable amount of astronomical knowledge among the ancient Chinese. We refer to the tragic fate of Hi and Ho, astronomers royal to one of the ancient emperors of that people. It was part of the duty of these men to carefully study the heavenly movements, and give timely warning of the approach of an eclipse or other remarkable phenomenon. But, neglecting this duty, they gave themselves up to drunkenness and riotous living. In consequence, an eclipse of the sun occurred without any notice being given; the religious rites due in such a case were not performed, and China was exposed to the anger of the gods. To appease their wrath, the unworthy astronomers were seized and summarily executed by royal command. Some historians have gone so far as to fix the date of this occurrence, which is variously placed at from 2128 to 2159 years before the Christian era. If this is correct, it is the earliest of which profane history has left us any record.

In the Hindoo astronomy we see the peculiarities of the contemplative Hindoo mind strongly reflected. Here the imagination revels in periods of time which, by comparison, dwarf even the measures of the celestial spaces made by modern astronomers. In this, and in perhaps other ancient systems, we find references to a supposed conjunction of all the planets 3102 years before the Christian era. Although we have every reason for believing that this conjunction was learned, not from any actual record of it, but by calculating back the position of the planets, yet the very fact that they were able to make this calculation shows that the motions of the planets must have been observed and recorded during many generations, either by the Hindoos themselves, or some other people from whom they acquired their knowledge. As a matter of fact, we now know from our modern tables that this conjunction was very far from being exact; but its error could not be certainly detected by the rude observations of the times in question.

Among a people so prone as the ancient Greeks to speculate upon the origin and nature of things, while neglecting the observation of natural phenomena, we cannot expect to find anything that can be considered a system of astronomy. But there are some ideas attributed to Pythagoras which are so frequently alluded to, and so closely connected with the astronomy of a subsequent age, that we may give them a passing mention. He is said to have taught that the heavenly bodies were set in a number of crystalline spheres, in the common centre of which the earth was placed. In the outer of these spheres were set the thousands of fixed stars which stud the firmament, while each of the seven planets had its own sphere. The transparency of each crystal sphere was perfect, so that the bodies set in each of the outer spheres were visible through all the inner ones. These spheres all rolled round on each other in a daily revolution, thus causing the rising and setting of the heavenly bodies. This rolling of the spheres on each other made a celestial music, the "music of the spheres," which filled the firmament, but was of too elevated a character to be heard by the ears of mortals.

It must be admitted that the idea of the stars being set in a hollow sphere of crystal, forming the vault of the firmament, was a very natural one. They seemed to revolve around the earth every day, for generation after generation, without the slightest change in their relative positions. If there were no solid connection between them, it does not seem possible that a thousand bodies could move around their vast circuit for such long periods of time without a single one of them varying its distance from one of the others. It is especially difficult to conceive how they could all move around the same axis. But when they are all set in a solid sphere, every one is made secure in its place. The planets could not be set in the same sphere, because they change their positions among the stars. This idea of the sphericity of the heavens held on to the minds of men with remarkable tenacity. The fundamental proposition of the system, both of Ptolemy and Copernicus, was that the universe is spherical, the latter seeking to prove the naturalness of the spherical form by the analogy of a drop of water, although the theory served him no purpose whatever. Faint traces of the idea are seen here and there in Kepler, with whom it vanished from the mind of the race, as the image of Santa Claus disappears from the mind of the growing child.

Pythagoras is also said to have taught in his esoteric lectures that the sun was the real centre of the celestial movements, and that the earth and planets moved around it, and it is this anticipation of the Copernican system which constitutes his greatest glory. But he never thought proper to make a public avowal of this doctrine, and even presented it to his disciples somewhat in the form of an hypothesis. It must also be admitted that the accounts of his system which have reached us are so vague and so filled with metaphysical speculation that it is questionable whether the frequent application of his name to the modern system is not more pedantic than justifiable.

The Greek astronomers of a later age not only rejected the vague speculations of their ancestors, but proved themselves the most careful observers of their time, and first made astronomy worthy the name of a science. From this Greek astronomy the astronomy of our own time may be considered as coming by direct descent. Still, were it not for the absence of historic records, we could probably trace back both their theories and their system of observation to the plains of Chaldea. The zodiac was mapped out and the constellations named many centuries before they commenced their observations, and these works marked quite an advanced stage of development. This prehistoric knowledge is, however, to be treated by the historian rather than the astronomer. If we confine ourselves to men whose names and whose labors have come down to us, we must concede to Hipparchus the honor of being the father of astronomy. Not only do his observations of the heavenly bodies appear to have been far more accurate than those of any of his predecessors, but he also determined the laws of the apparent motions of the planets, and prepared tables by which these motions could be calculated. Probably he was the first propounder of the theory of epicyclic motions of the planets, commonly called after the name of his successor, Ptolemy, who lived three centuries later.

Commencing with the time of Hipparchus, the general theory of the structure of the universe, or "system of the world," as it is frequently called, exhibits three great stages of development, each stage being marked by a system quite different from the other two in its fundamental principles. These are:

1. The so-called Ptolemaic system, which, however, really belongs to Hipparchus, or some more ancient astronomer. In this system the motion of the earth is ignored, and the apparent motions of the stars and planets around it are all regarded as real.

2. The Copernican system, in which it is shown that the sun is really the centre of the planetary motions, and that the earth is itself a planet, both turning on its axis and revolving round the sun.

3. The Newtonian system, in which all the celestial motions are explained by the one law of universal gravitation.

This natural order of development shows the order in which a knowledge of the structure of the universe can be most clearly presented to the mind of the general reader. We shall therefore explain this structure historically, devoting a separate chapter to each of the three stages of development which we have described. We commence with what is well known, or, at least, easily seen by every one who will look at the heavens with sufficient care. We imagine the observer out-of-doors on a starlit night, and show him how the heavenly bodies seem to move from hour to hour. Then, we show him what changes he will see in their aspects if he continues his watch through months and years. By combining the apparent motions thus learned, he forms for himself the ancient, or Ptolemaic, system of the world. Having this system clearly in mind, the passage to that of Copernicus is but a step. It consists only in showing that certain singular oscillations which the sun and planets seem to have in common are really due to a revolution of the earth around the sun, and that the apparent daily revolution of the celestial sphere arises from a rotation of the earth on its own axis. The laws of the true motions of the planets being perfected by Kepler, they are shown by Newton to be included in the one law of gravitation towards the sun. Such is the course of thought to which we first invite the reader.

CHAPTER I.

THE ANCIENT ASTRONOMY, OR THE APPARENT MOTIONS OF THE HEAVENLY BODIES.

§ 1. *The Celestial Sphere.*

It is a fact with which we are familiar from infancy, that all the heavenly bodies—sun, moon, and stars—seem to be set in an azure vault, which, rising high over our heads, curves down to the horizon on every side. Here the earth, on which it seems to rest, prevents our tracing it farther. But if the earth were out of the way, or were perfectly transparent, we could trace the vault downwards on every side to the point beneath our feet, and could see sun, moon, and stars in every direction. The celestial vault above us, with the corresponding one below us, would then form a complete sphere, in the centre of which the observer would seem to be placed. This has been known in all ages as the celestial sphere. The directions or apparent positions of the heavenly bodies, as well as their apparent motions, have always been defined by their situation and motions on this sphere. The fact that it is purely imaginary does not diminish its value as enabling us to form distinct ideas of the directions of the heavenly bodies from us.

It matters not how large we suppose this sphere, so long as we always suppose the observer to be in the centre of it, so that it shall surround him on all sides at an equal distance. But in the language and reasoning of exact astronomy it is always supposed to be infinite, as then the observer may conceive of himself as transported to any other point, even to one of the heavenly bodies themselves, and still be, for all practical purposes, in the centre of the sphere. In this case, however, the heavenly bodies are not considered as attached to the cir-

8   SYSTEM OF THE WORLD HISTORICALLY DEVELOPED.

cumference of the infinite sphere, but only as lying on the line of sight extending from the observer to some point of the sphere. Their relation to it may be easily understood by the observer conceiving himself to be luminous, and to throw out rays in every direction to the infinitely distant sphere. Then the apparent positions of the various heavenly bodies will be those in which their shadows strike the sphere. For instance, the observer standing on the earth and looking at the moon,

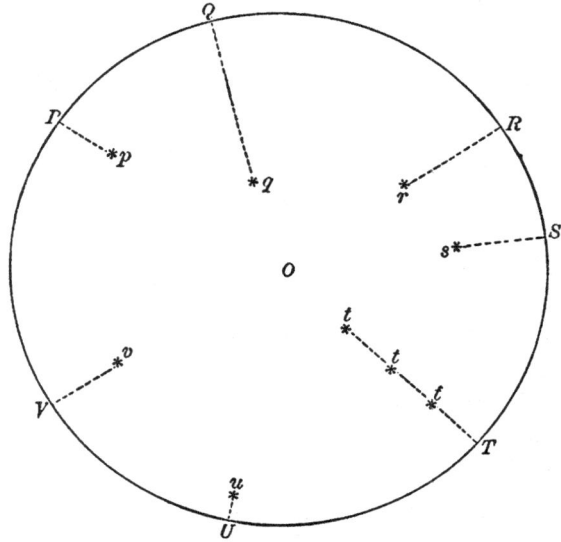

FIG. 1.—Section of the imaginary celestial sphere. The observer at *O*, looking at the stars or other bodies, marked *p, q, r, s, t, u, v*, will imagine them situated at *P, Q, R, S, T, U, V*, on the surface of the sphere, where they will appear projected along the straight *pP, qQ,* etc.

the shadow of the latter will strike the sphere at a point on a straight line drawn from the observer's eye through the centre of the moon, and continued till it meets the sphere. The point of meeting will represent the position of the moon as seen by the observer. Now, suppose the latter transported to the moon. Then, looking back at the earth, he will see it projected on the sphere in a point diametrically opposite to that in which he formerly saw the moon. To whatever planet he might transport himself, he would see the earth and the other planets projected on this imaginary sphere precisely as we always seem to see the heavenly bodies so projected.

This is all that is left of the old crystalline spheres of Pythagoras by modern astronomy. From being a solid which held all the stars, the sphere has become entirely immaterial, a mere conception of the mind, to enable it to define the directions in which the heavenly bodies are seen. By examining the figure it will be clear that all bodies which lie in the same straight line from the observer will appear on the same point of the sphere. For instance, bodies at the three points marked t will all be seen as if they were at T.

§ 2. *The Diurnal Motion.*

If we watch the heavenly bodies for a few hours we shall always find them in motion, those in the east rising upwards, those in the south moving towards the west, and those in the west sinking below the horizon. We know that this motion is only apparent, arising from the rotation of the earth on its axis; but as we wish, in this chapter, only to describe things as they appear, we may speak of the motion as real. A few days' watching will show that the whole celestial sphere seems to revolve, as on an axis, every day. It is to this revolution, carrying the sun alternately above and below the horizon, that the alternations of day and night are due. The nature and effects of this motion can best be studied by watching the apparent movement of the stars at night. We should soon learn from such a watch that there is one point in the heavens, or on the celestial sphere, which does not move at all. In our latitudes this point is situated in the north, between the zenith and the horizon, and is called the pole. Around this pole, as a fixed centre, all the heavenly bodies seem to revolve, each one moving in a circle, the size of which depends on the distance of the body from the pole. There is no star situated exactly at the pole, but there is one which, being situated little more than a degree distant, describes so small a circle that the unaided eye cannot see any change of place without making some exact and careful observation. This is therefore called the pole star. The pole star can nearly always be very readily found by means of the pointers, two stars of the constellation *Ursa Major*, the Great Bear, or, as it is familiarly called, the Dipper. By referring to the figure, the reader will readily find this constellation, by the dotted line from the pole and thence the pole star, which is near the centre of the map.

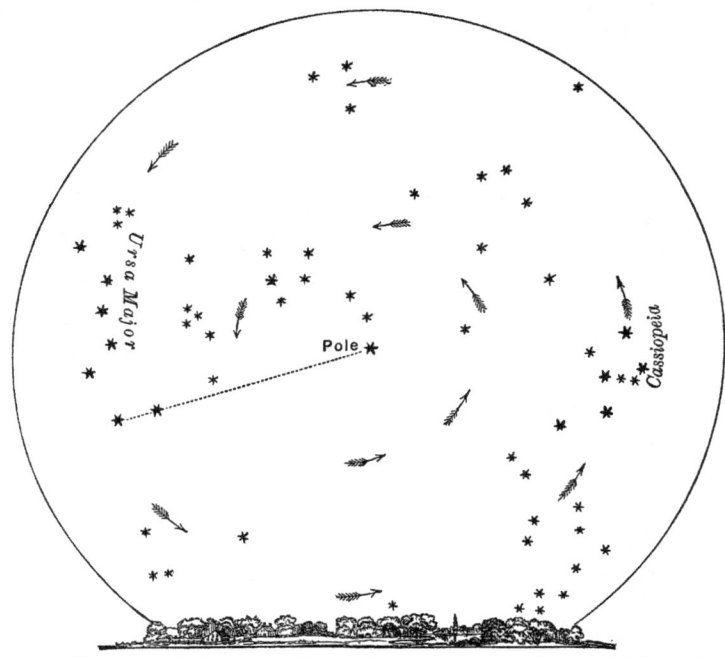

FIG. 2.—Map of the principal stars of the northern sky, showing the constellations which never set in latitude 40°, but revolve round the pole star every day in the direction shown by the arrows. The two lower stars of *Ursa Major*, on the left of the map, point to the pole star in the centre.

The altitude of the pole is equal to the latitude of the place. In the Middle States the latitude is generally not far from forty degrees; the pole is therefore a little nearer to the horizon than to the zenith. In Maine and Canada it is about halfway between these points, while in England and Northern Europe it is nearer the zenith.

THE DIURNAL MOTION.   11

Now, to see the effect of the diurnal motion near the pole, let us watch any star in the north between the pole and the horizon. We shall soon see that, instead of moving from east to west, as we are accustomed to see the heavenly bodies move, it really moves towards the east. After passing the north point, it begins to curve its course upwards, until, in the north-east, its motion is vertical. Then it turns gradually to the west, passing as far above the pole as it did below it, and, sinking down on the west of the pole, it again passes under it. The passage above the pole is called the upper culmination, and that below it the lower one. The course around the pole is shown by the arrows on Fig. 2. We cannot with the naked eye follow it all the way round, on account of the intervention of daylight; but by continuing our watch every clear night for a year, we should see it in every point of its course. A star following the course we have described never sets, but may be seen every clear night. If we imagine a circle drawn round the pole at such a distance as just to touch the horizon, all the stars situated within this circle will move in this way; this is therefore called the circle of perpetual apparition.

As we go away from the pole we shall find the stars moving in larger circles, passing higher up over the pole, and lower down below it, until we reach the circle of perpetual apparition, when they will just graze the horizon. Outside this circle every star must dip below the horizon for a greater or less time, depending on its distance. If it be only a few degrees outside, it will set in the north-west, or between north and north-west; and, after a few hours only, it will be seen to rise again between north and north-east, having done little more than graze the horizon. The possibility of a body rising so soon after having set does not always occur to those who live in moderate latitudes. In July, 1874, Coggia's comet set in the north-west about nine o'clock in the evening, and rose again about three o'clock in the morning; and some intelligent people who then saw it east of the pole supposed it could not be the same one that had set the evening before.

Passing outside the circle of perpetual apparition, we find

that the stars pass south of the zenith at their upper culmination, that they set more quickly, and that they are a longer time below the horizon. This may be seen in Fig. 3, the portion of the sphere to which we refer being between the celestial equator and the line LN. When we reach the equator one-half the course will be above and one-half below the hori-

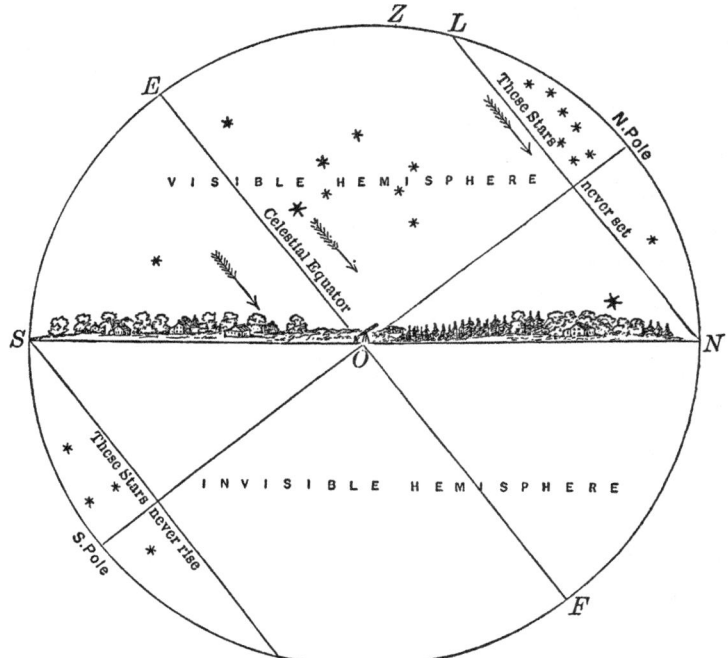

Fig. 3.—The celestial sphere and diurnal motion. S is the south horizon, N the north horizon, Z the zenith. The circle LN around the north pole contains the stars shown in Fig. 2; and the observer at O, in the centre of the sphere, looking to the north, sees the stars as they are depicted in that figure. The arrows show the direction of the diurnal motion in the west.

zon. South of the equator the circles described by the stars become smaller once more, and more than half their course is below the horizon. Near the south horizon the stars only show themselves above the horizon for a short time, while below it there is a circle of perpetual disappearance, the stars in which, to us, never rise at all. This circle is of the same magnitude with that of perpetual apparition, and the south pole is situated in its centre, just as the north pole is in the centre of the other.

If we travel southward we find that the north pole gradually sinks towards the horizon, while new stars come into view above the south horizon; consequently the circles of perpetual apparition and of perpetual disappearance both grow smaller. When we reach the earth's equator the south pole has risen to the south horizon, the north pole has sunk to the north horizon; the celestial equator passes from east to west directly overhead; and all the heavenly bodies in their diurnal revolutions describe circles of which one half is above and the other half below the horizon. These circles are all vertical.

South of the equator only the south pole is visible, the north one, which we see, being now below the horizon. Beyond the southern tropic the sun is north at noon, and, instead of moving from left to right, its course is from right to left.

The laws of the diurnal motion which we have described may be summed up as follows:

1. The celestial sphere, with the sun, moon, and stars, seems to revolve daily around an inclined axis passing through the point where we may chance to stand.

2. The upper end of this axis points (in this hemisphere) to the north pole; the other end passes into the earth, and points to the south pole, which is diametrically opposite, and therefore below the horizon.

3. All the fixed stars during this revolution move together, keeping at the same distance from each other, as if the revolving celestial sphere were solid, and they were set in it.

4. The circle drawn round the heavens half-way between the two poles being the celestial equator, all bodies north of this equator perform more than half their revolution above the horizon, while south of it less than half is above it.

§ 3. *Motion of the Sun among the Stars.*

The most obvious classification of the heavenly bodies which we see with the naked eye is that of sun, moon, and stars. But there is also this difference among the stars, that while the great mass of them preserve the same relative position on the celestial sphere, year after year and century after century, there are five which constantly change their positions relatively to the others. Their names are Mercury, Venus, Mars, Jupiter, and Saturn. These five, with the sun and moon, constitute the seven planets, or wandering stars, of the ancients, the motions of which are next to be described. Taking out the seven planets, the remaining heavenly bodies visible to the naked eye are termed *the Fixed Stars*, because they have no apparent motion, except the regular diurnal revolution described in the last section. But if we note the positions of the sun, moon, and planets among the stars for a number of successive nights, we shall find certain slow changes among them which we shall now describe, beginning with the sun. In studying this description, the reader must remember that we are not seeking for the apparent diurnal motion, but only certain much slower motions of the planets relative to the fixed stars, such as would be seen if the earth did not rotate on its axis.

If we observe, night after night, the exact hour and minute at which a star passes any point by its diurnal revolution, we shall find that passage to occur some four minutes earlier every evening than it did the evening before. The starry sphere therefore revolves, not in 24 hours, but in 23 hours 56 minutes. In consequence, if we note its position at the same hour night after night, we shall find it to be farther and farther to the west. Let us take, for example, the brightest star in the constellation Leo, represented on Map III., and commonly known as Regulus. If we watch it on the 22d of March, we shall find that it passes the meridian at ten o'clock in the evening. On April 22d it passes at eight o'clock, and at ten it is two hours west of the meridian. On the same day of May it passes at six, before sunset, so that it cannot be seen on the meridian at all. When it first becomes visible in the evening twilight, it will be an hour or more west of the meridian. In June it will be three hours west, and by the end of July it will set during twilight, and will soon be entirely lost in the rays of the sun. This shows that during the months in

MOTION OF THE SUN AMONG THE STARS. 15

question the sun has been approaching the star from the west, and in August has got so near it that it is no longer visible.

Carrying forward our computation, we find that on August 21st the star crosses the meridian at noon, and therefore at nearly the same time with the sun. In September it crosses at ten in the morning, while the sun is on the eastern side. The sun has therefore passed from the west to the east of the star, and the latter can be seen rising in the morning twilight before the sun. It constantly rises earlier and earlier, and therefore farther from the sun, until February, when it rises at sunset and sets at sunrise; and is therefore directly opposite the sun. In March the star would cross the meridian at ten o'clock once more, showing that in the course of a year the sun and star had resumed their first position. But, while the sun has risen and set 365 times, the star has risen and set 366 times, the sun having lost an entire revolution by the slow backward motion we have described.

If the stars were visible in the daytime (as they would be but for the atmosphere), the apparent motion of the sun among them could be seen in the course of a single day. For instance, if we could have seen Regulus rise on the morning of August 20th, 1876, we should have seen the sun a little south and west of it, the relative position of the sun being as shown by the circle numbered 1 in the figure. Watching the star all day, we should find that at sunset it was north from the sun, as from circle No. 2. The sun would during the day have moved nearly its own diameter. Next morning we should have seen that the sun had gone past the star into position 3, so that the latter would now rise before the former. By sunset it would have advanced to position 4, and so forth. The path which the sun describes among the stars in his annual revolution is called the ecliptic. It is marked down on Maps II., III., IV., and V., and the months in which the sun passes through each portion of the ecliptic are also indicated. A belt of the heavens, extending a few degrees on each side of

FIG. 4.—Motion of the sun past the star Regulus about August 26th of every year.

the ecliptic, is called the zodiac. The poles of the ecliptic are two opposite points, each in the centre of one of the two hemispheres into which the ecliptic divides the celestial sphere.

The determination of the solar motion around the ecliptic may be considered the birth of astronomical science. The prehistoric astronomers divided the ecliptic and zodiac into twelve parts, now familiarly known as the signs of the zodiac. This proceeding was probably suggested by the needs of agriculture, and of the chronological reckoning of years. A very little observation would show that the changes of the seasons are due to the variations in the meridian altitude of the sun, and in the length of the day; but it was only by a careful study of the position of the ecliptic, and the motion of the sun in it, that it could be learned how these variations in the daily course of the sun were brought about. This study showed that they were due to the fact that the ecliptic and equator did not coincide, but were inclined to each other at an angle of between twenty-three and twenty-four degrees. This inclination is known as the obliquity of the ecliptic. The two circles, equator and ecliptic, cross each other at two opposite points, the positions of which among the stars may be seen by reference to Maps II.-V. When the sun is at either of these points, it rises exactly in the east, and sets exactly in the west; one-half its diurnal course is above the horizon, and the other half below. The days and nights are therefore of equal length, from which the two points in question are called the *Equinoxes*.

The vernal equinox is on the right-hand edge of Map II. Leaving that equinox about March 21st, the sun crosses over the region represented by the map in the course of the next three months, working northward as it does so, until June 20th, when it is on the left-hand edge of the map, $23\frac{1}{2}°$ north of the equator. This point of the ecliptic is called the summer solstice, being that in which the sun attains its greatest northern declination. When near this solstice, it rises north of east, culminates at a high altitude (in our latitudes), and sets north of west. As explained in describing the diurnal motion of an

MOTION OF THE SUN AMONG THE STARS. 17

object north of the celestial equator, more than half the daily course of the sun is now above our horizon, so that our days are longer than our nights, while the great meridian altitude of the sun produces the heats of summer.

The portion of the ecliptic represented on Map II., commencing at the vernal equinox, where the sun crosses the equator, was divided by the early astronomers into the three signs of Aries, the Ram; Taurus, the Bull; and Gemini, the Twins. It will be seen that these signs no longer coincide with the constellations of the same name: this is owing to a change in the position of the equator, which will be described presently.

Turning to Map III., we see that during the three months, from June to September, the sun works downwards towards the equator, reaching it about September 20th. The point of crossing marks the autumnal equinox, found also on the right hand of Map IV. The days and nights are now once more of equal length.

During the next six months the sun is passing over the regions represented on Maps IV. and V., and is south of the equator, its greatest southern declination, or "the southern solstice," being reached about December 21st. More than half its daily course is then below the horizon, so that in our latitudes the nights are longer than the days, and the low noonday altitude of the sun gives rise to the colds of winter.

We have no historic record of this division of the zodiac into signs, and the ideas of the authors can only be inferred from collateral circumstances. It has been fancied that the names were suggested by the seasons, the agricultural operations, and so on. Thus the spring signs (Aries, the Ram; Taurus, the Bull; and Gemini, the Twins) are supposed to mark the bringing forth of young by the flocks and herds. Cancer, the Crab, marks the time when the sun, having attained its greatest declination, begins to go back towards the equator; and the crab having been supposed to move backwards, his name was given to this sign. Leo, the Lion, symbolizes the fierce heat of summer; and Virgo, the Virgin, gleaning corn, symbolizes the harvest. In Libra, the Balance, the day and night balance each other, being of equal length. Scorpius, the Scorpion, is supposed to have marked the presence of venomous reptiles in October; while Sagittarius, the Archer, symbolizes the season of hunting. The explanation of Capricornus, the Goat, is more fanciful, if possible, than that of Cancer. It was supposed that this animal, ascending the hill as he feeds, in order to reach the grass more easily, on reaching the top, turns back again, so that his name was used to mark the sign in which the sun, from going south, begins to return to the north. Aquarius, the Water-bearer, symbolizes the winter rains; and Pisces, the Fishes, the season of fishes.

All this is, however, mere conjecture; the only coincidences at all striking being Virgo and Libra. The names of the constellations were probably given to them several centuries, perhaps even thousands of years, before the Christian era; and in that case the zodiacal constellations would not have corresponded to the seasons we have indicated. An attempt has even been made to show that the names of the zodiacal constellations were intended to commemorate the twelve labors of Hercules; but this theory rests on no better foundation than the other.

The zodiacal constellations occupy quite unequal spaces in the heavens, as may be seen by inspection of the maps. In the beginning they were simply twelve houses for the sun, which that luminary occupied in the course of the year. Hipparchus found this system entirely insufficient for exact astronomy, and therefore divided the ecliptic and zodiac into twelve equal parts, of 30° each, called signs of the zodiac. He gave to these signs the names of the constellations most nearly corresponding to them. Commencing at the vernal equinox, the first arc of 30° was called the sign Aries, the second the sign Taurus, and so forth. The mode of reckoning positions on the ecliptic by signs was continued until the last century, but is no longer in use among professional astronomers, owing to its inconvenience. The whole ecliptic is now divided into 360°, like any other circle, the count commencing at the vernal equinox, and following the direction of the sun's motion all the way round to 360°.

§ 4. Precession of the Equinoxes.—The Solar Year.

By comparing his own observations with those of preceding astronomers, Hipparchus found that the equinoxes were slowly shifting their places among the stars, the change being at least a degree in a century towards the west. His successors determined it with greater exactness, and it is now known to be nearly a degree in seventy years. Careful study of the change shows that it is due mainly to a motion of the equator, which again arises from a change in the direction of the pole. The position of the ecliptic among the stars varies so slowly that the change can be seen only by the refined observations of modern times. In the explanation of the diurnal motion, it was stated that there was a certain point in the heavens around which all the heavenly bodies seem to perform a daily revolution. This point, the pole of the heavens, is marked on the centre of Map I., and is also in the centre of Fig. 2, page 10. It is little more than a degree distant from the pole star. Now, precession really consists in a very slow motion of this pole around the pole of the ecliptic, the rate of motion being such as to carry it all the way round in about 25,300 years. The exact time has never been calculated, and would not always be the same, owing to some small variations to which the motion is subject; but it will never differ much from this. There is a very slight motion to the ecliptic itself, and therefore to its pole; and this fact renders the motion of the pole of the equator around it somewhat complicated; but the curve described by the latter is very nearly a circle 46° in diameter. In the time of Hipparchus, our present pole star was 12° from the pole. The pole has been approaching it steadily ever since, and will continue to approach it till about the year 2100, when it will slowly pass by it at the distance of less than half a degree. The course of the pole during the next 12,000 years is laid down on the map, and it will be seen that at the end of that time it will be near the constellation Lyra. Since the equator is always 90° distant from the pole, there will be a corresponding motion to it, and hence to the point of its crossing the ecliptic. To show this, the position of the equator 2000 years ago, as well as its present position, is given on Map II.

The reader will, of course, understand that the various celestial movements of which we have spoken in this chapter are only apparent motions, and are due to the motion of the earth itself, as will be explained in the chapter on the Copernican system. The diurnal revolution of the celestial sphere is due to the rotation of the earth on its axis, while precession is really a change in the direction of that axis.

One important effect of precession is that one revolution of the sun among the stars does not accurately correspond to the return of the same seasons. The latter depend upon the position of the sun relative to the equinox, the time when the sun crosses the equator towards the north always marking the season of spring (in the northern hemisphere), no matter where the sun may be among the stars. If the equator did not move, the sun would always cross it at nearly the same point among the stars. But when, starting from the vernal equinox, it makes the circuit of the heavens, and returns to it again, the motion of the equator has been such that the sun crosses it 20 minutes before it reaches the same star. In one year, this difference is very small; but by its constant accumulation, at the rate of 20 minutes a year, it becomes very considerable after the lapse of centuries. We must, therefore, distinguish between the sidereal and the tropical year, the former being the period required for one revolution of the sun among the stars, the latter that required for his return to the same equinox, whence it is also called the equinoctial year. The exact lengths of these respective years are:

	Days.	Days.	Hours.	Min.	Sec.
Sidereal year	365.25636 =	365	6	9	9
Tropical year	365.24220 =	365	5	48	46

Since the recurrence of the seasons depends on the tropical year, the latter is the one to be used in forming the calendar, and for the purposes of civil life generally. Its true length is 11 minutes 14 seconds less than $365\frac{1}{4}$ days. Some results of this difference will be shown in explaining the calendar.

§ 5. *The Moon's Motion.*

Every one knows that the moon makes a revolution in the celestial sphere in about a month, and that during its revolution it presents a number of different phases, known as "new moon," "first quarter," "full moon," and so on, depending on its position relative to the sun. A study of these phases during a single revolution will make it clear that the moon is a globular dark body, illuminated by the light of the sun, a fact which has been evident to careful observers from the remotest antiquity. This may be illustrated by taking a large globe to represent the moon, painting one half white, to represent the half on which the sun shines, and the other half dark. Viewing it at a proper distance, and turning it into different positions, it will be found that the visible part of the white half may be made to imitate the various appearances of the moon.

As the sun makes a revolution around the celestial sphere in a year, so the moon makes a similar revolution among the stars in a little more than 27 days. This motion can be seen on any clear night between first quarter and full moon, if the moon happens to be near a bright star. If the position of the moon relatively to the star be noted from hour to hour, it will be found that she is constantly working towards the east by a distance equal to her own diameter in an hour. The following night she will be found from 12° to 14° east of the star, and will rise, cross the meridian, and set from half an hour to an hour later than she did the preceding night. At the end of 27 days 8 hours, she will be back in the same position among the stars in which she was first seen.

If, however, starting from one new moon, we count forwards this period, we shall find that the moon, although she has returned to the same position among the stars, has not got back to new moon again. The reason is that the sun has moved forwards, in virtue of his apparent annual motion, so far that it will require more than two days for the moon to overtake him. So, although the moon really revolves around the earth in 27⅓ days, the average interval between one new moon and the next is 29½ days.

A comparison of the phases of the moon with her direction will show that the sun is many times more distant than the moon. In Fig. 5, let E be the position of an observer on the earth, M the moon, and S the sun, illuminating one half of it. When the observer sees the moon in her first quarter—that is, when her disk appears exactly half illuminated—the angle at

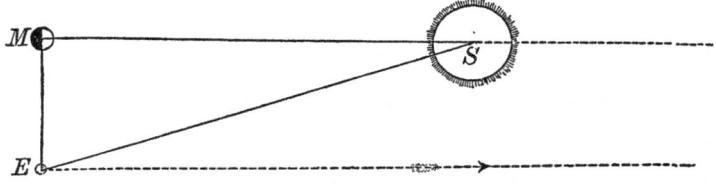

FIG. 5.—Showing the sun to be farther than the moon.

the moon, between the observer and the sun, must be a right angle. If the sun were only about four times as far as the moon, as in the figure, the observer, by measuring the angle SEM between the sun and moon, would find it to be 75°; and the nearer the sun, the smaller he would find it. But actual measurement would show it to be so near 90° that the difference would be imperceptible with ordinary instruments. Hence, the sun is really at the point where the dotted line and the line MS continued meet each other, which is many times the distance EM to the moon.

This idea was applied by Aristarchus, who flourished in the third century before Christ, preceding both Hipparchus and Ptolemy, to determine the distance of the sun, or, more exactly, how many times it exceeded the distance of the moon. He found, by measurement, that, in the position represented in the figure, the distance between the directions of the sun and moon was 87°, and that the sun was therefore something like twenty times as far as the moon. We now know that this result was twenty times too small, the angle being really so near 90° that Aristarchus could not determine the difference with certainty. In principle, the method is quite correct, and very ingenious, but it cannot be applied in practice. The one insuperable difficulty of the method arises from the impossibility of seeing when the moon is exactly half illuminated, the uncertainty arising from the inequalities in the lunar surface being greater than the whole angle to be measured.

Watching and mapping down the path of the moon among the stars, it is found not to be the same with that of the sun, being inclined to it about 5°. The paths cross each other in two opposite points of the heavens, called the moon's nodes. The path of the moon in the middle of the year 1877 is marked on star Maps II.–V. Referring to Map III., it will be seen that the descending node of the moon is in the constellation Leo, very near the star Regulus. Here the moon passes south of or below the ecliptic, and continues below it over the whole of Map IV. On Map V., it approaches the ecliptic again, crossing to the north of it in the constellation Aquarius, and continuing on that side till it reaches Regulus once more.

Such is the moon's path in July, 1877. But it is constantly changing in consequence of a motion of the nodes towards the west, amounting to more than a degree in every revolution. In order that the line drawn on the map may continue to represent the path of the moon, we must suppose it to slide along the ecliptic towards the right at the rate of about 20° a year, so that a slightly different path will be described in every monthly revolution. The path will always cross the ecliptic at the same angle, but the moon will not always pass over the same stars. In August, 1877, she will cross the ecliptic a little farther to the right (west), and will pass a little below Regulus. The change going on from month to month and from year to year, in a little less than ten years the ascending node will be found in Leo; and the other node, now in Leo, will have gone back to Aquarius. In a period of eighteen years and seven months, the nodes will have made a complete revolution, and the path of the moon will have resumed the position given on the map.

§ 6. *Eclipses of the Sun and Moon.*

The early inhabitants of the world were, no doubt, terrified by the occasional recurrence of eclipses many ages before there were astronomers to explain their causes. But the motions of the sun and moon could not be observed very long without the causes being seen. It was evident that if the moon should ever chance to pass between the earth and the sun, she must cut off some or all of his light. If the two bodies followed the same track in the heavens, there would be an eclipse of the sun every new moon; but, owing to the inclination of the two orbits, the moon will generally pass above or below the sun, and there will be no eclipse. If, however, the sun happens to be in the neighborhood of the moon's node when the moon passes, then there will be an eclipse. For an example, let us refer to Map III. We see that the sun passes the moon's descending node about August 25th, 1877, and is within 20° of this node from early in August till the middle of September. The moon passes the sun on August 8th and September 6th of that year, which are, therefore, the dates of new moon. At the first date, the moon passes so far to the north that, as seen from the centre of the earth, there is no eclipse at all; but in the northern part of Asia the moon would be seen to cut off a small portion of the sun.

While the moon is performing another circuit, the sun has moved so far past the node, that the moon passes south of it, and there is only a small eclipse, and that is visible only around the region of Cape Horn. Thus, there are two solar eclipses while the sun is passing this node in 1877, but both are very small. Indeed, every time the sun crosses a node, the moon is sure to cross his path, either before he reaches the node, or before he gets far enough from it to be out of the way. As he crosses both nodes in the course of the year, there must be at least two solar eclipses every year to some points of the earth's surface.

The cause of lunar eclipses might not have been so easy to guess as was that of solar ones; but a great number could

ECLIPSES OF THE SUN AND MOON. 25

not have been observed, and their times of occurrence recorded, without its being noticed that they always occurred at full moon, when the earth was opposite the sun. The idea that the earth cast a shadow, and that the moon passed into it, could then hardly fail to suggest itself; and we find, accordingly, that the earliest observers of the heavens were perfectly acquainted with the cause of lunar eclipses.

The reason why eclipses of the moon only occur occasionally is of the same general nature with that of the rare occurrence of solar eclipses. The centre of the earth's shadow is always, like the sun, in the ecliptic; and unless the moon happens to be very near the ecliptic, and therefore very near one of her nodes at the time of full moon, she will fail to strike the shadow, passing above or below it. Owing to the great magnitude of the sun, the earth's shadow is, at the distance of the moon, much smaller than the earth itself. The result of this is, that the moon must be decidedly nearer her node to produce a lunar than to produce a solar eclipse. Sometimes a whole year passes without there being any eclipse of the moon.

The nature of an eclipse will vary with the positions and apparent magnitudes of the sun and moon. Let us suppose, first, that, in a solar eclipse, the centre of the moon happens to pass exactly over the centre of the sun. Then, it is clear that if the apparent angular diameter of the moon exceed that of the sun, the latter will be entirely hidden from view. This is called a *total eclipse of the sun.* It is evident that such an eclipse can occur only when the observer is near the line joining the centres of the sun and moon. If, under the same circumstances, the apparent magnitude of the moon is less than that of the sun, it is evident that the whole of the latter cannot be covered, but a ring of light around his edge will still be visible. This is called an *annular eclipse.* If the moon does not pass centrally over the sun, then it can cover only a portion of the latter on one side or the other, and the eclipse is said to be partial. So with the moon: if the latter is only partially immersed in the earth's shadow, the eclipse of the moon is called *partial;* if she is totally immersed in it, so that no direct sunlight can reach her, the eclipse is said to be *total*. An an-

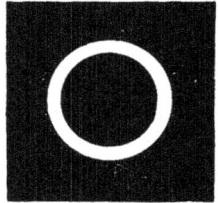

Fig. 6.—Annular eclipse of the sun.   Fig. 7.—Partial eclipse of the sun.

nular eclipse of the moon is impossible, because the earth's shadow always exceeds the diameter of the moon in breadth.

Some points respecting eclipses will be seen more clearly by reference to the accompanying figures, in which S represents the sun, E the earth, and M the moon. Referring to the first figure, it will be seen that an observer at either of the points marked O, or indeed anywhere outside the shaded portions, will see the whole of the sun, so that to him there will be no eclipse at all. Within the lightly shaded regions, marked PP, the sun will be partially eclipsed, and more so as the observer is near the centre. This region is called the penumbra.

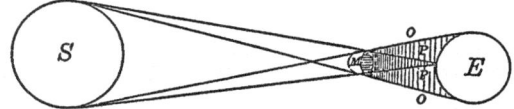

Fig. 8.—Eclipse of the sun, the shadow of the moon falling on the earth.

Within the darkest parts between the two letters P is a region where the sun is totally hidden by the moon. This is the shadow, and its form is that of a cone, with its base on the moon, and its point extending towards the earth. Now, it happens that the diameters of the sun and moon are very nearly proportional to their respective mean distances, so that the point of this shadow almost exactly reaches the surface of the earth. Indeed, so near is the adjustment, that the dark shadow sometimes reaches the earth, and sometimes does not, owing to the small changes in the distance of the sun and moon. When the shadow reaches the earth, it is comparatively very narrow, owing to its being so near its sharp point; but if an observer can station himself within it, he will see a total eclipse of the sun during the short time the shadow is passing over him. If the reader will study the figure, he will see why a total eclipse of the sun is so rare at any one place on the earth. The shadow, when it reaches the earth, is so near down to a point that its diameter is not generally more than a hundred miles; consequently, each total eclipse is visible only along a belt which may not average more than a hundred miles across.

In most eclipses, the shadow comes to a point before it reaches the earth; in this case, the apparent angular diameter of the moon is less than that of the sun, and there can be no total eclipse. But if an observer places himself in a line with the centre of the shadow, he will see an annular eclipse, the sun showing itself on all sides of the moon.

The next figure shows us the form of the earth's shadow.

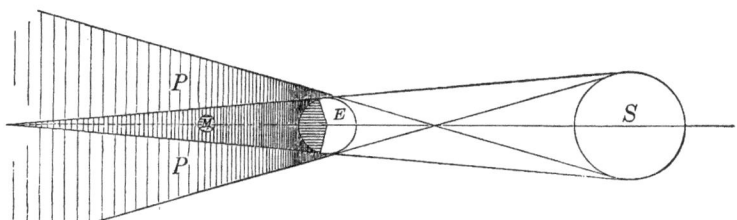

Fig. 9.—Eclipse of the moon, the latter being in the shadow of the earth.

The earth being much larger than the moon, its shadow extends far beyond it; and where it reaches the moon, it is always so much larger than the latter that she may be wholly immersed in it, as shown in the figure. Now, suppose the moon, in her course round the earth, to pass centrally through the shadow, and not above or below it, as she commonly does; then, when she entered the shaded region, marked P, which is called the penumbra, an observer on her surface would see a partial eclipse of the sun caused by the intervention of the earth. The time when this begins is given in the almanacs, being expressed by the words, "Moon enters penumbra." Some of the sunlight is then cut off from the moon, so that the latter is not so bright as usual; but the eye does not notice any loss of light until the moon almost reaches the dark shadow. As she enters the shadow, a portion of her surface seems to be cut off and to disappear entirely, and her visible portion continually grows smaller, until, in case of a total eclipse, her whole disk is immersed in the shadow. When this occurs, it is found that she is not entirely invisible, but still faintly shines with a lurid copper-colored light. This light is refracted into the shadow by the earth's atmosphere, and its amount may be greater or less, according to the quantity of clouds and vapor in the atmosphere around that belt of the earth which the sunlight must graze in order to reach the moon.

In about half of the lunar eclipses, the moon passes so far above or below the centre of the shadow that part of her body is in it, and part outside, at the time of greatest eclipse. This is called a *partial eclipse* of the moon. The magnitude of a partial eclipse, whether of the sun or moon, was measured by the older astronomers in digits. The diameter of the solar or lunar disk was divided into twelve equal parts, called digits; and the magnitude of the eclipse was said to be equal to the number of digits cut off by the shadow of the earth in case of a lunar eclipse, or by the moon in case of a solar eclipse. The most ancient astronomers were in the habit of measuring the digits by surface: when the moon was said to be eclipsed four digits, it meant that one-third of her surface, and not one-third her diameter, was eclipsed.

The duration of an eclipse varies between very wide limits, according to whether it is nearly central or the contrary. The duration of a solar eclipse depends upon the time required for the moon to pass over the distance from where she first comes into apparent contact with the sun's disk, until she separates from it again; and this, in the case of eclipses which are pretty large, may range between two and three hours. In a total eclipse, however, the apparent disk of the moon exceeds that of the sun by so small an amount, that it takes her but a short time to pass far enough to uncover some part of the sun's disk; the time is rarely more than five or six minutes, and sometimes only a few seconds. A total eclipse of the moon may, however, last nearly two hours, and the partial eclipses on each side of the total one may extend the whole duration of the eclipse to three or four hours.

Total eclipses of the sun afford very rare and highly prized opportunities for studying the operations going on around that luminary. Of these we shall speak in a subsequent chapter.

Returning, now, to the apparent motions of the sun and moon around the celestial sphere, we see that since the moon's orbit has two opposite nodes in which it crosses the ecliptic, and the sun passes through the entire course of the ecliptic in the course of the year, it follows that there are two periods in the course of a year during which the sun is near a node, and eclipses may occur. Roughly speaking, these periods are each about a month in duration, and we may call them seasons of eclipses. For instance, it will be seen on Map V. that the sun passes one node of the moon's orbit towards the end of February, 1877. A season of eclipses for that year is therefore February and the first half of March. Actually, there is a total eclipse of the moon on February 27th, and a very small eclipse of the sun on March 14th, of that year, visible only in Northern Asia.* From this time, the sun is so far from the node that there can be no eclipses until he approaches the other node in August. Then we have the two eclipses of the sun already mentioned, and, between them, a total eclipse of the moon on August 23d. Thus, in the year 1877, the first season of eclipses is in February and March, and the second in August and September.

We have said that the length of each eclipse season is about a month. To speak with greater accuracy, the average season for eclipses of the sun extends 18 days before and after the

---

* There is an extraordinary coincidence between this eclipse and that of August 8th of the same year, both being visible from nearly the same region in Central Siberia.

30   SYSTEM OF THE WORLD HISTORICALLY DEVELOPED.

sun's passage through the node, while that for lunar eclipses extends 11½ days on each side of the node. The total season is, therefore, 36 days for solar, and 23 days for lunar eclipses.

Owing to the constant motion of the moon's node already described, the season of eclipses will not be the same from year to year, but will occur, on the average, about 20 days earlier each year. We have seen that the sun passed the descending node of the moon marked on Map III. on August 24th, 1877; but during the year following the node will have moved so far to the west that the sun will again reach it on August 5th, 1878. The effect of this constant shifting of the nodes and seasons of eclipses is that in 1887 the August season will be shifted back to February, and the February season to August. The reader who wishes to find the middle of the eclipse seasons for twenty or thirty years can do so by starting from March 1st and August 24th, 1877, and subtracting 19¾ days for each subsequent year.

There is a relation between the motions of the sun and moon which materially assisted the early astronomers in the prediction of eclipses. We have said that the moon makes one revolution among the stars in about 27⅓ days. Since the node of the orbit is constantly moving back to meet the moon, as it were, she will return to her node in a little less than this period—namely, as shown by modern observations, in a mean interval of 27.21222 days. The sun, after passing any node of the orbit, will reach the same node again in 346.6201 days. The relation between these numbers is this: 242 returns of the moon to a node take very nearly the same time with 19 returns of the sun, the intervals being

  242 returns of the moon to her node.................... 6585.357 days;
   19 "   " sun to moon's node................. 6585.780 "

Consequently, if at any time the sun and moon should start out together from a node, they would, at the end of 6585 days, or 18 years and 11 days, be again found together very near the same node. During the interval, there would have been 223 new and full moons, but none so near the node as this. The exact time required for 223 lunations is 6585.3212 days; so that, in the case supposed, the 223d conjunction of the sun and moon would happen a little before they reached the node, their distance from it being, by calculation, a little less than one of their diameters, or, more exactly, 28'. If, instead of being exactly at the node, they are any given distance from it, say 3° east or west, then, in the same period, they will be again together within half a degree of the same distance from the node.

The period just found was called the Saros, and may be applied in this way: Let us note the exact time of the middle of any eclipse, either of the moon or of the sun; then let us count forwards 6585 days, 7 hours, 42 minutes, and we shall find another eclipse of very nearly the same kind. Reduced to years, the interval will be 18 years and 10 or 11 days, according to whether the 29th of February has intervened four or five times during the interval. This being true of every eclipse, if we record all the eclipses which occur during a period of 18 years, we shall find the same series after 10 or 11 days to begin over again; but the new series will not generally be visible at the same places with the old ones, or, at least, will not occur at the same time of day, since the middle will be nearly eight hours later. Not till the end of three periods will they recur near the same meridian; and then, owing to the period not being exact, the eclipse will not be precisely of the same magnitude, and, indeed, may fail entirely. Every successive recurrence of an eclipse at the end of the period being 28' farther back relatively to the node, the conjunction must, in process of time, be so far back from the node as not to produce an eclipse at all. During nearly every period it will be found that some eclipse fails, and that some new one enters in. A new eclipse of the moon thus entering will be a very small one indeed. At every successive recurrence of its period it will be larger, until, about its thirteenth recurrence, it will be total. It will be total for about twenty-two or twenty-three recurrences, when it will become partial once more, but on the opposite side of the moon from that on

32 SYSTEM OF THE WORLD HISTORICALLY DEVELOPED.

which it was first seen. There will then be about thirteen partial eclipses, each smaller than the last, until they fail entirely. The whole interval of time over which the recurrence of a lunar eclipse thus extends will be about 48 periods, or 865½ years. The solar eclipses, occurring farther from the node, will last yet longer, namely, from 65 to 70 periods, or over 1200 years.

As a recent example of the Saros, we may cite some total eclipses of the sun well known in recent times; for instance,

1842, July 8th, 1ʰ 8 A.M., total eclipse, observed in Europe;
1860, July 18th, 9ʰ A.M., total eclipse America and Spain;
1878, July 29th, 4ʰ 2 P.M., one visible in Colorado and on the Pacific Coast.

A yet more remarkable series of total eclipses of the sun occurs in the years 1850, 1868, 1886, etc., the dates being—

1850, August 7th, 4ʰ 4 P.M., in the Pacific Ocean;
1868, August 17th, 12ʰ P.M., in India;
1886, August 29th, 8ʰ A.M., in the Central Atlantic Ocean and Southern Africa;
1904, September 9th, noon, in South America.

This series is remarkable for the long duration of totality, amounting to some six minutes.

It must be understood that the various numbers we have given in this section are not accurate for all cases, because the motions both of the sun and moon are subject to certain small irregularities which may alter the times of eclipses by an hour or more. We have given only mean values, which are, however, always quite near the truth.

§ 7. *The Ptolemaic System.*

There is still extant a work which for fourteen centuries was a sort of astronomical Bible, from which nothing was taken, and to which nothing material in principle was added. This is the "Almagest" of Ptolemy, composed about the middle of the second century of our era. Nearly all we know of the ancient astronomy as a science is derived from it. Fragments of other ancient authors have come down to us, and most of the ancient writers make occasional allusions to astronomical phenomena or theories, from which various ideas respecting the ancient astronomy have been gleaned; but the work of Ptolemy is the only complete compendium which we possess. Although his system is in several important points erroneous, it yet represents the salient features of the apparent motions of the heavenly bodies with entire accuracy. Defective as it is when measured by our standard, it is a marvel of ingenuity and research when measured by the standard of the times.

The immediate object of the present chapter is to explain the apparent movements of the planets, which can be most easily done on the Ptolemaic system. But, on account of its historic interest, we shall begin with a brief sketch of the propositions on which the system rests, giving also Ptolemy's method of proving them. His fundamental doctrines are that the heavens are spherical in form, and all the heavenly motions spherical or in circles; that the earth is also spherical, and situated in the centre of the heavens, or celestial sphere, where it remains quiescent, and that it is in magnitude only a point when compared with the sphere of the stars. We shall give Ptolemy's views of these propositions, and his attempts to prove them, in their regular order.

1st. *The Heavenly Bodies move in Circles.*—Here Ptolemy refers principally to the diurnal motion, whereby every heavenly body is apparently carried around the earth, or, rather, around the pole of the heavens, in a circle every day. But all the ancient and mediæval astronomers down to the time of Kepler had a notion that, the circle being the most perfect plane figure, all the celestial motions must take place in circles; and as it was found that the motions were never uniform, they supposed these circles not to be centred on the earth. Where a single circle did not suffice to account for the motion, they introduced a combination of circular motions in a manner to be described presently.

2d. *The Earth is a Sphere.*—That the earth is rounded from east to west Ptolemy proves by the fact that the sun, moon, and stars do not rise and set at the same moment to all the inhabitants of the earth. The times at which eclipses of the moon are seen in different countries being compared, it is found that the farther the observer is west, the earlier is the hour after sunset. As the time is really the same everywhere, this shows that the sun sets later the farther we go to the west. Again, if the earth were not rounded from north to south, a star passing the meridian in the north or south horizon would always pass in the horizon, however far to the north or south the observer might travel. But it is found that when an observer travels towards the south, the stars in the north approach the horizon, and the circles of their diurnal motion cut below it, while new stars rise into view above the south horizon. This shows that the horizon itself changes its direction as the observer moves. Finally, from whatever direction we approach elevated objects from the sea, we see that their bases are first hidden from view by the curvature of the water, and gradually rise into view as we approach them.

3d. *The Earth is in the Centre of the Celestial Sphere.*— If the earth were displaced from the centre, there would be various irregularities in the apparent daily motion of the celestial sphere, the stars appearing to move faster on the side towards which the earth was situated. If it were displaced towards the east, we should be nearer the heavenly bodies when they are rising than when they are setting, and they would appear to move more rapidly in the east than in the west. The forenoons would therefore be shorter than the afternoons. Towards whatever side of the turning sphere it might be moved, the heavenly bodies would seem to move more rapidly on that side than on the other. No such irregularity being seen, but the diurnal motion taking place with perfect uniformity, the earth must be in the centre of motion.

4th. *The Earth has no Motion of Translation.*—Because if it had it would move away from the centre towards one side of the celestial sphere, and the diurnal revolution of the stars would cease to be uniform in all its parts. But the uniformity of motion just described being seen from year to year, the earth must preserve its position in the centre of the sphere.

It will be interesting to analyze these propositions of Ptolemy, to see what is true and what is false. The first proposition — that the heavenly bodies move in circles, or, as it is more literally expressed, that the heavens move spherically — is quite true, so far as the apparent diurnal motion is concerned. What Ptolemy did not know was that this motion is only apparent, arising from a rotation of the earth itself on its axis. The second proposition is perfectly correct, and Ptolemy's proofs that the earth is round are those still found in our school-books at the end of seventeen hundred years. Most curious, however, is the mixture of truth and falsehood in the third and fourth propositions, that the earth remains quiescent. We cannot denounce it as unqualifiedly false, because, in a certain sense, and indeed in the only sense in which there is any celestial sphere, the earth may be said to remain in the centre of the sphere. What Ptolemy did not see is that this sphere is only an ideal one, which the spectator carries with him wherever he goes. His demonstration that the centre of revolution of the sphere is in the earth is, in a certain sense, correct; but what he really proves is that the earth revolves on its own axis. He did not see that if the earth could carry the axis of revolution with it, his demonstration of the quiescence of the earth would fall to the ground.

Considerable insight into Ptolemy's views is gained by his answers to two objections against his system. The first is the vulgar and natural one, that it is paradoxical to suppose that a body like the earth could remain supported on nothing, and still be at rest. These objectors, he says, reason from what they see happen to small bodies around them, and not from what is proper to the universe at large. There is neither up nor down in the celestial spaces, for we cannot conceive of it in a sphere. What we call down is simply the direction of our feet towards the centre of the earth, the direction in which heavy bodies tend to fall. The earth itself is but a point in comparison with the celestial spaces, and is kept fixed by the forces exerted upon it on all sides by the universe, which is infinitely larger than it, and similar in all its parts.

This idea is as near an approach to that of universal gravitation as the science of the times would admit of.

He then says there are others who, admitting this reasoning, pretend that nothing hinders us from supposing that the heavens are immovable, and that the earth itself turns round its own axis once a day from west to east. It is certainly singular that one who had risen so far above the illusions of sense as to demonstrate to the world that the earth was round; that up and down were only relative; and that heavy bodies fell towards a centre, and not in some unchangeable direction, should not have seen the correctness of this view.

To refute the doctrine of the earth's rotation, he proceeds in a way the opposite of that which he took to refute those who thought the earth could not rest on nothing. He said of the latter that they regarded solely what was around them on the earth, and did not consider what was proper to the universe at large. To those who maintained the earth's rotation, he says, if we consider only the movements of the stars, there is nothing to oppose their doctrine, which he admits has the merit of simplicity; but in view of what passes around us and in the air, their doctrine is ridiculous. He then enters into a disquisition on the relative motion of light and heavy bodies, which is extremely obscure; but his conclusion is that if the earth really rotated with the enormous velocity necessary to carry it round in a day, the air would be left behind. If they say that the earth carries round the air with it, he replies that this could not be true of bodies floating in the air; and hence concludes that the doctrine of the earth's rotation is not tenable. It is clear, from this argument, that if Ptolemy and his contemporaries had devoted to experimental physics half the careful observation, research, and reasoning which we find in their astronomical studies, they could not have failed to establish the doctrine of the earth's rotation.

In the Ptolemaic system, all the celestial motions are represented by a series of circular motions. We have already explained the motions of the sun and moon among the stars, the first describing a complete circuit of the heavens from west to east in a year, and the second a similar circuit in a month. Though not entirely uniform, these movements are always forward. But it is not so with the five planets—Mercury, Venus, Mars, Jupiter, and Saturn. These move sometimes to the east and sometimes to the west, and are sometimes stationary.* On the whole, however, the easterly movements predominate; and the planets really oscillate around a certain mean point itself in regular motion towards the east. Let us take, for instance, the planet Jupiter. Suppose a certain fictitious Jupiter performing a circuit of the heavens among the stars every twelve years with a regular easterly motion, just as the sun performs such a circuit every year; then the real Jupiter will be found to oscillate, like a pendulum, on each side of the fictitious planet, but never swinging more than 12° from it. The time of each double oscillation is about thirteen months—that is, if on January 1st we find it passing the fictitious planet towards the west, it will continue its westerly swing about three months, when it will gradually stop, and return with a somewhat slower motion to the fictitious planet again, passing to the east of it the middle of July. The easterly swing will continue till about the end of October, when it will return towards the west. The westerly or backward motion is called *retrograde*, and the easterly motion *direct*. Between the two is a point at which the planet appears stationary once more. The westerly motions are called *retrograde* because they are in the opposite direction both to the motion of the sun among the stars, and to the average direction in which all the planets move. It was seen by Hipparchus, who lived three centuries before Ptolemy, that this oscillating motion could be represented by supposing the real Jupiter to describe a circular orbit around the fictitious Jupiter once in a year. This orbit is called the epicycle, and thus we have the celebrated epicyclic theory of the planetary motions laid down in the "Almagest." The movement of the planet on this theory can be seen by

---

* It may not be amiss to remind the reader once more that we here leave the diurnal motion of the stars entirely out of sight, and consider only the motions of the planets relative to the stars.

Fig. 10. *E* is the earth, around which the fictitious Jupiter moves in the dotted circle, 1, 2, 3, 4, etc. To form the epicycle in which the real planet moves, we must suppose an arm to be constantly turning round the fictitious planet once a year, on the end of which Jupiter is carried. This arm will then be in the successive positions, 1 1', 2 2', 3 3', etc., represented by the light dotted lines. Drawing a line through the successive positions 1', 2', 3', etc., of the real Jupiter, we shall have a series of loops representing its apparent orbit.

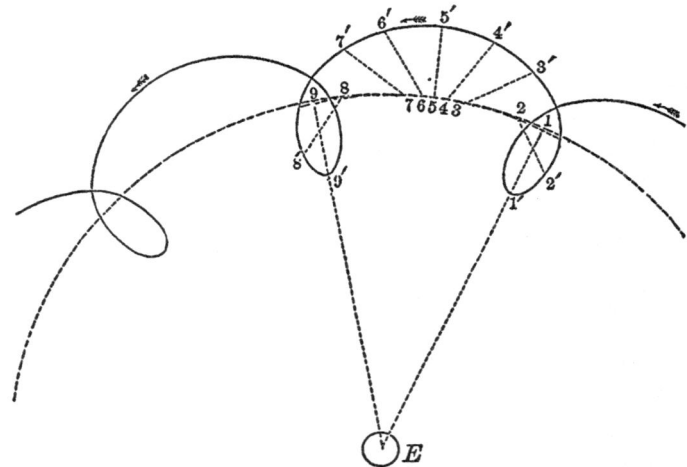

FIG. 10.—Showing the apparent orbit of a planet, regarding the earth as at rest.

It will be seen that although it requires only a year for the arm carrying the real Jupiter to perform a complete revolution and return to its primitive direction, it requires about thirteen months to form a complete loop, because, owing to the motion of the fictitious planet in its orbit, the arm must move more than a complete revolution to finish the loop. For instance, referring again to Fig. 10, comparing the positions 1 1' and 8 8', it will be seen that the arm, being in the same direction, has performed a complete revolution; but, owing to the curvature of the orbit, it does not reach the middle of the second loop until it attains the position 9 9'.

THE PTOLEMAIC SYSTEM. 39

The planets of which the radius of the epicycle makes an annual revolution in this way are Mars, Jupiter, and Saturn. The complete apparent orbits of the last two planets are shown in the next figure, taken from Arago. By the radius of the epicycle we mean the imaginary revolving arm which, turning round the fictitious planet, carries the real planet at its

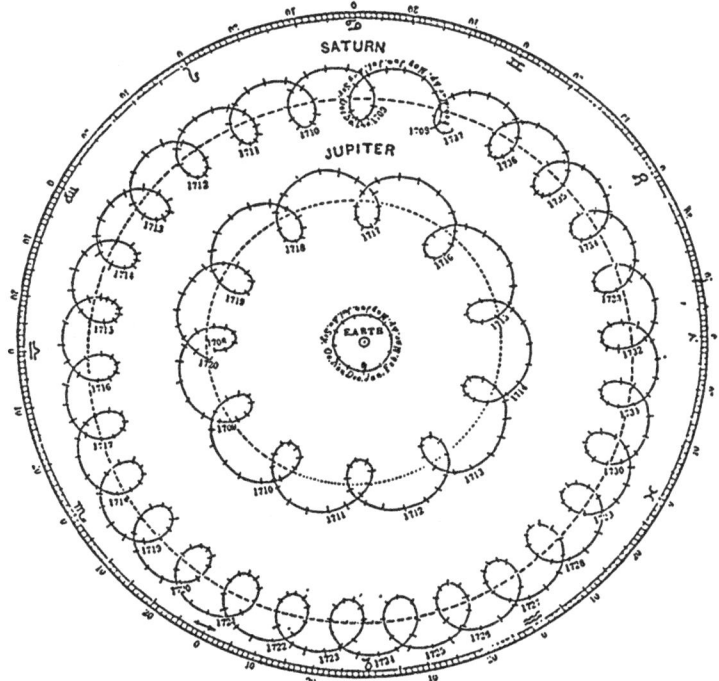

FIG. 11.—Apparent orbits of Jupiter and Saturn, 1708-1737, after Cassini.

end. The law of revolution of this arm is, that whenever the planet is opposite the sun, the arm points towards the earth, as in the positions 1 1', 9 9', in which cases the sun will be on the side of the earth opposite the planet; while, whenever the planet is in conjunction with the sun, the arm points from the earth. This fact was well known to the ancient astronomers, and their calculations of the motions of the planets were all founded upon it; but they do not seem to have noticed the very important corollary from it, that the direction of the radius of the epicycle of Mars, Jupiter, and Saturn is always the same with that of the sun from the earth. Had they done so, they could hardly have failed to see that the epicycles could be abolished entirely by supposing that it was the earth which moved round the sun, and not the sun round the earth.

The peculiarity of the planets Mercury and Venus is that the fictitious centres around which they oscillate are always in the direction of the sun, or, as we now know, the sun himself is the centre of their motions. They are never seen more than a limited distance from that luminary, Venus oscillating about 45° on each side of the sun, and Mercury from 16° to 29°. It is said that the ancient Egyptians really did make the sun the centre of the motion of these two planets; and it is difficult to see how any one could have failed to do so after learning the laws of their oscillation. Yet Ptolemy rejected this system, placing their orbits between the earth and sun without assigning any good reason for the course.

The arrangement of the planets on the Ptolemaic system is shown in Fig. 12. The nearest planet is the moon, of which the ancient astronomers actually succeeded in roughly measuring the distance. The remaining planets are arranged in the same order with their real distance from the sun, except that the latter takes the place assigned to the earth in the modern system. Thus we have the following order:

        The Moon,
        Mercury,
        Venus,
        The Sun,
        Mars,
        Jupiter,
        Saturn.

Outside of Saturn was the sphere of the fixed stars.

This order of the planets must have been a matter of opinion rather than of demonstration, it being correctly judged by the ancient astronomers that those which seemed to move

THE PTOLEMAIC SYSTEM.

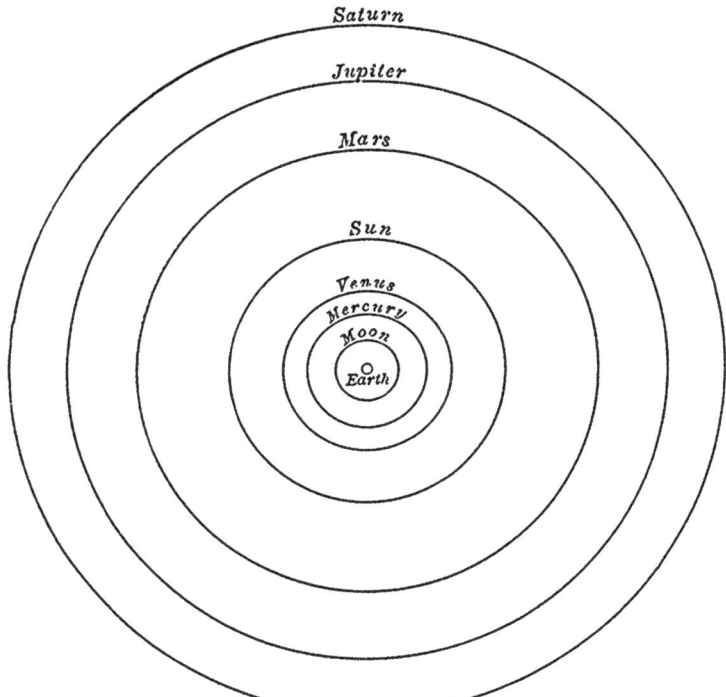

FIG. 12.—Arrangement of the seven planets in the Ptolemaic system. The orbits, as marked, are those of the fictitious planets, the real planets being supposed to describe a series of loops.

more slowly were the more distant. This system made it quite certain that the moon was the nearest planet, and Mars, Jupiter, and Saturn, in their order, the most distant ones. But the relative positions of the Sun, Mercury, and Venus were more in doubt, since they all performed a revolution round the celestial sphere in a year. So, while Ptolemy, as we have just said, placed Mercury and Venus between the earth and the sun, Plato placed them beyond the sun, the order being, Moon, Sun, Mercury, Venus, Mars, Jupiter, Saturn.

Hipparchus and Ptolemy made a series of investigations respecting the times of revolution of the planets, and the inequalities of their motions, of which it is worth while to give a brief

42 SYSTEM OF THE WORLD HISTORICALLY DEVELOPED.

summary. The former was no doubt an abler astronomer than Ptolemy; but as he was, so far as we know, the first accurate observer of the celestial motions, he could not make a sufficiently long series of observations to determine all the periods of the planets. Ptolemy had the advantage of being able to combine his own observations with those of Hipparchus, three centuries earlier.

Imperfect though their means of observation were, these observers found that the easterly movements of the planets among the stars were none of them uniform. This held true not only of the sun and moon, but of the fictitious planets already described. Hence they invented the eccentric, and supposed the motions to be really circular and uniform, but in circles not centred in the earth. In Fig. 13, let E be the earth, and C the centre around which the planet really revolves. Then, when the planet is passing the point P, which is nearest the earth, its angular motion would seem more rapid than the average, because in general the angular velocity of a moving body is greater the nearer the observer is to it, while when passing A it will seem to be more slow than the average. The angular velocity being always greatest in one point of the orbit, and least in a point directly opposite, changing regularly from the maximum to the minimum, the general features of the movement are correctly represented by the eccentric. By comparing the angular velocities in different points of the orbit, Hipparchus and Ptolemy were able to determine the supposed distance of the earth from the centre, or rather the proportion of this distance to the distance of the planet. The distance thus determined is double its true amount. The point P is called the Perigee,

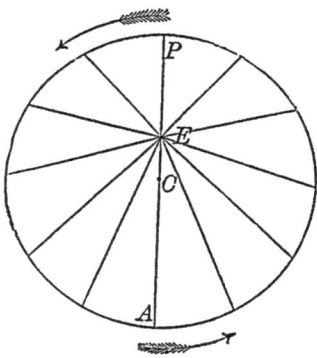

FIG. 13.—The eccentric. Shows how the ancients represented the unequal apparent velocities of the planets when their real motion was supposed uniform, by placing the earth away from the centre of motion, at E.

and *A* the Apogee. The distance *CE* from the earth to the centre of motion is the eccentricity. As there was no way of determining the absolute dimensions of the orbit, it was necessary to take the ratio of *CE* to the radius of the orbit *CP* or *CE* for the eccentricity.*

In determining the motions of the moon, Hipparchus and Ptolemy depended almost entirely on observations of lunar eclipses. The first of these, it is said, was observed at Babylon in the first year of Mardocempad, between the 29th and 30th days of the Egyptian month Thoth. It commenced a little more than an hour after the moon rose, and was total. The date, in our reckoning, was B.C. 720, March 19th. The series of eclipses extended from this date to that of Ptolemy himself, who lived between eight and nine centuries later. If the observations of these eclipses had been a little more precise, they would still be of great value to us in fixing the mean motion of the moon. As it is, we can now calculate the circumstances of an ancient eclipse from our modern tables of the sun and moon almost as accurately as any of the ancient astronomers could observe it.

Notwithstanding the extremely imperfect character of the observations, both Hipparchus and Ptolemy made discoveries respecting the peculiarities of the moon's motions which show a most surprising depth of research. By comparing the intervals between eclipses, they found that her motion was not uniform, but that, like the sun, she moved faster in some parts of her orbit than in others. To account for this, they supposed her orbit eccentric, like that of the sun; that is, the earth, instead of being in the centre of the circular orbit of the moon, was supposed to be displaced by about a tenth part the whole distance of that body. So far the orbit of the moon was like that of the sun and the fictitious planets, except that its eccentricity was greater. But a long series of observations showed

---

* Compared with the modern theory of the elliptic motion, approximately treated, the distance *CE* is double the eccentricity of the ellipse. One-half the apparent inequality is really caused by the orbit being at various distances from the earth or sun, but the other half is real.

that the perigee and apogee did not, as in the case of the sun and planets, remain in the same points of the orbit, but moved forwards at such a rate as to carry them round the heavens in nine years; that is, supposing Fig. 13 to represent the orbit of the moon, the centre of the circle C revolved round the earth in nine years, and the orbit changed its position accordingly.

It was also found by Ptolemy, by measuring the apparent angle between the moon and sun in various points of the orbit of the former, that there was yet another inequality in her motion. This has received the name of the evection. In consequence of this inequality, the moon oscillates more than a degree on each side of her position as calculated from the eccentric, in a period not differing much from her revolution round the earth. To represent this motion, Ptolemy had to introduce a small additional epicycle, as in the case of the planets, only the radius was so small that there was no looping of the orbit. In consequence, his theory of the moon's motion was quite complicated; yet he managed to represent this motion, within the limits of the errors of his observations, by a combination of circular motions, and thus saved the favorite theory of the times, that all the celestial motions were circular and uniform.

§ 8. *The Calendar.*

One of the earliest purposes of the study of the celestial motions was that of finding a convenient measurement of time. This application of astronomy, being of great antiquity, having been transmitted to us without any fundamental alteration, and depending on the apparent motions of the sun and moon, which we have studied in this chapter, is naturally considered in connection with the ancient astronomy.

The astronomical divisions of time are the day, the month, and the year. The week is not such a division, because it does not correspond to any astronomical cycle, although, as we shall presently see, a certain astronomical signification was said to have been given to it by the ancient astrologers. Of these divisions the day is the most well-marked and striking through-

THE CALENDAR. 45

out the habitable portion of the globe. Had a people lived at or near the poles, it would have been less striking than the year. But wherever man existed, there was a regular alternation of day and night, with a corresponding alternation in his physical condition, both occurring with such regularity and uniformity as to furnish in all ages the most definite unit of time. For merely chronological purposes the day would have been the only unit of time theoretically necessary; for if mankind had begun at some early age to number every day by counting from 1 forwards without limit, and had every historical event been recorded in connection with the number of the day on which it happened, there would have been far less uncertainty about dates than now exists. But keeping count of such large numbers as would have accumulated in the lapse of centuries would have been very inconvenient, and a simple count of time by days has never been used for the purposes of civil life through any greater period than a single month.

Next to the day, the most definite and striking division of time is the year. The natural year is that measured by the return of the seasons. All the operations of agriculture are so intimately dependent on this recurrence, that man must have begun to make use of it for measuring time long before he had fully studied the astronomical cause on which it depends. The years in the lifetime of any one generation not being too numerous to be easily reckoned, the year was found to answer every purpose of measuring long intervals of time.

The number of days in the year is, however, too great to be conveniently kept count of; an intermediate measure was therefore necessary. This was suggested by the motion and phases of the moon. The "new moon" being seen to emerge from the sun's rays at intervals of about 30 days, a measure of very convenient length was found, to which a permanent interest was attached by the religious rites connected with the reappearance of the moon.

The week is a division of time entirely disconnected with the month and year, the employment of which dates from the Mosaic dispensation. The old astrologers divided the seven days of the week among the seven planets, not in the order of their distance from the sun, but in one shown by the following figure. If we go round the circle in the direction of the hands of a watch, we shall find the names of the seven planets of the ancient astronomy, in the order of their supposed distances;* while, if we follow the lines drawn in the circle from side to side, we shall have the days of the week in their order.

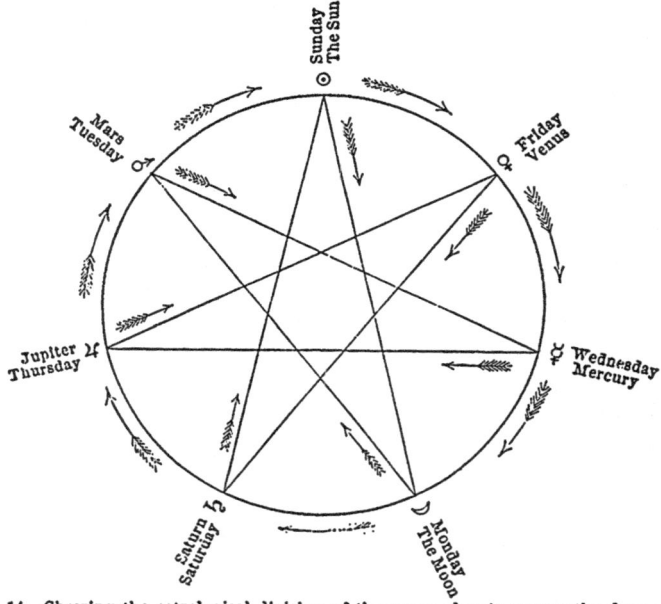

Fig. 14.—Showing the astrological division of the seven planets among the days of the week.

If the lunar month had been an exact number of days, say 30, and the year an exact number of months, as 12, there would have been no difficulty in the use of these cycles for the measurement of time. But the former is several hours less than 30 days, while the latter is nearly 12½ lunar months. In the attempt to combine these measures, the ancient calen-

---

* See pages 40, 41.

THE CALENDAR. 47

dars were thrown into a confusion which made them very perplexing, and which we see to this day in the irregular lengths of our months. To describe all the devices which we know to have been used for remedying these difficulties would be very tedious; we shall therefore confine ourselves to their general nature.

The lunar month, or the mean interval between successive new moons, is very nearly $29\frac{1}{2}$ days. In counting months by the moon, it was therefore common to make their length 29 and 30 days, alternately. But the period of $29\frac{1}{2}$ days is really about three-quarters of an hour too short. In the course of three years the count will therefore be a day in error, and it will be necessary to add a day to one of the months. When lunar months were used, the year, comprising 12 such months, would consist of only 354 days, and would therefore be 11 days too short. Nevertheless, such a year was used both by the Greeks and Romans, and is still used by the Mahometans; the Romans, however, in the calendar of Numa, adding 22 or 23 days to every alternate year by inserting the intercalary month *Mercedonius* between the 23d and 24th of February.

The irregularity and inconvenience of reckoning by lunar months caused them to be very generally abandoned, the only reason for their retention being religious observances due at the time of new moon, which, among the Jews and other ancient nations, were regarded as of the highest importance. Accordingly, we find the Egyptians counting by months of 30 days each, and making every year consist of 12 such months and five additional days, making 365 days in all. As the true length of the year was known to be about six hours greater than this, the equinox would occur six hours later every year, and a month later after the lapse of 120 years. After the lapse of 1460 years, according to the calculations of the time, each season would have made a complete course through the twelve months, and would then have returned once more at the same time of year as in the beginning. This was termed the *Sothic Period;* but the error of each year being estimated a little too great, as we now know, the true length of the period would have been about 1500 years.

The confusion in the Greek year was partly remedied through the discovery by Meton of the cycle which has since borne his name. This cycle consists of 19 solar years, during which the moon changes 235 times. The error of this cycle is very small, as may be seen from the following periods, computed from modern data:

	Days.	Hours.	Min.
235 lunations require in the mean............	6939	16	31
19 true solar years (tropical)...............	6939	14	27
19 Julian years of 365¼ days...............	6939	18	0

Hence, if we take 235 lunar months, and divide them up as nearly evenly as is convenient into 19 years, the mean length of these years will be near enough right for all the purposes of civil reckoning. The years of each cycle were numbered from 1 to 19, and the number of the year was called the Golden Number, from its having been ordered to be inscribed on the monuments in letters of gold.

The Golden Number is still used in our church calendars for finding the date of Easter Sunday. This is the solitary religious festival which, in Christian countries, depends on the motion of the moon. The nominal rule for determining Easter is that it is the Sunday following the first new moon which occurs after the 21st of March. The dates of the new moon correspond to the Metonic Cycle; that is, after the lapse of 19 years they recur on or about the same day of the year. Consequently, if we make a list of the dates on which the Paschal new moon occurs, we shall find no two dates to be the same for nineteen successive years; but the twentieth will occur on the same day with the first, or, at most, only one day different, and then the whole series will be repeated. Consequently, the Golden Number for the year shows, with sufficient exactness for ecclesiastical purposes, on what day, or how many days after the equinox, the Paschal new moon occurs. The church calculations of Easter Sunday are, however, founded upon very old tables of the moon, so that if we fixed it by the actual moon, we should often find the calendar feast a week in error.

The basis of the calendars now employed throughout Christendom was laid by Julius Cæsar. Previous to his time, the Roman calendar was in a state of great confusion, the nominal length of the year depending very largely on the caprice of the ruler for the time being. It was, however, very well known that the real length of the solar year was about $365\frac{1}{4}$ days; and, in order that the calendar year might have the same mean length, it was prescribed that the ordinary year should consist of 365 days, but that one day should be added to every fourth year. The lengths of the months, as we now have them, were finally arranged by the immediate successors of Cæsar.

The Julian calendar continued unaltered for about sixteen centuries; and if the true length of the tropical year had been $365\frac{1}{4}$ days, it would have been in use still. But, as we have seen, this period is about $11\frac{1}{4}$ minutes longer than the solar year, a quantity which, repeated every year, amounts to an entire day in 128 years. Consequently, in the sixteenth century, the equinoxes occurred 11 or 12 days sooner than they should have occurred according to the calendar, or on the 10th instead of the 21st of March. To restore them to their original position in the year, or, more exactly, to their position at the time of the Council of Nice, was the object of the Gregorian reformation of the calendar, so called after Pope Gregory XIII., by whom it was directed. The change consisted of two parts:

1. The 5th of October, 1582, according to the Julian calendar, was called the 15th, the count being thus advanced 10 days, and the equinoxes made once more to occur about March 21st and September 21st.

2. The closing year of each century, 1600, 1700, etc., instead of being each a leap-year, as in the Julian calendar, should be such only when the number of the century was divisible by 4. While 1600, 2000, 2400, etc., were to be leap-years, as before, 1700, 1800, 1900, 2100, etc., were to be reduced to 365 days each.

50 SYSTEM OF THE WORLD HISTORICALLY DEVELOPED.

This change in the calendar was soon adopted by all Catholic countries, and, more slowly, by Protestant ones—England, among the latter, holding out for more than a century, but finally entering into the change in 1752. In Russia it was never adopted at all, the Julian calendar being still continued in that country. Consequently, the Russian reckoning is now 12 days behind ours, the 10 days' difference during the sixteenth and seventeenth centuries being increased by the days dropped from the years 1700 and 1800 in the new reckoning.

The length of the mean Gregorian year is $365^d\ 5^h\ 49^m\ 12^s$; while that of the tropical year, according to the best astronomical determination, is $365^d\ 5^h\ 48^m\ 46^s$. The former is, therefore, still 26 seconds too long, an error which will not amount to an entire day for more than 3000 years. If there were any object in having the calendar and the astronomical years in exact coincidence, the Gregorian year would be accurate enough for all practical purposes during many centuries. In fact, however, it is difficult to show what practical object is to be attained by seeking for any such coincidence. It is important that summer and winter, seed-time and harvest, shall occur at the same time of the year through several successive generations; but it is not of the slightest importance that they should occur at the same time now that they did 5000 years ago, nor would it cause any difficulty to our descendants of 5000 years hence if the equinox should occur in the middle of February, as would be the case should the Julian calendar have been continued.

The change of calendar met with much popular opposition, and it may hereafter be conceded that in this instance the common sense of the people was more nearly right than the wisdom of the learned. An additional complication was introduced into the reckoning of time without any other real object than that of making Easter come at the right time. As the end of the century approaches, the question of making 1900 a leap-year, as usual, will no doubt be discussed, and it is possible that some concerted action may be taken on the part of leading nations looking to a return to the old mode of reckoning.

CHAPTER II.

THE COPERNICAN SYSTEM, OR THE TRUE MOTIONS OF THE HEAVENLY BODIES.

§ 1. *Copernicus.*

IN the first section of the preceding chapter we described the apparent diurnal motion of the heavens, whereby all the heavenly bodies appear to be carried round in circles, thus performing a revolution every day. Any observer of this motion who should suppose the earth to be flat, and the direction we call downward everywhere the same, would necessarily regard it as real. A very little knowledge of geometry would, however, show him that the appearance might be accounted for by supposing the earth to revolve. The seemingly fatal objection against this view would be that, if such were the case, the surface of the earth could not remain level, and every thing would slide away from its position. But it was impossible for men to navigate the ocean without perceiving the rotundity of its surface, and we have no record of a time when it was not known that the earth was round. We have seen that Ptolemy not only was acquainted with the true figure of the earth, but knew that in magnitude it was so much smaller than the celestial spaces, or sphere of the heavens, as to be only a point in comparison. He had, therefore, all the knowledge necessary to enable him to see that the moving body was much more likely to be the earth than to be the sphere of the heavens. Nevertheless, he rejected the theory on obscure physical grounds, as shown in the last chapter, the untenability of which would have been proved him by a few very simple physical experiments. And although it is known that the doctrine of the earth's motion was sustained by others in his age, notably by Timocharis, yet the weight of his authority was so great as not only to override all their arguments, but to carry his views through fourteen centuries of the intellectual history of man.

The history of astronomy during these centuries offers hardly anything of interest to the general reader. There was no telescope to explore the heavens, and no genius arose of sufficient force to unravel the maze of their mechanism. It was mainly through the Arabs that any systematic knowledge of the science was preserved for the use of posterity. The astronomers of this people invented improved methods of observing the positions of the heavenly bodies, and were thus able to make improved tables of their motions. They measured the obliquity of the ecliptic, and calculated eclipses of the sun and moon with greater precision than the ancient Greeks could do. The predictions of the science thus gradually increased in accuracy, but no positive step was taken in the direction of discovering the true nature of the apparent movements of the heavens.

The honor of first proving to the world what the true theory of the celestial motions is belongs almost exclusively to Copernicus. It is true that we have some reason to believe that Pythagoras taught that the sun, and not the earth, was the centre of motion, and that he was, therefore, the first to solve the great problem. But he did not teach this doctrine publicly, and the very vague statements of his private teachings on this point which have been handed down to us are so mixed up with the speculations which the Greek philosophers combined with their views of nature, that it is hard to say with precision whether Pythagoras had or had not fully seized the truth. It is certain that no modern would receive the credit of any discovery without giving more convincing proofs of the correctness of his views than we have any reason to suppose that Pythagoras gave to his disciples.

The great merit of Copernicus, and the basis of his claim to the discovery in question, is that he was not satisfied with a mere statement of his views, but devoted a large part of the labor of a life to their demonstration, and thus placed them in such a light as to render their ultimate acceptance inevitable.

Apart from all questions of the truth or falsity of his theory, the great work in which it was developed, "*De Revolutionibus Orbium Cœlestium*" would deservedly rank as the most important compendium of astronomy which had appeared since Ptolemy. Few books have been more completely the labor of a lifetime than this. Copernicus was born at Thorn, in Prussia, in 1473, twenty years before the discovery of America, but studied at the University of Cracow. He became an ecclesiastical dignitary, holding the rank of canon during a large portion of his life, and finding ample leisure in this position to pursue his favorite studies. He is said to have conceived of the true system of the world as early as 1507. He devoted the years of his middle life to the observations and computations necessary to the perfection of his system, and communicated his views to a few friends, but long refused to publish them, fearing the popular prejudice which might thus be excited. In 1540, a brief statement of them was published by his friend Rheticus; and, as this was favorably received, he soon consented to the publication of his great work. The first printed copy was placed in his hands only a few hours before his death, which occurred in May, 1543.

The fundamental principles of the Copernican system are embodied in two distinct propositions, which have to be proved separately, and one of which might have been true without the other being so. They are as follows:

1. The diurnal revolution of the heavens is only an apparent motion, caused by a diurnal revolution of the earth on an axis passing through its centre.

2. The earth is one of the planets, all of which revolve round the sun as the centre of motion. The true centre of the celestial motions is therefore not the earth, but the sun. For this reason the Copernican system is frequently spoken of in historical discussions as the "heliocentric theory."

The first proposition is the one with the proof of which Copernicus begins. He explains how an apparent motion may result from a real motion of the person seeing, as well as from a motion of the object seen, and thus shows that the diurnal motion may be accounted for just as well by a revolution of the earth as by one of the heavens. To sailors on a ship sailing on a smooth sea, the ship, and every thing in it, seems to be at rest and the shore to be in motion. Which, then, is more likely to be in motion, the earth or the whole universe outside of it? In whatever proportion the heavens are greater than the earth, in the same proportion must their motion be more rapid to carry them round in twenty-four hours. Ptolemy himself shows that the heavens were so immense that the earth was but a point in comparison, and, for any thing that is known, they may extend into infinity. Then we should require an infinite velocity of revolution. Therefore, it is far more likely that it is this comparative point that turns, and that the universe is fixed, than the reverse.

The second principle of the Copernican system—that the apparent annual motion of the sun among the stars, described in § 3 of the preceding chapter, is really due to an annual revolution of the earth around the sun—rests upon a very beautiful result of the laws of relative motion. This movement of the earth explains not only this apparent revolution of the sun, but the apparent epicyclic motion of the planets described in treating of the Ptolemaic system.

In Fig. 15, let S represent the sun, $ABCD$ the orbit of the earth around it, and the figures 1, 2, 3, 4, 5, 6, six successive positions of the earth. These positions would be about two weeks apart. Also, let $EFGH$ represent the apparent sphere of the fixed stars. Then, an observer at 1, viewing the sun in the direction $1S$, will see him as if he were in the celestial sphere at the point $1'$, because, having no conception of the actual distance, the sun will appear to him as if actually among the stars at $1'$ which lie in the same straight line with him. When the earth, with the observer on it, reaches 2, he will see the sun in the direction $2S2'$, that is, as if among the stars in $2'$. That is, during the two weeks' interval, the sun will apparently have moved among the stars by an angle equal to the actual angular motion of the earth around the sun. So, as the earth passes through the successive positions 3, 4, 5, 6, the sun will appear in the positions 3′, 4′, 5′, 6′, and the motion of the earth continuing all the way round its orbit, the sun will appear to move through the entire circle *EFGH*. Thus we have, as a result of the annual motion of the earth around the sun, the annual motion of the sun around the celestial sphere already described in the third section of the preceding chapter.

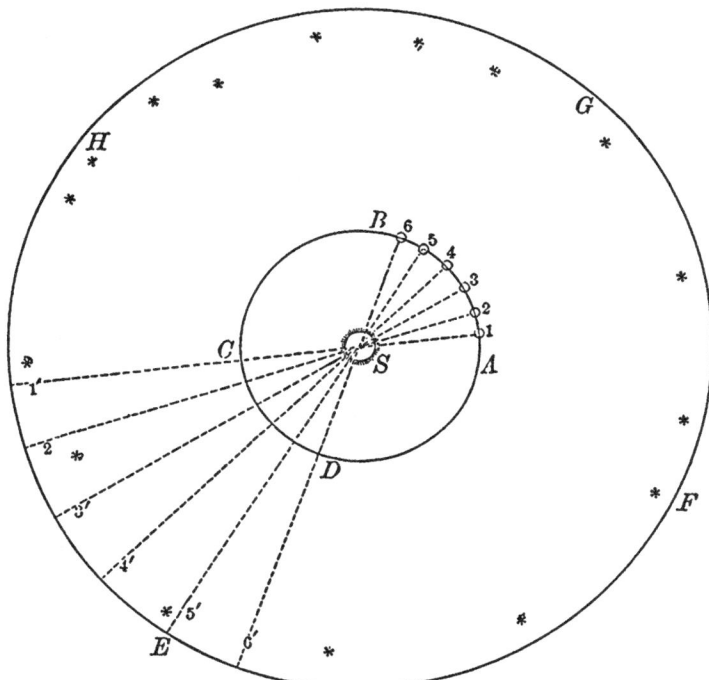

Fig. 15.—Apparent annual motion of the sun explained.

Let us now see how this same motion abolishes the complicated system of epicycles by which the ancient astronomers represented the planetary motions. A theorem on which this explanation rests is this: *If an observer in unconscious motion sees an object at rest, that object will seem to him to be moving in a direction opposite to his own, and with an equal velocity.* A familiar instance of this is the apparent motion

56 SYSTEM OF THE WORLD HISTORICALLY DEVELOPED.

of objects on shore to passengers on a steamer. In Fig. 16, let us suppose an observer on the earth carried around the sun S in the orbit $ABCDEF$, but imagining himself at rest in the centre of motion S. Suppose that he observes the apparent motion of the planet P, which is really at rest. How will the planet appear to move? To show this, we represent apparent directions and motions by dotted lines. Let us begin with the observer at A, from which position he really sees the planet in the direction and distance AP. But, imagining himself at S, he thinks he sees the planet at the point a, the distance and direction of which Sa is the same with AP. As he passes unconsciously from A to B, the planet seems to him to move past from a to b in the opposite direction; and, still thinking himself at rest in S, he sees the planet in b, the line Sb being equal and parallel to BP. As he recedes from the planet through the arc BCD, the

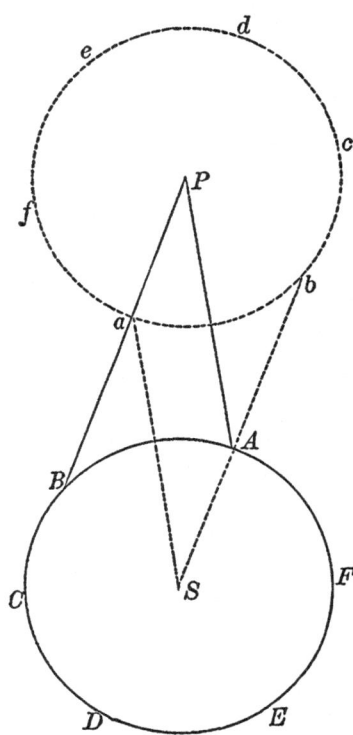

FIG. 16.—Showing how the apparent epicyclic motion of the planets is accounted for by the motion of the earth round the sun.

planet seems to recede from him through bcd. While he moves from left to right through DE, the planet seems to move from right to left through de. Finally, as he approaches the planet through the arc EFA, the planet will seem to approach him through efa, and when he gets back to A he will locate the planet at a, as in the beginning. Thus, in consequence of the motion of the observer around the circle $ABCDEF$, the planet, though really at rest, will seem to him to move through a corresponding circle, *abcdef*. If there are a number of planets, they will all seem to describe corresponding circles of the same magnitude.

If the planet *P*, instead of being at rest, is in motion, the apparent circular motion will be combined with the forward motion of the planet, and the latter will now describe a circle around a centre which is in motion. Thus we have the apparent motion of the planets around a moving centre, as already described in the Ptolemaic system. We have said, in § 7 of the preceding chapter, that by this system the motions of the planets are represented by supposing a fictitious planet to revolve around the heavens with a regular motion, while the real planet revolves around this fictitious one as a centre once a year. Here, *the progressive motion of the fictitious planet is* (*in the case of the outer planets Mars, Jupiter, and Saturn*) *the motion of the real planet around the sun, while the circle which the real planet describes around this moving centre is only an apparent motion due to the observer being carried around the sun on the earth.* If the reader will compare the epicyclic motion of Ptolemy, represented in Figs. 10 and 11 with the motion explained in Fig. 16, he will find that they correspond in every particular. In the case of the inner planets, Mercury and Venus, which never recede far from the sun, the epicyclic motion by which they seem to vibrate from one side of the sun to the other is due to their orbital motion around the sun, while the progressive motion with which they follow the sun is due to the revolution of the earth around the sun.

We may now see clearly how the retrograde motion and stationary phases of the planets are explained on the Copernican system. The earth and all the planets are really moving round the sun in a direction which we call east on the celestial sphere. When the earth and an outer planet are on the same side of the sun, they are moving in the same direction; but the earth is moving faster than the planet. Hence, to an observer on the earth, the planet seems to be moving west, though its real motion is east. As the earth passes to the opposite side of the sun from the planet, it changes its motion to a direction the opposite of that of the planet, and thus the westerly motion of the latter appears to be increased by the whole motion of the earth.* Between these two motions there is a point at which the planet does not seem to move at all. This is called the stationary point. If the planet we consider is not an outer, but an inner one, Mercury or Venus, and we view it when between us and the sun, its motion to us is reversed, because we see it from the side opposite the sun. Hence it seems to move west to us, and it is retrograde. The earth is indeed moving in the same real direction; but since the planet moves faster than the earth, its retrograde motion seems to predominate. As the planet passes round in its orbit, it first appears stationary, and then, passing to the opposite side of the sun, it seems direct.

Let us now dwell for a moment on some considerations which will enable us to do justice to the Ptolemaic system, as it is called, by seeing how necessary a step it was in the evolution of the true theory of the universe. The great merit of that system consisted in the analysis of the seemingly complicated motions of the planets into a combination of two circular motions, the one that of a fictitious planet around the celestial sphere, the other that of the real planet around the fictitious one. Without that separation, the constant oscillations of the planets back and forth could not have suggested any idea whatever, except that of a motion too complicated to be explained on mechanical principles. But when, leaving out of sight the regular forward motion of the mean or fictitious planet, the attention was directed to the epicyclic motion alone, one could not fail to see the remarkable correspondence between this latter motion and the apparent annual motion of the sun. Seeing this, it took a very small step to see that

---

* It must not be forgotten that the direction *east* in the heavens is a curved direction, as it were, and is opposite on opposite sides of the sun or celestial sphere. For instance, the motions of the stars as they rise and as they set are opposite, but both are considered west.

the sun, and not the earth, was the centre of planetary motion. Then nothing but the illusions of sense remained to prevent the acceptance of the theory that the earth was itself a planet moving round the sun, and that both the annual motion of the sun and the epicyclic motion of the planets were not real, but apparent motions, due to the motion of the earth itself; and in no other way than this could the heliocentric theory have been developed.

The Copernican system affords the means of determining the proportions of the solar system, or the relative distances of the several planets, with great accuracy. That is, if we take as our measuring-rod the distance of the earth from the sun, we can determine how many lengths of this rod, or what fractional parts of its length, will give the distance of each planet, although the length of the rod itself may remain unknown. This determination rests on the principle that the apparent circle or epicycle described by the planet in Fig. 16 is of the same magnitude with the actual orbit described by the earth around the sun. Hence, the nearer the observer is to this circle, the larger it will appear. The apparent epicycle described by Neptune is rather less than two degrees in radius; that is, the true planet Neptune is seen to swing a little less than two degrees on each side of its mean position in consequence of the annual motion of the earth round the sun. This shows that the orbit of the earth, as seen from Neptune, subtends an angle of only two degrees. On the other hand, the planet Mars generally swings more than 40° on each side; sometimes, indeed, more than 45°. From this a trigonometrical calculation shows that its mean distance is only about half as much again as that of the earth; and the fact that the apparent swing is variable shows the distance to be different at different times.

As it will be of interest to see how nearly Copernicus was able to determine the distances of the planets, we present his results in the following table, together with what we now know to be the true numbers. The numbers given are decimal fractions, expressing the least and greatest distance of each planet from the sun, the distance of the earth being taken as unity.*

Planets.	Least Distance.		Greatest Distance.	
	Copernicus.	Modern.	Copernicus.	Modern.
Mercury...........	0.326	0.308	0.405	0.467
Venus..............	0.709	0.718	0.730	0.728
Mars................	1.373	1.382	1.666	1.666
Jupiter............	5.453	5.454	4.980	4.952
Saturn.............	9.76	10.07	8.66	9.00

Considering the extremely imperfect means of observation which the times afforded, these results of Copernicus come very near the truth. The greatest proportional deviation is in the case of Mercury, the most difficult of all the planets to observe, even to the present day. It is said that Copernicus died without ever seeing this planet.

The eccentricities of the orbits were represented by Copernicus in a way which agrees exactly with the modern formulæ when only a rough approximation is sought for. Like Ptolemy, he supposed the orbits of the planets not to be centred on the sun, but to be displaced by a small quantity termed the *eccentricity*. But it had long been known that the theory of uniform motion in an eccentric circle, though it might make the irregularities in the planet's angular motion come out all right, would make the changes of distance double their true value. He therefore took for the eccentricity a mean between that which would satisfy the motion in longitude, and that which would give the changes of distance, and added a small epicycle of one-third this eccentricity; and, by supposing the planet to make two revolutions in this epicycle for every revolution around the sun, he represented both irregularities.†

---

* I have deduced these numbers from the tables given in Book V. of "De Revolutionibus Orbium Cœlestium." They are probably the most accurate that Copernicus was able to obtain.

† The mathematical form of this theory of Copernicus is as follows: Putting

The work of Copernicus was the greatest step ever taken in astronomy. But he still took little more than the single step of showing what apparent motions in the heavens were real, and what were due to the motion of the observer. Not only was his work in other respects founded on that of Ptolemy, but he had many of the notions of the ancient philosophy respecting the fitness of things. Like Ptolemy, he thought the heavens as well as the earth to be spherical, and all the celestial motions to be circular, or composed of circles. He argues against Ptolemy's objections to the theory of the earth's motion, that that philosopher treats of it as if it were an enforced or violent motion, entirely forgetting that if it exists it must be a natural motion, the laws of which are altogether different from those of violent motion. Thus, part of his argument was really without scientific foundation, though his conclusion was correct. Still, Copernicus did about all that could have been done under the circumstances. His hypothesis of a small epicycle one-third the eccentricity represented the motions of the planets around the sun with all the exactness that observation then admitted of, while, in the absence of any knowledge of the laws of motion, it was impossible to frame any dynamical basis for the motions of the planets.

§ 2. *Obliquity of the Ecliptic; Seasons, etc.; on the Copernican System.*

We have next to explain the relations of the ecliptic and equator on the new system. Since, on this system, the celestial sphere does not revolve at all, what is the significance of the pole and axis around which it seems to revolve? The

---

e for his eccentricity, and g for the mean anomaly of the planet, he represented its rectangular coördinates in the form

$$x = a\,(\cos. g - e + \tfrac{1}{2}e \cos. 2g),$$
$$y = a\,(\sin. g + \tfrac{1}{2}e \sin. 2g);$$

while the approximate modern formulæ of the elliptic motion are—

$$x = a\,(\cos. g - \tfrac{3}{2}e + \tfrac{1}{2}e \cos. 2g),$$
$$y = a\,(\sin. g + \tfrac{1}{2}e \sin. 2g),$$

which agree exactly when we put $e = \tfrac{3}{2}e$.

answer is, that the celestial poles are the points among the stars towards which the axis of the earth is directed. Here the stars are supposed to be infinitely distant, and the axis of the earth to be continued in an infinite straight line to meet them. Since this point appears to the unassisted sight to be the same during the entire year, it follows that as the earth moves round the sun, its axis keeps pointing in the same absolute direction, as will be shown in Fig. 18. But in the preceding chapter we showed that there is a slow but constant change in the position of the pole among the stars, called precession, which the ancient astronomers discovered by studying observations extend-

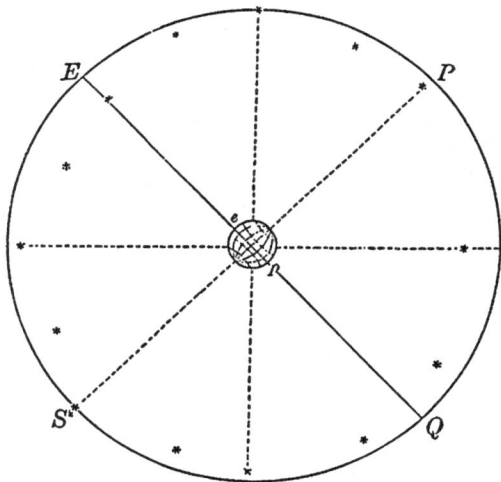

FIG. 17.—Relation of the terrestrial and celestial poles and equators.

ing through several centuries, and this shows that on the Copernican system the direction of the earth's axis is slowly changing.

To conceive of the celestial equator on the Copernican system, we must imagine the globular earth to be divided into two hemispheres by a plane intersecting the earth around its equator, and continued out on all sides till it reaches the celestial sphere. This may, perhaps, be better understood by referring to Fig. 17, representing the earth in the centre of the

OBLIQUITY OF THE ECLIPTIC.

imaginary celestial sphere. The dotted lines passing from the poles of the earth to the points P and S mark the poles of that sphere. It is evident that as the earth turns on this axis, the celestial sphere, no matter how great it may seem to be, will appear to turn on the same axis in the opposite direction. Again, ep being the earth's equator, dividing it into two equal parts, we have only to imagine it to be extended to E and Q, all round the celestial sphere, to cut the latter into two equal parts.

Let us next examine more closely the relation of the earth to the sun. We have already shown that as the earth moves around the sun, the latter seems to move around the celestial sphere, and the circle in which he seems to move is called the ecliptic. But the ecliptic and the celestial equator are inclined to each other by an angle of about $23\frac{1}{2}°$. This shows that the axis of the earth is not perpendicular to its orbit, but

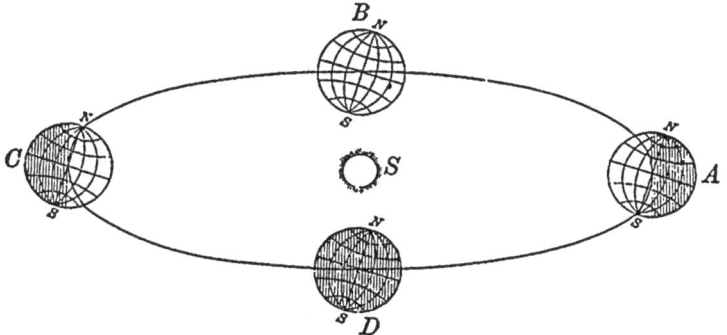

FIG. 18.—Causes of changes of seasons on the Copernican system.

is inclined $23\frac{1}{2}°$ to that perpendicular, as shown in Fig. 18, which represents the annual course of the earth round the sun. It is of necessity drawn on a very incongruous scale, because the distance of the sun from the earth being nearly 12,000 diameters of the latter and 110 that of the sun, both bodies would be almost invisible if they were not greatly magnified in the figure. A difficulty which may suggest itself is, that the present figure represents the earth as moving away from its position in the centre of the sphere. There are two ways of avoiding this difficulty. One is to suppose that the observer carries the imaginary celestial sphere with him as he is carried around the sun; the other is to consider the sphere as nearly infinite in diameter. The latter is probably the easiest mode of conception for the general reader. He must, therefore, in the last figure suppose the sphere to extend out to the fixed stars, which are so distant that the whole orbit of the earth is but a point in comparison; and the different points of the sphere towards which the poles and the equator of the earth point, as the latter moves round the sun, are so far as to appear always the same. It now requires but an elementary idea of the geometry of the sphere to see that these two great circles of the celestial sphere—the ecliptic, around which the sun seems to move, and the equator, which is everywhere equally distant from the points in which the earth's axis intersects the sphere—will appear inclined to each other by the same angle by which the earth's axis deviates from the perpendicular to the ecliptic.

Next, we have to see how the changes of the seasons, the equinoxes, etc., are explained on the Copernican theory. In the last figure the earth is represented in four different positions of its annual orbit around the sun. In the position A, the south pole is inclined $23\frac{1}{2}°$ towards the sun, while the north pole, and the whole region within the arctic circle, is enveloped in darkness. Hence, in this position, the sun neither rises to the inhabitants of the arctic zone, nor sets to those of the antarctic zone. Outside of these zones, he rises and sets, and the relative lengths of day and night at any place can be estimated by studying the circles around which that place is carried by the diurnal turning of the earth on its axis. To facilitate this, we present on the following page a magnified picture of the earth at A, showing more fully the hemisphere in which it is day and that in which it is night. The seven nearly horizontal lines on the globe are examples of the circles in question. We see that a point on the arctic circle just grazes the dividing-line between light and darkness once in its revolution, or once a day; that is, the sun just shows himself in the horizon once a day. Of the next circle towards the south about two-thirds is in the dark, and one-third in the light hemisphere. This shows that the days are about twice as long as the nights. This circle is near that around which London is carried by the diurnal revolution of the earth on its axis. As we go south, we see that the proportion of light on the diurnal circles constantly increases, while that of darkness diminishes, until we reach the equator, where they are equal.

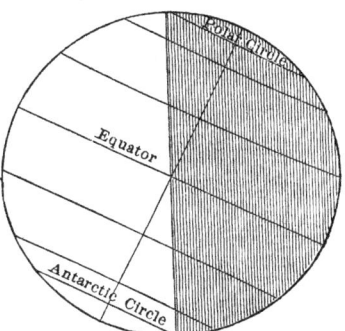

Fig. 19.—Enlarged view of the earth in the position A of the preceding figure, showing winter in the northern hemisphere, and summer in the southern.

When we pass into the southern hemisphere, we see the light covering more than half of each circle, the proportion of light to darkness constantly increasing, at the same rate that the opposite proportion would increase in going to the north. When we reach the antarctic circle, the whole circle is in the light hemisphere, the observer just grazing the dividing-line at midnight. Inside of that circle the observer is in sunlight all the time, so that the sun does not set at all. We see, then, that at the equator the days and nights are always of the same length, and that the inequality increases as we approach either pole.

We now go on three months to the position *B*, which the earth occupies in March. Here the plane of the terrestrial equator being continued, passes directly through the sun; the latter, therefore, seems to be in the celestial equator. All the diurnal circles are here one-half in the illuminated, and one-half in the unilluminated hemisphere, the latter being invisible in the figure, through its being behind the earth. The days and nights are, therefore, of equal length all over the globe, if we call it night whenever the sun is geometrically below the horizon. In the position *C*, which the earth takes in June, everything is the same as in position *A*, except that effects are reversed in the two hemispheres. The northern hemisphere now has the longest days, and the southern one the longest nights. At *D*, which the earth reaches in September, the days and nights are equal once more, for the same reason as in *B*. Thus, all the seemingly complicated phenomena which we have described in the preceding chapter are completely explained in the simplest way on the new system. We have next to see how the details of the system were filled in by the immediate successors of Copernicus.

§ 3. *Tycho Brahe.*

We have said that no great advance could be made upon the Copernican system, without either a better knowledge of the laws of motion or more exact observations of the positions of the heavenly bodies. It was in the latter direction that the advance was first made. The leader was Tycho Brahe, who was born in 1546, three years after the death of Copernicus. His attention was first directed to the study of astronomy by an eclipse of the sun on August 21st, 1560, which was total in some parts of Europe. Astonished that such a phenomenon could be predicted, he devoted himself to a study of the methods of observation and calculation by which the prediction was made. In 1576 the King of Denmark founded the celebrated Observatory of Uraniberg, at which Tycho spent twenty years, assiduously engaged in observations of the positions of the heavenly bodies with the best instruments that could then be made. This was just before the invention of the telescope, so that the astronomer could not avail himself of that powerful instrument. Consequently, his observations were superseded by the improved ones of the centuries following, and their celebrity and importance are principally due to their having afforded Kepler the means of discovering his celebrated laws of planetary motion.

As a theoretical astronomer, Tycho was unfortunate. He rejected the Copernican system, for a reason which, in his day, had some force, namely, the incredible distance at which it was necessary to suppose the fixed stars to be situated if that system were accepted. We have shown how, on the Copernican system, the outer planets seem to describe an annual revolution in an epicycle, in consequence of the annual revolution of the earth around the sun. The fixed stars, which are situated outside the solar system, must appear to move in the same way, if the system be correct. But no observations, whether of Tycho or his predecessors, had shown any such motion. To this the friends of Copernicus could only reply that the distance of the fixed stars must be so great that the motion could not be seen. Since a vibration of three or four minutes of arc might have been detected by Tycho, it would be necessary to suppose the stellar sphere at least a thousand times the distance of the sun, and a hundred times that of Saturn, then the outermost known planet. That a space so vast should intervene between the orbit of Saturn and the fixed stars seemed entirely incredible: to the philosophers of the day it was an axiom that nature would not permit the waste of space here implied. At the same time, the proofs given by Copernicus that the sun was the centre of the planetary motions were too strong to be overthrown. Tycho, therefore, adopted a system which was a compound of the Ptolemaic and the Copernican; he supposed the five planets to move around the sun as the centre of their motions, while the sun was itself in motion, describing an annual orbit around the earth, which remained at rest in the centre of the universe.

Perhaps it is fortunate for the reception of the Copernican system that the astronomical instruments of Tycho were not equal to those of the beginning of the present century. Had he found that there was no annual parallax among the stars amounting to a second of arc, and therefore that, if Copernicus was right, the stars must be at least 200,000 times the distance of the sun, the astronomical world might have stood aghast at the idea, and concluded that, after all, Ptolemy must be right, and Copernicus wrong.

Tycho never elaborated his system, and it is hard to say how he would have answered the numerous objections to it.

He never had any disciples of eminence, except among the ecclesiastics; in fact, the invention of the telescope did away with the last remaining doubts of the correctness of the Copernican system before a new one would have had time to gain a foothold.

§ 4. *Kepler.—His Laws of Planetary Motion.*

Kepler was born in 1571, in Würtemberg. He was for a while the assistant of Tycho Brahe in his calculations, but was too clear-sighted to adopt the curious system of his master. Seeing the truth of the Copernican system, he set himself to determine the true laws of the motion of the planets around the sun. We have seen that even Copernicus had adopted the ancient theory, that all the celestial motions are compounded of uniform circular motions, and had thus been obliged to introduce a small epicycle to account for the irregularities of the motion. The observations of Tycho were so much more accurate than those of his predecessors, that they showed Kepler the insufficiency of this theory to represent the true motions of the planets around the sun. The planet most favorable for this investigation was Mars, being at the same time one of the nearest to the earth, and one of which the orbit was most eccentric. The only way in which Kepler could proceed in his investigation was to make various hypotheses respecting the orbit in which the planet moved, and its velocity in various points of its orbit, and from these hypotheses to calculate the positions and motions of the planet as seen from the earth, and then compare with observations, to see whether the observed and calculated positions agreed. As our modern tables of logarithms by which such calculations are immensely abridged were not then in existence, each trial of an hypothesis cost Kepler an immense amount of labor. Finding that the form of the orbit was certainly not circular, but elliptical, he was led to try the effect of placing the sun in the focus of the ellipse. Then, the motion of the planet would be satisfied if its velocity were made variable, being greater the nearer it was to the sun. Thus he was at length led to the first two of his three celebrated laws of planetary motion, which are as follows:

1. *The orbit of each planet is an ellipse, having the sun in one focus.*

2. *As the planet moves round the sun, its radius-vector (or the line joining it to the sun) passes over equal areas in equal times.*

To explain these laws, let PA (Fig. 20) be the ellipse in which the planet moves. Then the sun will not be in the cen-

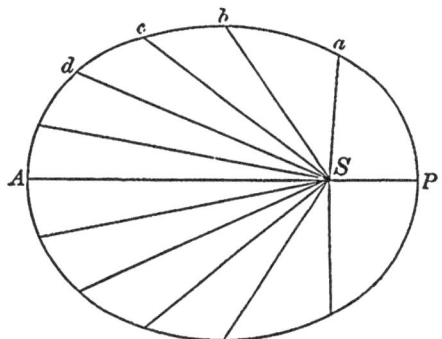

FIG. 20.—Illustrating Kepler's first two laws of planetary motion.

tre of the ellipse, but in one focus, say at S, the other focus being empty. When the planet is at P, it is at the point nearest the sun; this point is therefore called the *perihelion*. As it passes round to the other side of the sun, it continues to recede from him till it reaches the point A, when it attains its greatest distance. This point is the *aphelion*. Then it begins to approach the sun again, and continues to do so till it reaches P once more, when it again begins to repeat the same orbit. It thus describes the same ellipse over and over.

Now, suppose that, starting from P, we mark the position of the planet in its orbit at the end of any equal intervals of time, say 30 days, 60 days, 90 days, 120 days, and so on. Let a, b, c, d be the first four of these positions between each of which the planet has required 30 days to move. Draw lines from each of the five positions of the planet, beginning at P, to the sun at *S*. We shall thus have four triangular spaces, over each of which the radius-vector of the planet has swept in 30 days. The first of Kepler's laws means that the areas of all of these spaces will be equal.

The old theory that the motions of the heavenly bodies must be circular and uniform, or, at least, composed of circular and uniform motions, was thus done away with forever. The ellipse took the place of the circle, and a variable motion the place of a uniform one.

Another law of planetary motion, not less important than these two, was afterwards discovered by Kepler. Copernicus knew, what had been surmised by the ancient astronomers, that the more distant the planet, the longer it took it to perform its course around the sun, and this not merely because it had farther to go, but because its motion was really slower. For instance, Saturn is about 9½ times as far as the earth, and if it moved as fast as the earth, it would perform its revolution in 9½ years; but it actually requires between 29 and 30 years. It does not, therefore, move one-third so fast as the earth, although it has nine times as far to go. Copernicus, however, never detected any relation between the distances and the periods of revolution. Kepler found it to be as follows:

Third law of planetary motion. The square of the time of revolution of each planet is proportional to the cube of its mean distance from the sun.

This law is shown in the following table, which gives (1) the mean distance of each planet known to Kepler, expressed in astronomical units, each unit being the mean distance of

Planets.	(1) Distance.	(2) Cube of Distance.	(3) Period (Years).	(4) Square of Period.
Mercury	0.387	0.058	0.241	0.058
Venus	0.723	0.378	0.615	0.378
Earth	1.000	1.000	1.000	1.001
Mars	1.524	3.540	1.881	3.538
Jupiter	5.203	140.8	11.86	140.66
Saturn	9.539	868.0	29.46	867.9

the earth from the sun; (2) the cube of this quantity; (3) the time of revolution in years; and (4) the square of this time.

The remarkable agreement between the second and fourth columns will be noticed.

§ 5. *From Kepler to Newton.*

So far as the determination of the laws of planetary motion from observation was concerned, we might almost say that Kepler left nothing to be done. Given the position and magnitude of the elliptic orbit in which any planet moved, and the point of the orbit in which it was found at any date, and it became possible to calculate the position of the planet in all future time. More than that science could not do. It is true that the places of the planet thus predicted were not found to agree exactly with observation; and had Kepler had at his command observations as accurate as those of the present day, he would have found that his laws could not be made to perfectly represent the motion of the planets. Not only would the elliptic orbit have been found to vary its position from century to century, but the planets would have been found to deviate from it, first in one direction and then in the other, while the areas described by the radius-vector would have been sometimes larger and sometimes smaller. Why should a planet move in an elliptic orbit? Why should its radius-vector describe areas proportional to the time? Why should there be that exact relation between their distances and times of revolutions? Until these questions were answered, it would have been impossible to say why the planets deviated from Kepler's laws; and they were questions which it was impossible to answer until the general laws of motion, unknown in Kepler's time, were fully understood.

The first important step in the discovery of these laws was taken by Galileo, the great contemporary of Kepler, one of the inventors of the telescope, and the first who ever pointed that instrument at the heavens. From a scientific point of view, as inventor of the telescope, founder of the science of dynamics, teacher and upholder of the Copernican system, and

72   SYSTEM OF THE WORLD HISTORICALLY DEVELOPED.

sufferer at the hands of the Inquisition, for promulgating what he knew to be the truth, Galileo is perhaps the most interesting character of his time. If any serious doubt could remain of the correctness of the Copernican system, it was removed by the discoveries made by the telescope. The phases of Venus showed that she was a dark globular body, like the earth, and that she really revolved around the sun. In Jupiter and his satellites, the solar system, as described by Copernicus, was repeated on a small scale with a fidelity which could not fail to strike the thinking observer. There was no longer any opposition to the new doctrines from any source entitled to respect. The Inquisition forbade their promulgation as absolute truths, but were perfectly willing that they should be used as *hypotheses*, and rather encouraged men of science in the idea of investigating the interesting mathematical problems to which the explanation of the celestial motions by the Copernican system might give rise. The only restriction was that they must stop short of asserting or arguing the hypotheses to be a reality. As this assertion was implicitly contained in several places in the great work of Copernicus, they condemned this work in its original form, and ordered its revision.* Probably the decree of the Inquisition was entirely without effect in stopping the reception of the Copernican system outside of Italy and Spain.

It will be seen, from what has been said, that the next step to be taken in the direction of explaining the celestial motions must be the discovery of some general cause of those motions, or, at least, their reduction to some general law. The first attempt to do this was made by Descartes in his celebrated theory of vortices, which for some time disputed the field with Newton's theory of gravitation. This philosopher supposed the sun to be immersed in a vast mass of fluid, extending indefinitely in every direction. The sun, by its rotation, set the

---

* The order for this revision was made at the time of condemning Galileo's work, but I am not aware that it was ever executed. An edition of Copernicus, revised to satisfy the Inquisition, would certainly be an interesting work to the astronomical bibliopole at the present time.

parts of the fluid next to it in rotation; these communicated their motions to the parts still farther out, and so on, until the whole mass was set in rotation like a whirlpool. The planets were carried around in this ethereal whirlpool. The more distant planets moved more slowly because the ether was less affected by the rotation of the sun the more distant it was from him. In the great vortex of the solar system were smaller ones, each planet being the centre of one; and thus the satellites, floating in the ether, were carried round their primaries. Had Descartes been able to show that the parts of his vortex must move in ellipses having the sun in one focus, that they must describe equal areas in equal times, and that the velocity must diminish as we recede from the sun, according to Kepler's third law, his theory would so far have been satisfactory. Failing in this, it cannot be regarded as an advance in science, but rather as a step backwards. Yet, the great eminence of the philosopher and the number of his disciples secured a wide currency for his theory, and we find it supported by no less an authority than John Bernoulli.

After Galileo, the man who, perhaps, did most to prepare the way for gravitation was Huyghens. As a mathematician, a mechanician, and an observer, he stood in the first rank. He discovered the laws of centrifugal force, and if he had simply applied these laws to the solar system, he would have been led to the result that the planets are held in their orbits by a force varying as the inverse square of their distance from the sun. Having found this, the road to the theory of gravitation could hardly have been missed. But the great discovery seemed to require a mind freshly formed for the occasion.

CHAPTER III.

UNIVERSAL GRAVITATION.

§ 1. *Newton.—Discovery of Gravitation.*

THE real significance of Newton's great discovery of universal gravitation is fully appreciated by but few. Gravitation is generally thought of as a mysterious force, acting only between the heavenly bodies, and first discovered by Newton. Had gravitation itself been discovered by Newton as some new principle to account for the motions of the planets, it would not have been so admirable a discovery as that which he actually made. Gravitation, in a somewhat limited sphere, is known to all men. It is simply the force which causes all heavy bodies to fall, or to tend towards the centre of the earth. Every one who had ever seen a stone fall, or felt it to be heavy, knew of the existence of gravitation. What Newton did was to show that the motions of the planets were determined by a universal force, of which the force which caused the apple to fall was one of the manifestations, and thus to deprive the celestial motions of all the mystery in which they had formerly been enshrouded. To his predecessors, the continuous motion of the planets in circles or ellipses was something so completely unlike any motion seen on the surface of the earth, that they could not imagine it to be governed by the same laws; and, knowing of no law to limit the planetary motions, the idea of the heavenly bodies moving in a manner which set all the laws of terrestrial motion at defiance was to them in no way incredible.

The idea of a cosmical force emanating from the sun or the earth, and causing the celestial motions, did not originate with Newton. We have seen that even Ptolemy had an idea of a force which, always directed towards the centre of the earth, or, which was to him the same thing, towards the centre of the universe, not only caused heavy bodies to fall, but bound the whole universe together. Kepler also maintained that the force which moved the planets resided in, and emanated from, the sun. But neither Ptolemy nor Kepler could give any adequate explanation of the force on the basis of laws seen in action around us; nor was it possible to form any conception of its true nature without a knowledge of the general laws of motion and force, to which neither of these philosophers ever attained.

The great misapprehension which possessed the minds of nearly all mankind till the time of Galileo was, that the continuous action of some force was necessary to keep a moving body in motion. That Kepler himself was fully possessed of this notion is shown by the fact that he conceived a force acting only in the direction of the sun to be insufficient for keeping up the planetary motions, and to require to be supplemented by some force which should constantly push the planet ahead. The latter force, he conceived, might arise from the rotation of the sun on his axis. It is hard to say who was the first clearly to see and announce that this notion was entirely incorrect, and that a body once set in motion, and acted on by no force, would move forwards forever—so gradually did the great truth dawn on the minds of men. It must have been obvious to Leonardo da Vinci; it was implicitly contained in Galileo's law of falling bodies, and in Huyghens's theory of central forces; yet neither of these philosophers seems to have clearly and completely expressed it. We can hardly be far wrong in saying that Newton was the first who clearly laid down this law in connection with the correlated laws which cluster around it. The basis of Newton's discovery were these three laws of motion:

First law. *A body once set in motion and acted on by no force will move forwards in a straight line and with a uniform velocity forever.*

Second law. *If a moving body be acted on by any force, its deviation from the motion defined in the first law will be in the direction of the force, and proportional to it.*

Third law. *Action and reaction are equal, and in opposite directions; that is, whenever any one body exerts a force on a second one, the latter exerts a similar force on the first, only in the opposite direction.*

The first of these laws is the fundamental one. The circumstance which impeded its discovery, and set man astray for many centuries, was that there was no body on the earth's surface acted on by no force, and therefore no example of a body moving in a continuous straight line. Every body on which an experiment could be made was at least acted on by the gravitation of the earth—that is, by its own weight—and, in consequence, soon fell to the earth. Other forces which impeded its motion were friction and the resistance of the air. It needed research of a different kind from what the predecessors of Galileo had given to physical problems to show that, but for these forces, the body would move in a straight line without hinderance.

We are now prepared to understand the very straightforward and simple way in which Newton ascended from what he saw on the earth to the great principle with which his name is associated. We see that there is a force acting all over the earth by which all bodies are drawn towards the earth's centre. This force extends without sensible diminution, not only to the tops of the highest buildings, but of the highest mountains. How much higher does it extend? Why should it not extend to the moon? If it does, the moon would tend to drop to the earth, just as a stone thrown from the hand does. Such being the case, why should not this simple force of gravity be the force which keeps the moon in her orbit, and prevents her from flying off in a straight line under the first law of motion? To answer this question, it was necessary to calculate what force was requisite to retain the moon in her orbit, and to compare it with gravity. It was at that time well known to astronomers that the distance of the moon was sixty semidiameters of the earth. Newton at first supposed the earth to be less than 7000 miles in diameter, and consequently his calculations failed to lead him to the right result. This was in 1665, when he was only twenty-three years of age. He laid aside his calculations for nearly twenty years, when, learning that the measures of Picard, in France, showed the earth to be one-sixth larger than he had supposed, he again took up the subject. He now found that the deflection of the orbit of the moon from a straight line was such as to amount to a fall of sixteen feet in one minute, the same distance which a body falls at the surface of the earth in one second. The distance fallen being as the square of the time, it followed that the force of gravity at the surface of the earth was 3600 times as great as the force which held the moon in her orbit. This number was the square of 60, which expresses the number of times the moon is more distant than we are from the centre of the earth. Hence, *the force which holds the moon in her orbit is the same as that which makes a stone fall, only diminished in the inverse square of the distance from the centre of the earth.*

To the mathematician the passage from the gravitation of an apple to that of the moon is quite simple; but the non-mathematical reader may not, at first sight, see how the moon can be constantly falling towards the earth without ever becoming any nearer. The following illustration will make the matter clear: any one can understand the law of falling bodies, by which a body falls sixteen feet the first second, three times that distance the next, five times the third, and so on. If, in place of falling, the body be projected horizontally, like a cannon-ball, for example, it will fall sixteen feet out of the straight line in which it is projected during the first second, three times that distance the next, and so on, the same as if dropped from a state of rest. In the annexed figure, let AB represent a portion of the curved surface of the earth, and AD a straight line horizontal at A, or the line along which an observer at A would sight if he set a small telescope in a horizontal position. Then, owing to the curvature of the earth, the surface will fall away from this line of sight at the rate of about eight inches in the first mile, twenty-four inches more in the second mile, and so on. In five miles the fall will amount to sixteen feet. In ten miles, in addition to this sixteen feet, three times that amount will be added, and so on, the law being the same

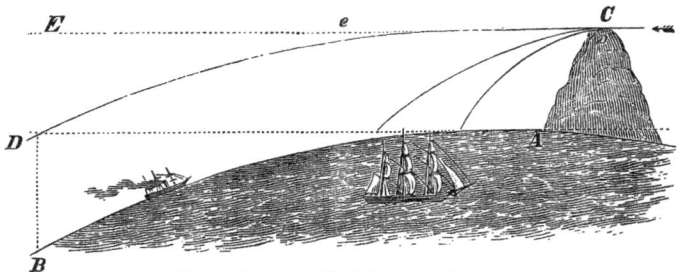

Fig. 21.—Illustrating the fall of the moon towards the earth.

with that of a falling body. Now, let AC be a high steep mountain, from the summit of which a cannon-ball is fired in the horizontal direction CE. The greater the velocity with which the shot is fired, the farther it will go before it reaches the ground. Suppose, at length, that we should fire it with a velocity of five miles a second, and that it should meet with no resistance from the air. Suppose e to be the point on the line five miles from C. Since it would reach this point in one second, it follows, from the law of falling bodies just cited, that it will have dropped sixteen feet below e. But we have just seen that the earth itself curves away sixteen feet at this distance. Hence, the shot is no nearer the earth than when it was fired. During the next second, while the ball would go to E, it would fall forty-eight feet more, or sixty-four feet in all. But here, again, the earth has still been rounding off, so the distance DB is sixty-four feet. Hence, the ball is still no nearer the earth than when it was fired, although it has been dropping away from the line in which it was fired exactly like a falling body. Moreover, meeting with no resistance, it is still going on with undiminished velocity; and, just as it has been falling for two seconds without getting any nearer the earth, so it can get no nearer in the third second, nor in the fourth, nor in any subsequent second; but the earth will constantly curve away as fast as the ball can drop. Thus the latter will pass clear round the earth, and come back to the first point C, from which it started, in the direction of the arrow, without any loss of velocity. The time of revolution will be about an hour and twenty-four minutes, and the ball will thus keep on revolving round the earth in this space of time. In other words, the ball will be a satellite of the earth, just like the moon, only much nearer, and revolving much faster.

Our next step is to extend gravitation to other bodies than the earth. The planets move around the sun as the moon does around the earth, and must, therefore, be acted on by a force directed towards the sun. This force can be no other than the gravitation of the sun itself. A very simple calculation from Kepler's third law shows that the force with which each planet thus gravitates towards the sun is inversely as the square of the mean distance of the planet.

Only one more step is necessary. What sort of an orbit will a planet describe if acted on by a force directed towards the sun, and inversely as the square of the distance? A very simple demonstration will show that, no matter what the law of force, if it be constantly directed towards the sun, the radius-vector of the planet will sweep over equal areas in equal times. And, conversely, it cannot sweep over equal areas in equal times if the force acts in any other direction than that of the sun. Hence it follows, from Kepler's second law, that the force is directed towards the sun itself.

The problem of determining what form of orbit would be described was one with which very few mathematicians of that day were able to grapple. Newton succeeded in proving, by a rigorous demonstration, that the orbit would be an ellipse, a parabola, or a hyperbola, according to circumstances, having the sun in one of its foci, which, in the case of the ellipse, was Kepler's first law. Thus, all mystery disappeared from the celestial motions, and the planets were shown to be simply heavy bodies moving according to the same laws we see acting all around us, only under entirely different circumstances. All three of Kepler's laws were expressed in the single law of gravitation towards the sun, with a force acting inversely as the square of the distance.

Very beautiful is the explanation which gravity gives of Kepler's third law. We have seen that if we take the cubes of the mean distances of the several planets, and divide them by the square of the times of revolution, the quotient will be the same for each planet of the system. If we proceed in the same way with the satellites of Jupiter, cubing the distance of each satellite from Jupiter, and dividing the cube by the square of the time of revolution, the quotient will be the same for each satellite, but will not be the same as for the planets. This quotient, in fact, is proportional to the mass or weight of the central body. In the case of the planets it is 1050 times as great as in the case of the satellites of Jupiter. This shows that the sun is 1050 times as heavy as Jupiter. We thus have a very convenient way of "weighing" such of the planets as have satellites, by measuring the orbits of the satellites, and determining the times of their revolution. But the weight is not thus expressed in tons, but only in fractions of the mass of the sun.

The law, however, is not yet complete. The attraction between the sun and planets must, by the third law of motion, be mutual. If the earth attracts the moon, she must, if the law be a general one, attract the planets also, and the planets must attract each other, and thus alter their motions around the sun. Now, it is known from observation that the planets do not move in exact accordance with Kepler's laws. The final question, then, arises whether the attraction of the planets on each other fully and exactly accounts for the deviations. This question Newton could answer only in an imperfect way, the problem being too intricate for his mathematics. He was able to show that the attraction of the sun would cause inequalities in the motion of the moon of the same nature as those observed, but he could not calculate their exact amount. Still, the general correspondence of his theory with the motions of the heavens was so striking that there ought not to be any doubt of its truth. Very remarkable, therefore, is it to see the French Academy of Sciences, as late as 1732—more than forty years later—awarding a prize to John Bernoulli, the celebrated mathematician, for a paper in which the motions of the planets were explained on the theory of vortices. It should not be inferred from this that that justly celebrated body still considered that theory to be correct; but we may infer that they still considered it an open question whether the theory of gravitation was correct.

To express Newton's theory with completeness, it is not sufficient to say simply that the sun, earth, and planets attract each other. Divide matter as finely as we may, we find it still possessing the power of attraction, because it has weight. Since the earth attracts the smallest particles, they must, by the third law of motion, attract the earth with equal force. Hence we conclude that the power of attraction resides, not in the earth as a whole, but in each individual particle of the matter composing it; that is, the attraction of the earth upon a stone is simply the sum total of the attractions between the stone and all the particles composing the earth.

There is no known limit to the distance to which the attraction of gravitation extends. The attraction of the sun upon the most distant known planets, Uranus and Neptune, shows not the slightest variation from the law of Newton. But, owing to the rapid diminution with the distance to which the law of the inverse square gives rise when we take distances so immense as those which separate us from the fixed stars, the gravitation even of the sun is so small that a million years would be required for it to produce any important effect. We are thus led to the law of universal gravitation, expressed as follows:

Every particle of matter in the universe attracts every other particle with a force directly as their masses, and inversely as the square of the distance which separates them.

§ 2. *Gravitation of Small Masses.—Density of the Earth.*

To make perfect the proof that gravity does really reside in each particle of matter, it was desirable to show, by actual experiment, that isolated masses did really attract each other, as required by Newton's law. This experiment has been made in various ways with entire success, the object, however, being not to prove the existence of the attraction, but to measure the mean density of the earth, which admits of being thus determined. The attraction of a sphere upon a point at its surface is shown, mathematically, to be the same as if the entire mass of the sphere were concentrated in its centre. It is, therefore, directly as the total amount of matter in the sphere, that is, its weight, and inversely as the square of its radius. Let us, then, compare the attraction of two spheres of the same material, of which the diameter of the one is double that of the other. The larger will have eight times the bulk, and therefore eight times the mass, of the smaller. But against this is the disadvantage that a particle on its surface is twice as far from its centre as in the case of the smaller sphere, which causes a diminution of one-fourth. Consequently, it will attract such a particle with double the force that the smaller sphere will; that is, the attractions are directly as the diameters of the spheres, if the densities are equal. If the densities are not equal, the attraction is proportional to the product of the density into the diameter.

The diameter of the earth is, in round numbers, forty millions of feet. Consequently, the attraction of a sphere of the same mean density as the earth, but one foot in diameter, will be $\frac{1}{40\,000\,000}$ part the attraction of the earth; that is, $\frac{1}{40\,000\,000}$ the weight of the body attracted. Consequently, if we should measure the attraction of such a sphere of lead, and find that it was just $\frac{1}{40\,000\,000}$ that of the weight of the body attracted, we would conclude that the mean density of the earth was equal to that of lead. But the attraction is actually found to be nearly twice as great as this; consequently, a leaden sphere is nearly twice as dense as the average of the matter composing the earth. Such a determination of the density of the earth is known as the Cavendish experiment, from the name of the physicist who first executed it.

The method in which a task seemingly so hopeless as measuring a minute force like this is accomplished is shown in the following figures. It consists primarily of a torsion balance;

that is, a very light rod, *e*, with a weight at each end, suspended horizontally by a fine fibre of silk. In order to protect it against currents of air, it must be completely enclosed in a case. In Fig. 22, the balance *eb* is suspended from the end

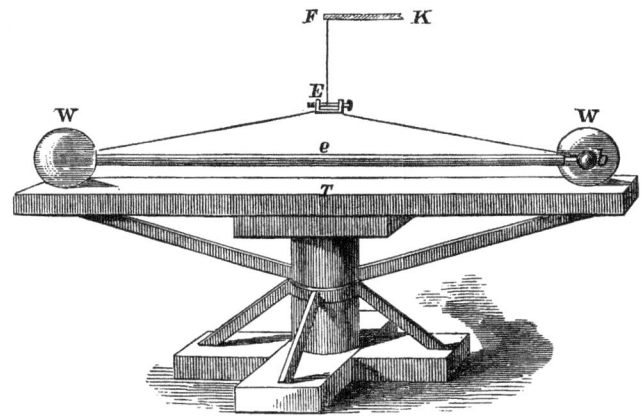

FIG. 22.—Baily's apparatus for determining the density of the earth by the Cavendish experiment. The left-hand ball *b* is hidden behind the weight *W*.

of the arm *KF* by the fine fibre of silk, *FE*. The weights to be attracted are at the two ends, *bb*. When thus suspended, the balance will swing round in a horizontal direction, twisting the silk fibre, by a very small force. The attracting masses consist of a pair of leaden balls, *WW*, as large as the experimenter can procure and manage, which are supported on the turn-table, *T*. In Fig. 23, a view of the apparatus from above is given, showing the relative positions of the leaden balls, and the suspended weights which they are to attract. It will be seen that in the position in which the weights are represented in the figure their attraction tends to make the torsion balance turn in the direction opposite that of the hands of a watch. The effect of placing the leaden balls in this position is, that the balance begins to turn as described, and, being carried by its momentum beyond the position of equilibrium, at length comes to rest by the twisting of the silk thread by which it is suspended, and then is carried part of the way back to its original position. It makes several vibrations, each requiring some minutes, and at length comes to rest in a position different from its original one. The attracting balls are then placed in the reverse position, corresponding to the

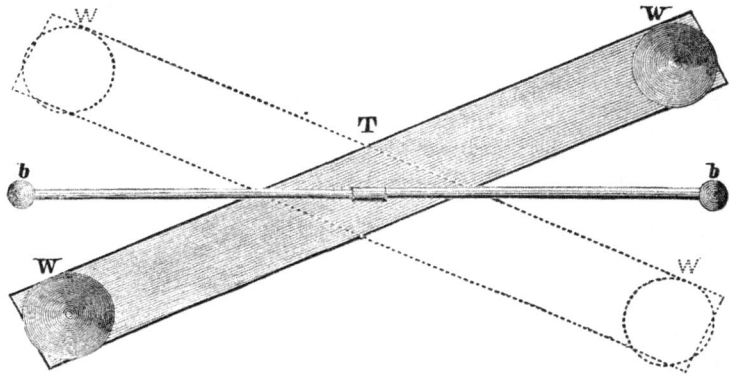

Fig. 23.—View of Baily's apparatus from above.

dotted lines, so that they tend to make the balance swing in the opposite direction, and the motions of the balance are again determined. These motions are noted by a small microscope, viewed through the enclosure in which the whole apparatus is placed, and from these motions the attractions of the balls can be computed.

Since this experiment was first made by Cavendish, it has been repeated by several other physicists; first by Professor Reich, of Freiberg, in 1838, and again by Francis Baily, Esq., of London. The latter repetition forms one of the most elaborate and exhaustive series of experiments ever made; we have therefore chosen Baily's apparatus for the purpose of illustration. The results for the mean density of the earth obtained by these several experiments are:

Cavendish (his own result)	5.48
" (Hutton's revision)	5.32
Reich	5.44
Baily	5.66*

* Memoirs of the Royal Astronomical Society, vol. xix.

DENSITY OF THE EARTH. 85

The same problem has been attacked by attempting to determine the attraction of mountains, or portions of the crust of the earth. In fact, the first attempt of the sort ever made was by Maskelyne, Astronomer Royal of England from 1766 to 1811, who determined the attraction of the mountain Schehallien, in Scotland, by observing its effect on the plumb-line. The principle of this is very clear: on whichever side of a steep isolated mountain we hang a plumb-line, the attraction of the mountain will cause it to incline towards it, the direction of gravity, or the apparent vertical, being changed from AB (Fig. 24) to AE, and from CD to CG. The density of the earth thus obtained was 4.71, a quantity much smaller than that afterwards given by the leaden balls. But this method is necessarily extremely uncertain, owing to the fact that the earth immediately beneath the mountain will probably not be of the same density as at a distance from it, and it is impossible to determine and allow for this difference.

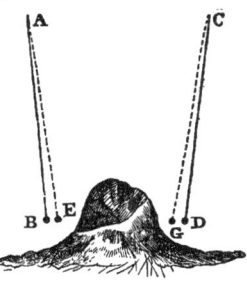

FIG. 24.

A third method is to determine the diminution of gravity as we descend into the earth. We have said that the attraction of the earth upon a point outside of it is the same as if the whole mass of the earth were concentrated in its centre. Hence, as we rise above the surface of the earth, thus receding from the centre, the force of gravity diminishes. If this force all resided in the centre of the earth, it would continue to increase as we go below the surface. But such is not the case, because, once inside the earth, we have matter round and above us the attraction of which tends to lessen the gravity towards the centre. If we could actually reach the centre, the attraction would be nothing, because a point there would be equally attracted in every direction. If the density of the earth were uniform, the force of gravity would diminish with perfect uniformity from the surface to the centre. If the density increases as we approach the centre, the diminution of gravity will be less rapid.* A determination of the density of the earth by the diminution of gravity in a mine was made by Professor Airy, at the Harton Colliery, in Wales, in 1855. His result was 6.56. This method is subject to uncertainty, from the difficulty of determining the density of that portion of the earth the attraction of which causes the gravity of bodies in the bottom of the mine to be diminished.

§ 3. *Figure of the Earth.*

If the earth did not revolve, the mutual attraction of all its parts would tend to make it assume a spherical form. If the cohesion of the solid parts prevented the spherical form from being accurately assumed, nevertheless the surface of the ocean, or of any fluid covering the earth, would assume that form. If, now, we set such a spherical earth in rotation around an axis, a centrifugal force will be generated towards the equatorial regions, which will cause the ocean to move from the poles towards the equator, so that the surface will tend to assume the form of an oblate spheroid, the longest diameter passing through the equator, and the shortest through the poles. A computation of the centrifugal force at the equator shows it to be $\frac{1}{290}$ the force of gravity itself. Consequently, the oblateness ought to be easily measurable in geodetic operations. Yet another result was that, in consequence of the centrifugal force at the equator, bodies would be lighter, and a clock regulated to northern latitudes would lose time when taken thither.

This last result accorded with the experience of Richer, sent by the French Academy to Cayenne, in 1672, to make observations on Mars. After that, to deny the oblate figure of the earth was not so much to deny Newton's theory of gravity

---

* The general law which regulates the force of gravity within the earth is this: The total attraction of the shell of earth, which is outside the attracted point extending all around the globe, is nothing, while the remainder of the globe, being a sphere with the point on its surface, attracts as if it were all concentrated at the centre. But this presupposes that the whole earth is composed of spherical layers, each of uniform density, which is not strictly the case.

as to deny that mechanical forces produced their natural effect in changing the form of the surface of the ocean. Nevertheless, the French astronomers long refused their assent, because the geodetic operations they had undertaken in France seemed to indicate that the earth was elongated rather than flattened in the direction of the poles. The real cause of this result was, that the distance measured in France was so short that the effect of the earth's ellipticity was entirely masked by the unavoidable errors of the measures, yet it long delayed the entire acceptance of the Newtonian theory by the French astronomers. We must, however, give the latter, or, speaking of them individually, their successors of the next generation, the credit of taking the most thorough measures to settle the question. Their government sent one expedition to Peru, to measure the length of a degree of latitude at the equator, and another to Lapland, to measure one as near as possible to the pole. The result was entirely in accord with the theory of Newton, and gave it a confirmation which had in the mean time become entirely unnecessary.

Newton was unable to determine the exact figure which the earth ought to assume under the influence of its own attraction and the centrifugal force of rotation, though he could see that its meridian lines would be curves not very different from an ellipse. The complication of the problem arises from the fact that, as the earth changes its form in consequence of the rotation, the direction and force of attraction at the various points of its surface change also; and this, in its turn, leads to a different figure. It was not until the middle of the last century that the problem of the form of a rotating fluid mass was solved, and the answer found to be an ellipsoid.

The figure of the earth is, however, not an exact ellipsoid, there being two causes of deviation. (When we speak of the figure or dimensions of the earth, we mean those of the ocean as they would be if the ocean covered the entire earth.) One cause of deviation is that the density of the earth increases as we approach its centre. The other cause is that there are great irregularities in the density of its superficial portions.

In consequence of this, the real figure of the water-line is full of small deviations, which are rendered very evident by the refined determinations of modern times, and which are very troublesome to all who are engaged in exact geodetic operations.

§ 4. *Precession of the Equinoxes.*

Yet another mysterious phenomenon which gravity completely explained was that of the precession of the equinoxes. We have already described this as a slow change in the position of the pole of the celestial sphere among the stars, leading to a corresponding change in the position of the celestial equator. But the Copernican theory shows the celestial pole to be purely fictitious, because the heavens do not revolve at all, but the earth. The pole of the celestial sphere is only that point of the heavens towards which the axis of the earth points. Hence, when we come to the Copernican system, we see that precession must be in the earth, and not in the heavens, and must consist simply in a change in the direction of the earth's axis, in virtue of which it describes a circle in the heavens in about 25,800 years. This effect was traced by Newton to the attraction of the sun and moon on the protuberance produced, as just described, by the centrifugal force at the earth's equator. In the present case the effect is much the same as if the earth, being itself spherical, were enveloped by a huge ring extending round its equator. In Fig. 25 let

FIG. 25.

AB represent this ring revolving around the sun, S; the centrifugal force at its centre, c, will then balance the attraction of the sun at the same point. But the point A being nearer the sun, his attraction will be greater than at c, and the centrifugal force will be less, so that there will be a surplus force pulling A towards the sun. At B, on the other hand, the attractive force of the sun is less, and the centrifugal force is greater. Consequently, there is a surplus force tending to draw B from the sun. The ring being oblique towards the sun, the effect of these surplus forces would be to make the ring turn round at c until the line AB pointed towards the sun. The spherical earth being fastened in the ring, as just supposed, would very slowly be turned round with the ring, so that its equator would be directed towards the sun. But this effect is prevented by the earth's rotation on its axis, which makes it act like a gyroscope, or like a spinning-top. Instead of being brought down towards the sun, a very slow motion, at right angles to this direction, is produced, and thus we have the motion of precession. The nature of this motion may be best seen by Fig. 17, where the north pole of the earth is represented as constantly inclined to the right of the observer as the earth moves round the sun, so that the solstices are at A and C, and the equinoxes at B and D. The effect of the attraction of the sun and moon on the protuberance at the equator is, that in 6500 years the axis of the earth will incline towards the observer of the picture, with nearly the inclination of 23°; so that the solstices will be at B and D, and the equinoxes at A and C. In 6500 years more the north pole will be pointed towards the left instead of the right, as in the figure; in 6500 more it will be directed from the observer; and, finally, at the end of a fourth period it will be once more near its present position.

The effects we have described would not occur if the plane of the ring, AB, passed through the sun, because then the forces which draw A towards the sun and B from it, would act directly against each other, and so destroy each other's effect. Now, this is the case twice a year, namely, when the sun is on the equator. Therefore, the motion of precession is not uniform, but is much greater than the average in June and December, when the sun's declination is greatest; and is less in March and September, when the sun is on the plane of the equator. Moreover, in December the earth is nearer the sun than in June, and the force greater, so that we have still another inequality from this cause.

Precession is not produced by the sun alone. The moon is a yet more powerful agent in producing it, its smaller mass being more than compensated by its greater proximity to us.* The same causes which make the action of the sun variable make that of the moon variable also, and we have the additional cause that, owing to the revolution of the moon's node, the inclination of the moon's orbit to the plane of the earth's equator is subject to an oscillation having a period of 18.6 years, producing an inequality of this same period in the precession. The several inequalities in the precession which we have described are known as *nutation of the earth's axis*, and are all accurately computed and laid down in astronomical tables.

§ 5. *The Tides.*

It has been known to seafaring nations from a remote antiquity that there was a singular connection between the ebb and flow of the tides, and the diurnal motion of the moon. Cæsar's description of his passages across the English Channel shows that he was acquainted with the law. In describing the motion of the moon, it was shown that, owing to her revolution in a monthly orbit, she rises, passes the meridian, and sets about fifty minutes later every day. The tides ebb and flow twice a day, but the corresponding tide is always later than the day before, by the same amount, on the average, that the moon is later. Hence, at any one place, the tides always occur when the moon is near the same point of her apparent diurnal course.

---

* This may need some explanation, as the attractive force of the sun upon the earth is more than a hundred times that of the moon. The force which produces precession is proportional to the *difference* of the attractions on the two sides of the earth, or on A and B in Fig. 25, and this difference is greater in the case of the moon's attraction. In fact, it varies inversely as the cube of the distance of the attracting body.

THE TIDES. 91

The cause of this ebb and flow of the sea, and its relation to the moon, was a mystery until gravitation showed it to be due to the attraction of the moon on the waters of the ocean. The reason why there are two tides a day will appear by studying the case of the moon's revolution around the earth. Let M be the moon, E the earth, and EM the line joining their centres. Now, strictly speaking, the earth does not revolve around the moon, any more than the moon around the earth; but, by the principle of action and reaction, both move around their common centre of gravity. The earth being eighty times as heavy as the moon, this centre is situated within the former, about three-fourths of the way from its centre to its surface, at the point G in the figure. The manner in which

FIG. 26.—Attraction of the moon tending to produce tides.

the moon produces the tides is much the same as that in which precession is produced. Near the centre of the earth, E, the centrifugal force of the earth's monthly rotation around G, and the attraction of the moon, counterbalance each other, so that a point there has no disposition to move under the influence of these combined forces. As we pass from E to D, the part of the earth's surface opposite the moon, the centrifugal force around G keeps increasing, owing to our greater distance from the centre, while the attraction of the moon diminishes. Hence, at D the centrifugal force predominates, and tends to throw the waters of the ocean out, as shown in the figure. Again, as we pass from the centre E to C, the centrifugal force constantly diminishes till we reach the centre of revolution, G, when it vanishes, and, beyond G, begins to act in the opposite direction. Hence, at C the attraction of the moon and the small centrifugal force around G both combine to throw the waters of the ocean out in the direction of the moon. Thus, there is a force causing the waters to rise at D and C, and therefore to fall at A and B; and there are, therefore, two tides to each apparent diurnal revolution of the moon.

If the waters everywhere yielded immediately to the attractive force of the moon, it would always be high-water when the moon was on the meridian, low-water when she was rising or setting, and high-water again when she was in the middle of that portion of her course which is under the horizon. But, owing to the inertia of the water, some time is necessary for so slight a force to set it in motion, and, once in motion, it continues so after the force has ceased, and until it has acted some time in the opposite direction. Therefore, if the motion of the water were unimpeded, it would not be high-water until some hours after the moon had passed the meridian. Yet another circumstance interferes with the free motion of the water — namely, the islands and continents. These deflect the tidal wave from its course in such a way that it may, in some cases, be many hours behind its time, or even a whole day. Sometimes two waves may meet each other, and raise an extraordinarily high tide. At other times the tides may have to run up a long bay, where the motion of a long mass of water will cause an enormous tide to be raised. In the Bay of Fundy both of these causes are combined. A tidal wave coming up the Atlantic coast meets the ocean wave from the east, and, entering the bay with their combined force, the water at the head of it is forced up to the height of sixty or seventy feet, on the principle seen in the hydraulic ram.

The sun produces a tide as well as the moon, the force which it exerts on the two sides of the earth being the same, which, acting on the equatorial protuberance of the earth, produces precession. The tide-producing force of the sun is about $\frac{4}{10}$ of that of the moon. At new and full moon the two bodies unite their forces, and the result is that the ebb and flow are greater than the average, and we have the "spring-tides." When the moon is in her first or third quarter, the

INEQUALITIES IN THE MOTIONS OF THE PLANETS. 93

two forces act against each other; the tide-producing force is the difference of the two, the ebb and flow are less than the average, and we have the "neap-tides."

§ 6. *Inequalities in the Motions of the Planets produced by their Mutual Attraction.*

The profoundest question growing out of the theory of gravitation is whether all the inequalities in the motion of the moon and planets admit of being calculated from their mutual attraction. This question can be completely answered only by actually making the calculation, and seeing whether the resulting motion of each planet agrees exactly with that observed. The problem of computing the motion of each planet under the influence of the attraction of all the others is, however, one of such complexity that no complete and perfect solution has ever been found. Stated in its most general form, it is as follows: Any number of planets of which the masses are known are projected into space, their positions, velocities, and directions of motion all being given at some one moment. They are then left to their mutual attractions, according to the law of gravitation. It is required to find general algebraic formulæ by which their position at any time whatever shall be determined. In this general form, no approximation to an entire solution has ever been found. But the orbits described by the planets around the sun, and by the satellites around their primaries, are nearly circular; and this circumstance affords the means of computing the theoretical place of the planet as accurately as we please, provided the necessary labor can be bestowed upon the work.

What makes the problem so complex is that the forces which act upon the planets are dependent on their motions, and these again are determined by the forces which act on them. If the planets did not attract each other at all, the problem could be perfectly solved, because they would then all move in ellipses, in exact accordance with Kepler's laws. Supposing them to move in ellipses, their positions and distances at any time could be expressed in algebraic formulæ;

and their attractions on each other could be expressed in the same way. But, owing to these very attractions, they do not move in ellipses, and therefore the formulæ thus found will not be strictly correct. To put the difficulty into a nut-shell, the geometer cannot strictly determine the motion of the planet until he knows the attractions of all the other planets on it, and he cannot determine these without first knowing the position of the planet, that is, without having solved his problem.

The question how to surmount these difficulties has, to a greater or less extent, occupied the attention of all great mathematicians from the time of Newton till now; and although complete success has not attended their efforts, yet the marvellous accuracy with which sun, moon, and planets move in their prescribed orbits, and the certainty with which the laws of variation of those orbits through countless ages past and to come have been laid down, show that their labor has not been in vain. Newton could attack the problem only in a geometrical way; he laid down diagrams, and showed in what way the forces acted in various parts of the orbits of the two planets, or in various positions of the sun and moon. He was thus enabled to show how the attraction of the sun upon the moon changes the orbit of the latter around the earth, and causes its nodes to revolve from east to west, as observations had shown them to do, and to calculate roughly one or two of the inequalities in the motion of the moon in her orbit.

When the Continental mathematicians were fully convinced of the correctness of Newton's theory, they immediately attacked the problem of planetary motion with an energy and talent which placed them ahead of the rest of the world. They saw the entire insufficiency of Newton's geometrical method, and the necessity of having the forces which moved the planets expressed by the algebraic method, and, by adopting this system, were enabled to go far ahead both of Newton and his countrymen. The last half of the last century was the Golden Age of mathematical astronomy. Five illustrious names of this period outshine all others: Clairaut, D'Alembert, Euler, Lagrange, and Laplace, all, except Euler,

INEQUALITIES IN THE MOTIONS OF THE PLANETS. 95

French by birth or adoption. The great works which closed it were the "Mécanique Céleste" of Laplace, and the "Mécanique Analytique" of Lagrange, which embody the substance of all that was then known of the subject, and form the basis of nearly everything that has since been achieved. We shall briefly mention some of the results of these works, and those of their successors which may interest the non-mathematical reader.

Perhaps the most striking of these results is that of the secular variations of the planetary orbits. Copernicus and Kepler had found, by comparing the planetary orbits as observed by themselves with those of Ptolemy, that the forms and positions of those orbits were subject to a slow change from century to century. The immediate successors of Newton were able to trace this change to the mutual action of the planets, and thus arose the important question, Will it continue forever? For, should it do so, it would end in the ultimate subversion of the solar system, and the destruction of all life on our globe. The orbit of the earth, as well as of the other planets, would become so eccentric that, approaching near the sun at one time, and receding far from it at another, the vicissitudes of temperature would be insupportable. Lagrange, however, was enabled to show by a mathematical demonstration that these changes were due to a regular system of oscillations extending throughout the whole planetary system, the periods of which were so immensely long that only a progressive motion could be perceived during all the time that men had observed the planets. The number of these combined oscillations is equal to that of the planets, and their periods range from 50,000 years all the way up to 2,000,000—"Great clocks of eternity, which beat ages as ours beat seconds." In consequence of these oscillations, the perihelia of the planets will turn in every direction, and the orbits will vary in eccentricity, but will never become so eccentric as to disturb the regularity of the system. About 18,000 years ago, the eccentricity of the earth's orbit was about .019; it has been diminishing ever since. and will continue to diminish for 25,000 years to come, when it will be more nearly a circle than any orbit of our system now is.

Some of the questions growing out of the moon's motion are not completely settled yet. Early in the last century it was found by Halley, from a comparison of ancient eclipses with modern observations of the moon, that our satellite was accelerating her motion around the earth. She was, in fact, about a degree ahead of where she ought to have been had her motion been uniform from the time of Hipparchus and Ptolemy. The existence of this acceleration was fully established in the time of Lagrange and Laplace, and was to them a source of great perplexity, because they had conceived themselves to have shown mathematically that the mutual attractions of the planets or satellites could never accelerate or retard their mean motions in their orbits, and thus the motion of the moon seemed to be affected by some other force than gravitation. After several vain attempts to account for the motion, it was found by Laplace that, in consequence of the secular diminution of the eccentricity of the earth's orbit, the action of the sun on the moon was progressively changing in such a manner as to accelerate its motion. Computing the amount of the acceleration, he found it to be about 10 seconds in a century, and its action on the moon being like that of gravity on a falling body, the total effect would increase as the square of the time; that is, while in one century the moon would be 10 seconds ahead, in two centuries she would be 40 seconds ahead, in three centuries 90 seconds, and so on.

This result agreed so well with the observed acceleration, as determined by a comparison of ancient eclipses with modern data, that no one doubted its correctness till long after the time of Laplace. But, in 1853, Mr. J. C. Adams, of England, celebrated as one of the two mathematicians who had calculated the position of Neptune from the motions of Uranus, undertook to recompute the effect of the variation of the earth's eccentricity on the mean motion of the moon. He was surprised to find that, carrying his process farther than Laplace had done, the effect in question was reduced from 10 seconds,

INEQUALITIES IN THE MOTION OF THE MOON.

the result of Laplace, to 6 seconds. On the other hand, the further examination of ancient and modern observations seemed to show that the acceleration as given by them was even greater than that found by Laplace, being more nearly 12 seconds than 10 seconds; that is, it was twice as great as that computed by Mr. Adams from the theory of gravitation.

The announcement of this result by Mr. Adams was at first received with surprise and incredulity, and led to one of the most remarkable of scientific discussions. Three of the great astronomical mathematicians of the day—Hansen, Plana, and De Pontécoulant—disputed the correctness of Mr. Adams's result, and maintained that that of Laplace was not affected with any such error as Mr. Adams had found. In fact, Hansen, by a method entirely different from that of his predecessors, had found a result of 12 seconds, which was yet larger than that of Laplace. On the other hand, Delaunay, of Paris, by a new and ingenious method of his own, found a result agreeing exactly with Mr. Adams's. Thus, the five leading experts of the day were divided into two parties on a purely mathematical question, and several years were required to settle the dispute. The majority had on their side not only the facts of observation, so far as they went, but the authority of Laplace; and, if the question could have been settled either by observation or by authority, they must have carried the day. But the problem was altogether one of pure mathematics, depending on the computation of the effect which the gravitation of the sun ought to produce on the motion of the moon. Both parties were agreed as to the data, and but one correct result was possible, so that an ultimate decision could be reached only by calculation.

The decision of such a question could not long be delayed. There was really no agreement among the majority as to what the supposed error of Mr. Adams consisted in, or what the exact mathematical expression for the moon's acceleration was. On the other hand, Mr. Adams showed conclusively that the methods of De Pontécoulant and Plana were fallacious; and the more profoundly the question was examined, the more evident it became that he was right. Mr. Cayley made a computation of the result by a new method, and Delaunay by yet another method, and both agreed with Mr. Adams's. Although their antagonists never formally surrendered, they tacitly abandoned the field, leaving Delaunay and Adams in its undisturbed possession.*

Now, however, there was a discrepancy between the theoretical and observed acceleration, the cause of which was to be investigated. A possible cause happened to be already known: the friction of the tidal wave must constantly retard the diurnal motion of the earth on its axis, though it is impossible to say how much this retardation may amount to. The consequence would be that the *day* would gradually, but unceasingly, increase in length, and our count of time, depending on the day, would be always getting too slow. The moon would, therefore, appear to be going faster, when really it was only the earth which was moving more slowly. So long as theory had agreed with the observed acceleration of the moon, there had been no need to invoke this cause; but, now that there was a discrepancy, it afforded the most plausible explanation. The amount of retardation necessary to account for the excess of the apparent acceleration over that computed is about ten seconds in a century; that is, we must suppose that the diurnal rotation of the earth, at the end of one hundred years, is ten seconds behind what it would have been if it had rotated uniformly at the rate it had at the beginning of the century. This change is so minute that there is no way of detecting it except by celestial observations; and we are not yet in a position to pronounce upon it with certainty.

The secular acceleration is not the only variation in the moon's mean motion which has perplexed the mathematicians. About the close of the last century, it was found by Laplace that the moon had, for a number of years, been falling behind

---

* The writer has reason to believe it an historical fact that Hansen, on revising his own calculations, and including terms he at first supposed to be insensible, found that he would be led substantially to the result of Adams, although he never made any formal publication of this fact.

her calculated place, a result which seemed to show that there was some oscillation of long period which had been overlooked. He made two conjectural explanations of this inequality, but both were disproved by subsequent investigators. The question, therefore, remained without any satisfactory solution till 1846, when Hansen announced that the attraction of Venus produced two inequalities of long period in the moon's motion, which had been previously overlooked, and that these fully accounted for the observed deviations of the moon's position. These terms were recomputed by Delaunay, and he found for one of them a result agreeing very well with Hansen's. But the second came out so small that it could never be detected from observations, so that here was another mathematical discrepancy. There was not room, however, for much discussion this time. Hansen himself admitted that he had been unable to determine the amount of this inequality in a satisfactory manner from the theory of gravitation, and had therefore made it agree with observation, an empirical process which a mathematician would never adopt if he could avoid it. Even if observations were thus satisfied, doubt would still remain. But it has lately been found that this empirical term of Hansen's no longer agrees with observation, and that it does not satisfactorily agree with observations before 1700. In consequence, there are still slow changes in the motion of our satellite which gravitation has not yet accounted for. We are, apparently, forced to the conclusion either that the motion of the moon is influenced by some other cause than the gravitation of the other heavenly bodies, or that these inequalities are only apparent, being really due to small changes in the earth's axial rotation, and in the consequent length of the day. If we admit the latter explanation, it will follow that the earth's rotation is influenced by some other cause than the tidal friction; and that, instead of decreasing uniformly, it varies from time to time in an irregular manner. The observed inequalities in the motion of the moon may be fully accounted for by changes in the earth's rotation, amounting in the aggregate to half a minute or so of time—changes which could be detected by a perfect clock kept going for a number of years. But, as it takes many years for these changes to occur, no clock yet made will detect them.

Yet another change not entirely accounted for on the theory of gravitation occurs in the motion of the planet Mercury. From a discussion of all the observed transits of this planet across the disk of the sun, Leverrier has found that the motion of the perihelion of Mercury is about 40 seconds in a century greater than that computed from the gravitation of the other planets. This he attributes to the action of a group of small planets between Mercury and the sun. In this form, however, the explanation is not entirely satisfactory. In the first place, it seems hardly possible that such a group of planets could exist without being detected during total eclipses of the sun, if not at other times. In the next place, granting them to exist, they must produce a secular variation in the position of the orbit of Mercury, whereas this variation seems to agree exactly with theory. Leverrier explains this by supposing the group of asteroids to be in the same plane with the orbit of Mercury, but it is exceedingly improbable that such a group would be found in this plane. There is, however, an allied explanation which is at least worthy of consideration. The phenomenon of the zodiacal light, to be described hereafter, shows that there is an immense disk of matter of some kind surrounding the sun, and extending out to the orbit of the earth, where it gradually fades away. The nature of this matter is entirely unknown, but it may consist of a swarm of minute particles, revolving round the sun, and reflecting its light, like planets. If the total mass of these particles is equal to that of a very small planet, say a tenth the mass of the earth, it would cause the observed motion of the perihelion of Mercury. The evidence on this subject will be considered more fully in treating of Mercury.

With the exceptions just described, all the motions in the solar system, so far as known, agree perfectly with the results of the theory of gravitation. The little imperfections which still exist in the astronomical tables seem to proceed mainly from errors in the data from which the mathematician must start in computing the motion of any planet. The time of revolution of a planet, the eccentricity of its orbit, the position of its perihelion, and its place in the orbit at a given time, can none of them be computed from the theory of gravitation, but must be derived from observations alone. If the observations were absolutely perfect, results of any degree of accuracy could be obtained from them; but the imperfections of all instruments, and even of the human sight itself, prevent observations from attaining the degree of precision sought after by the theoretical astronomer, and make the considerations of "errors of observation" as well as of "errors of the tables" constantly necessary.

§ 7. *Relation of the Planets to the Stars.*

In Chapter I., § 3, it was stated that the heavenly bodies belong to two classes, the one comprising a vast multitude of stars, which always preserved their relative positions, as if they were set in a sphere of crystal, while the others moved, each in its own orbit, according to laws which have been described. We now know that these moving bodies, or planets, form a sort of family by themselves, known as the Solar System. This system consists of the sun as its centre, with a number of primary planets revolving around it, and satellites, or secondary planets, revolving around them. Before the invention of the telescope but six primary planets were known, including the earth, and one satellite, the moon. By the aid of that instrument, two great primary planets, outside the orbit of Saturn, and an immense swarm of smaller ones between the orbits of Mars and Jupiter, have been discovered; while the four outer planets — Jupiter, Saturn, Uranus, and Neptune — are each the centre of motion of one or more satellites. The sun is distinguished from the planets, not only by his immense mass, which is several hundred times that of all the other bodies of his system combined, but by the fact that he shines by his own light, while the planets and satellites are dark bodies, shining only by reflecting the light of the sun.

A remarkable symmetry of structure is seen in this system, in that all the large planets and all the satellites revolve in orbits which are nearly circular, and, the satellites of the two outer planets excepted, nearly in the same plane. This family of planets are all bound together, and kept each in its respective orbit, by the law of gravitation, the action of which is of such a nature that each planet may make countless revolutions without the structure of the system undergoing any change.

Turning our attention from this system to the thousands of fixed stars which stud the heavens, the first thing to be considered is their enormous distance asunder, compared with the dimensions of the solar system, though the latter are themselves inconceivably great. To give an idea of the relative distances, suppose a voyager through the celestial spaces could travel from the sun to the outermost planet of our system in twenty-four hours. So enormous would be his velocity, that it would carry him across the Atlantic Ocean, from New York to Liverpool, in less than a tenth of a second of the clock. Starting from the sun with this velocity, he would cross the orbits of the inner planets in rapid succession, and the outer ones more slowly, until, at the end of a single day, he would reach the confines of our system, crossing the orbit of Neptune. But, though he passed eight planets the first day, he would pass none the next, for he would have to journey eighteen or twenty years, without diminution of speed, before he would reach the nearest star, and would then have to continue his journey as far again before he could reach another. All the planets of our system would have vanished in the distance, in the course of the first three days, and the sun would be but an insignificant star in the firmament. The conclusion is, that our sun is one of an enormous number of self-luminous bodies scattered at such distances that years would be required to traverse the space between them, even when the voyager went at the rate we have supposed. The solar and the stellar systems thus offer us two distinct fields of inquiry, into which we shall enter after describing the instruments and methods by which they are investigated.

PART II.—PRACTICAL ASTRONOMY.

INTRODUCTORY REMARKS.

SHOULD the reader ask what Practical Astronomy is, the best answer might be given him by a statement of one of its operations, showing how eminently practical our science is. "Place an astronomer on board a ship; blindfold him; carry him by any route to any ocean on the globe, whether under the tropics or in one of the frigid zones; land him on the wildest rock that can be found; remove his bandage, and give him a chronometer regulated to Greenwich or Washington time, a transit instrument with the proper appliances, and the necessary books and tables, and in a single clear night he can tell his position within a hundred yards by observations of the stars." This, from a utilitarian point of view, is one of the most important operations of Practical Astronomy. When we travel into regions little known, whether on the ocean or on the Western plains, or when we wish to make a map of a country, we have no way of finding our position by reference to terrestrial objects. Our only course is to observe the heavens, and find in what point the zenith of our place intersects the celestial sphere at some moment of Greenwich or Washington time, and then the problem is at once solved. The instruments and methods by which this is done may also be applied to celestial measurements, and thus we have the art and science of Practical Astronomy. To speak more generally, Practical Astronomy consists in the description and investigation of the instruments and methods employed by astronomers in the work of exploring and measuring the heavens, and of determining positions on the earth by observations of the heavenly bodies. The general construction of these instruments, and the leading principles which underlie their use and employment, can be explained with the aid of a few technical terms which we shall define as we have occasion for them.

The instruments employed by the ancients in celestial observations were so few and simple that we may dispose of them very briefly. The only ones we need mention at present are the gnomon and the astrolabe, or armillary sphere. The former was little more than a large sun-dial of the simplest construction, by which the altitude and position of the sun were determined from the length and direction of the shadow of an upright pillar. If the sun were a point to the sight, this method would admit of considerable accuracy, because the shadow would then be sharply defined. In fact, however, owing to the apparent size of the solar disk, the shadow of any object at the distance of a few feet becomes ill defined, shading off so gradually that it is hard to say where it ends. No approach to accuracy can therefore be attained by the gnomon.

Notwithstanding the rudeness of this instrument, it seems to have been the one universally employed by the ancients for the determination of the times when the sun reached the equinoxes and solstices. The day when the shadow was shortest marked the summer solstice, and a comparison of the length of the shadow with the height of the style gave, by a trigonometric calculation, the altitude of the sun. The day when the shadow was longest marked the winter solstice; and the day when the altitude of the sun was midway between the altitudes at the two solstices marked the equinoxes. Thus this rude instrument served the purpose of determining the length of the year with an accuracy sufficient for the purposes of daily life. But so immensely superior are our modern methods in accuracy, that the astronomer can to-day compute the position of the sun at any hour of any day 2000 years ago with far greater accuracy than it could have been observed with a gnomon.

INTRODUCTORY REMARKS.

The armillary sphere consisted of a combination of three circles, one of which could be set in the plane of the equator or the ecliptic; that is, an arm moving around this circle would always point towards some part of the equator or the ecliptic, according to the way the instrument was set. The circle in question, being divided into degrees, served the purpose of measuring the angular distance of any two bodies in or near the ecliptic, as the sun and moon, or a star and planet. It was by such measures that Hipparchus and Ptolemy were able to determine the larger inequalities in the motions of the sun, moon, and planets.

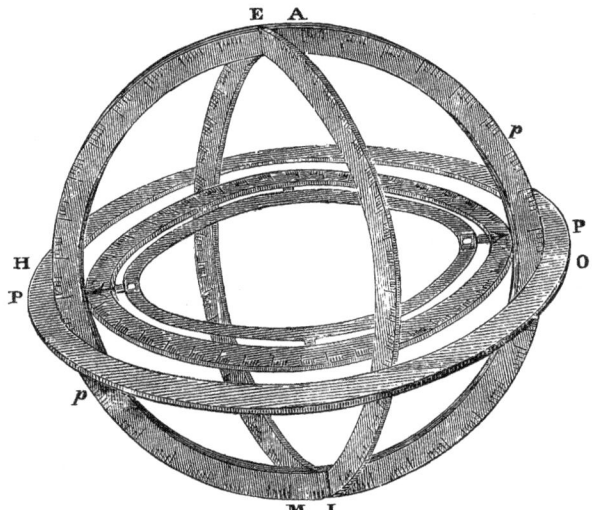

FIG. 27.—Armillary sphere, as described by Ptolemy, and used by him and by Hipparchus. The circle EI is set in the plane of the ecliptic, the line PP being directed towards its pole. The circle A*p*M*p* passes through the poles of both the ecliptic and the equator. The inner pair of circles turn on the axis PP, and are furnished with sights which may be directed on the object to be observed. The latitude and longitude of the object are then read off by the position of the circles.

CHAPTER I.

THE TELESCOPE.

§ 1. *The First Telescopes.*

THE telescope is so essential a part of every instrument intended for astronomical measurement, that, apart from its own importance, it must claim the first place in any description of astronomical instruments. The question, Who made the first telescope? was long discussed, and, perhaps, will never be conclusively settled. If the question were merely, Who is entitled to the credit of the invention under the rules according to which scientific credit is now awarded? we conceive that the answer must be, Galileo. The first publisher of a result or discovery, supposing such result or discovery to be honestly his own, now takes the place of the first inventor; and there is little doubt that Galileo was the first one to show the world how to make a telescope. But Galileo himself says that it was through hearing that some one in France or Holland had made an instrument which magnified distant objects, and brought them nearer to the view, that he was led to inquire how such a result could be reached. He seems to have obtained from others the idea that the instrument was possible, but no hint as to how it was made.

As a historic fact, however, there is no serious question that the telescope originated in Holland; but the desire of the inventors, or of the authorities, or both, to profit by the possession of an instrument of such extraordinary powers, prevented the knowledge of its construction from spreading abroad. The honor of being the originator has been claimed for three men, each of whom has had his partisans. Their names are Metius,

Lipperhey, and Jansen; the last two being spectacle-makers in the town of Middleburg, and the first a professor of mathematics.

The claims of Jansen were sustained by Peter Borelli, author of a small book* on the subject, and on the strength of his authority Jansen was long held to be the true inventor. His story was that Jansen had shown a telescope sixteen inches long to Prince Maurice and the Archduke Albert, who, perceiving the importance of the invention in war, offered him money to keep it a secret. If this story be true, it would be interesting to know on what terms Jansen was induced to sell out his right to immortality. But Borelli's case rests on the testimony of two or three old men who had known Jansen in their youth, taken forty-five or fifty years after the occurrence of the events, when Jansen had long been dead, and has therefore never been considered as fully proved.

About 1830, documentary evidence was discovered which showed that Hans Lipperhey, whom Borelli claims to have been a second inventor of the telescope, made application to the States-general of Holland, on November 2d, 1608, for a patent for an instrument to see with at a distance. About the same time a similar application was made by James Metius. The Government refused a patent to Lipperhey, on the ground that the invention was already known elsewhere, but ordered several instruments from him, and enjoined him to keep their construction a secret.

It will be seen from this that the historic question, Who made the first telescope? does not admit of being easily answered; but that the powers of the instrument were well known in Holland in 1608 seems to be shown by the refusal of a patent to Lipperhey. The efforts made in that country to keep the knowledge of the construction a secret were so far successful that we must go from Holland to Italy to find how that knowledge first became public property. About six months after the petitions of Lipperhey and Metius, Galileo

---

* "De Vero Telescopii Inventore," The Hague, 1655.

was in Venice on a visit, and there received a letter from Paris, in which the invention was mentioned. He at once set himself to the reinvention of the instrument, and was so successful that in a few days he exhibited a telescope magnifying three times, to the astonished authorities of the city. Returning to his home in Florence, he made other and larger ones, which revealed to him the spots on the sun, the phases of Venus, the mountains on the moon, the satellites of Jupiter, the seeming handles of Saturn, and some of the myriads of stars, separately invisible to the naked eye, whose combined light forms the milky-way. But the largest of these instruments magnified only about thirty times, and was so imperfect in construction as to be far from showing as much as could be seen with a modern telescope of that power. The Galilean telescope was, in fact, of the simplest construction, consisting of the combination of a pair of lenses, of which the larger was convex and the smaller concave, as shown in the following figure:

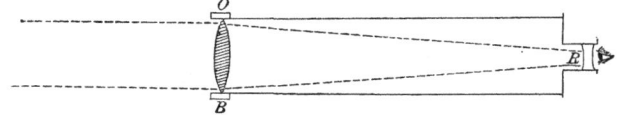

FIG. 23.—The Galilean telescope. The dotted lines show the course of the rays through the lenses.

The distance of the lenses was such that the rays of light from a star passing through the large convex lens, or object-glass, OB, met the concave lens, R, before reaching the focus. The position of this concave lens was such that the rays should emerge from it nearly parallel. This form of telescope is still used in opera-glasses, because it can be made shorter than any other.

The improvements in the telescope since Galileo can be best understood if we give a brief statement of the principles on which all modern telescopes are constructed. The properties of every such instrument depend on the power possessed by a lens or by a concave mirror of forming an image of any distant object in its focus. This is done in the

THE FIRST TELESCOPES. 109

case of the lens by refracting the light which passes through it, and in the case of the mirror by reflecting back the rays which strike it. In order to form an image of a point, it is necessary that a portion of the rays of light which emanate from the point shall be collected and made to converge to some other point. For instance, in the following figure, the

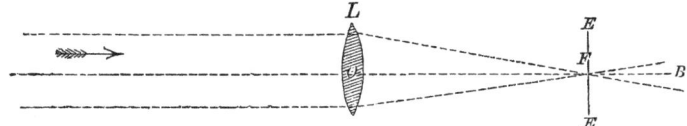

FIG. 29.—Formation of an image by a lens.

nearly parallel rays emanating from a distant point in the direction from which the arrow is coming strike the lens, L, and as they pass through it are bent out of their course, and made to converge to a point, F. Continuing their course, they diverge from F *exactly as if F itself were a luminous point*, a cone of light being formed with its apex at F. An observer placing his eye within this cone of rays, and looking at F, will there seem to see a shining point, although really there is nothing there. This apparent shining point is, in the language of astronomy, called *the image of the real point*. The distance, OF, is called *the focal length of the lens*.

If, instead of a simple point, we have an object of some apparent magnitude, as the moon, a house, or a tree, then the light from each point of the object will be brought to a corresponding point near F. To find where this corresponding point is, we have only to draw a line from each point of an object through the centre of the lens, and continue it as far as the focus. Each point of the object will then have its own point in the image. These points, or images, will be spread out over the surface, EFE, which is called the focal plane, and will make up a representation, or image, of the entire object on a small scale, but in a reversed position, exactly as in the camera of a photographer. An eye at B within the cone of rays will then see all or a part of the object reversed in the focal plane. The image thus formed may be viewed by the eye as if it were a real object; and as a minute object may be viewed by a magnifying lens, so such a lens may be used to view and magnify the image formed in the focal plane. In the large lens of long focus to form the image in the focal plane, and the small lens to view and magnify this image, we have the two essential parts of a refracting telescope. The former lens is called the *objective*, or *object-glass*, and the latter the *eye-piece, eye-lens*, or *ocular*.

The magnifying power of a telescope depends upon the relative focal lengths of the objective and ocular. The greater the focal length of the former—that is, the greater the distance OF—the larger the image will be; and the less the focal length of the eye-lens, the nearer the eye can be brought to the image, and the more the latter will be magnified. The magnifying power is found by dividing the focal length of the objective by that of the eye-lens. For instance, if the focal length of an objective were 36 inches, and that of the eye-lens were three-quarters of an inch, the quotient of these numbers would be 48, which would be the magnifying power. If the focal lengths of these lenses were equal, the telescope would not magnify at all. By simply turning a telescope end for end, and looking in at the objective, we have a reversed telescope, which diminishes objects in the same proportion that it magnifies them when not reversed.

From the foregoing rule it follows that we can, theoretically, make any telescope magnify as much as we please, by simply using a sufficiently small eye-lens. If, for instance, we wish our telescope of 36 inches focal length to magnify 3600 times, we have only to apply to it an eye-lens of $\frac{1}{100}$ of an inch focal length. But, in attempting to do this, a difficulty arises with which astronomers have always had to contend, and which has its origin in the imperfection of the image formed by the object-glass. No lens will bring all the rays of light to absolutely the same focus. When light passes through a prism, the various colors are refracted unequally, red being refracted the least, and violet the most. It is the same when light is refracted by a lens, and the consequence is that

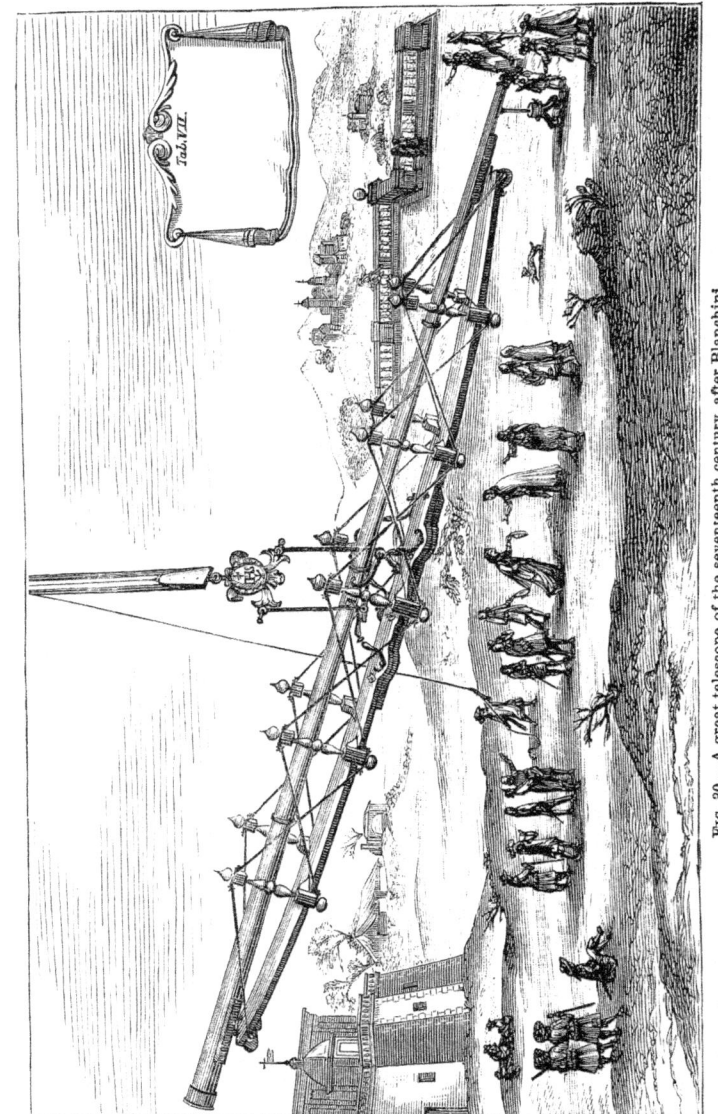

Fig. 30.—A great telescope of the seventeenth century, after Bianchini.

the red rays will be brought to the farthest focus, and the violet to the nearest, while the intermediate colors will be scattered between. As all the light is not brought to the same focus, it is impossible to get any accurate image of a star or other object at which the telescope is pointed, the eye seeing only a confused mixture of images of various colors. When a sufficiently low magnifying power is used, the confusion will be slight, the edges of the object being indistinct, and made up of colored fringes. When the magnifying power is increased, the object will indeed look larger, but these confused fringes will look larger in the same proportion; so that the observer will see no more than before. This separation of the light in a telescope is termed *chromatic aberration.*

Such was the difficulty which the successors of Galileo encountered in attempting to improve the telescope, and which they found it impossible to obviate. They found, however, that they could diminish it by increasing the length of the telescope, and the consequent size of the confused image. If they made an object-glass of any fixed diameter, say six inches, they found that the image was no more confused when the focal length was sixty feet than when it was six, and the same eye-lens could therefore be used in both cases. But the image in the focus of the first was ten times as large as in the second, and thus using the same eye-lens would give ten times the magnifying power. Huyghens, Cassini, Hevelius, and other astronomers of the latter part of the seventeenth century, made telescopes a hundred feet or upwards in length. Some astronomers then had to dispense with a tube entirely; the objective being mounted by Cassini on the top of a long pole, while the ocular was moved along near the ground. Hevelius kept his objective and ocular connected by a long rod which replaced the tube. Very complicated and ingenious arrangements were sometimes used in managing these huge instruments, of which we give one specimen, taken from the work of Blanchini, "*Hesperi et Phosphori Nova Phœnomena,*" in which that astronomer describes his celebrated observations on the rotation of Venus.

§ 2. *The Achromatic Telescope.*

A century and a half elapsed from the time when Galileo showed his first telescope to the authorities of Venice before any method of destroying the chromatic aberration of a lens was discovered. It is to Dollond, an English optician, that the practical construction of the achromatic telescope is due, although the principle on which it depends was first published by Euler, the German mathematician. The invention of Dollond consists in the combination of a convex and concave lens of two kinds of glass in such a way that their aberrations shall counteract each other. How this is effected will be best seen by taking the case of refraction by a prism, where the same principle comes into play. The separation of the light into its prismatic colors is here termed *dispersion*. Suppose, now, that we take two prisms of glass, *ABC* and *ACD*, (Fig. 31), and join them in the manner shown in the figure. If a

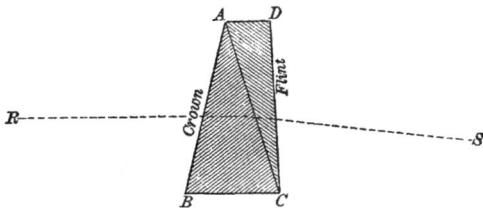

FIG. 31.—Refraction through a compound prism.

ray, *RS*, pass through the two, their actions on it will tend to counteract each other, owing to the opposite directions in which their angles are turned, and the ray will be refracted only by the difference of the refractive powers, and dispersed by the difference of the dispersive powers. If the dispersive powers are equal, there will be no dispersion at all, the ray passing through without any separation of its colors. If the two prisms are made of the same kind of glass, their dispersive powers can be made equal only by making them of the same angle, and then their refractive powers will be equal also, and the ray will pass through without any refraction. As our ob-

THE ACHROMATIC TELESCOPE. 115

ject is to have refraction without dispersion, a combination of prisms of the same kind of glass cannot effect it.

The problem which is now presented to us is, Can we make two prisms of different kinds of glass such that their dispersive powers shall be equal, but their refractive powers unequal? The researches of Euler and Dollond answered this question in the affirmative by showing that the dispersive power of dense flint-glass is double that of crown-glass, while its refractive power is nearly the same. Consequently, if we make the prism ABC of crown glass, and the prism ACD of flint, the angle of the flint at C being half that of the crown at A, the two opposite dispersions will neutralize each other, and the rays will pass through without being broken up into the separate colors. But the crown prism, with double the angle, will have a more powerful refractive power than the flint; so that, by combining the two, we shall have *refraction* without *dispersion*, which solves the problem.

The manner in which this principle is applied to the construction of an object-glass is this: a convex lens of crown is combined with a concave lens of flint of about half the curvature. No exact rule respecting the ratio of the two curvatures can be given, because the refractive powers of different specimens of glass differ greatly, and the proper ratio must, therefore, be found by trial in each case. Having found it, the two lenses will then have equal aberrations, but in opposite directions, while the crown refracting more powerfully than the flint, the rays will be brought to a focus at a distance a little more than double the focal distance of the former. A combination of this sort is called an *achromatic objective*. Some of the earlier achromatic objectives were made of three lenses, a double concave lens of flint glass being fitted between two double convex ones of crown. At present, however, but two lenses are used, the forms of which, as used in the smaller European telescopes, and in all the telescopes of Mr. Alvan Clark, are shown in Fig. 32. The crown-glass is here a double convex lens, and the

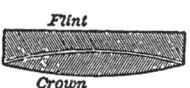

FIG. 32.—Section of an achromatic objective.

curvatures of the two faces are equal. The curvature of the inside face of the flint is the same as that of the crown, so that the two faces fit accurately together, while the outer face is nearly flat. If the dispersive power of the flint were just double that of the crown, this face would have to be flat to produce achromatism; but this is not generally the case. The fact is that, as no two specimens of glass made at different meltings have exactly the same refractive and dispersive powers, the optician, in making a telescope, must find the ratios of dispersion of his two glasses, and then give the outer face of his flint such a degree of curvature as to neutralize the dispersion of his crown glass. Usually, this face will have to be slightly concave.

When the inner faces of the glasses are thus made to fit, it is not uncommon to join the glasses together with a transparent balsam, in order to diminish the loss of light in passing through the glass. Whenever light falls upon transparent glass, between three and four per cent. of it is reflected back, and when, after passing through, it leaves again, about the same amount is reflected back into the glass. Consequently, about seven per cent. of the light is lost in passing through each lens. But when the two lenses are joined with balsam or castor-oil, the reflection from the second surface of the flint and the first surface of the crown is greatly diminished, and a loss of perhaps six per cent. of the light is avoided.*

As larger and more perfect achromatic telescopes were made, a new source of aberration was discovered, no practical method of correcting which is yet known. It arises from the fact that flint glass, as compared with crown, disperses the blue end of the spectrum more than the red end. If we make

---

* When there is no balsam, another inconvenience sometimes arises from a double reflection of light from the inner surfaces of the glass. Of the light reflected back from the first surface of the crown, four per cent. is again reflected from the second surface of the flint, and sent down to the focus of the telescope with the direct rays. If there be the slightest misplacement of one of the lenses, the reflected rays will come to a different focus from the direct ones, and every bright star will seem to have a small companion star along-side of it.

lenses of flint and crown having equal dispersive power, we shall find that the red end is longest in the crown-glass spectrum, and the blue end in the flint-glass spectrum. The consequence is that when we join a pair of prisms in reversed positions, as shown in Fig. 31, the two dispersions cannot be made to destroy each other entirely. Instead of the refracted light being all joined in one white ray, the spectrum will be folded over, as it were, the red and indigo ends being joined together, the faint violet light extending out by itself, while the yellow and green are joined at the opposite end. This end will, therefore, be of a yellowish green, while the other end is purple.

The spectrum thus formed by the combination of a flint and crown prism is termed the *secondary spectrum*. It is very much shorter than the ordinary spectra formed by either the crown or the flint glass, and a large portion of the light is condensed near the yellowish-green end. The effect of it is that the refracting telescope is not perfectly achromatic, though very nearly so. In a small telescope the defect is hardly noticeable, the only drawback being that a bright star or other object is seen surrounded by a blue or violet areole, formed by the indigo rays thrown out by the flint-glass. If the eye-piece is pushed in, so that the star is seen, not as a point, but as a small disk, the centre of this disk will be green or yellow, while the border will be reddish purple. But, in the immense refractors of two-feet aperture or upwards, of which a number have been produced of late years, the secondary aberration constitutes the most serious optical defect; and it is a defect which, arising from the properties of glass itself, no art can diminish. The difficulty may be lessened in the same way that the chromatic aberration was lessened in the older telescopes, namely, by increasing the length of the instrument. In doing this, however, with glasses of such large size, engineering difficulties are encountered which soon become insurmountable. We must, therefore, consider that, in the great refractors of recent times, the limit of optical power for such instruments has been very nearly attained.

The eye-piece of a telescope, as well as its objective, consists of two glasses. A single lens will, indeed, answer all the purposes of seeing an object in the centre of the field of view, but the field itself will be narrow and indistinct at the edges. An additional lens, termed the field-lens, is therefore placed very near the image, for the purpose of refracting the outer rays into the proper direction to form a distinct image with the aid of the eye-lens. In Fig. 33 such an eye-piece is represented, in which the field-lens is between the image and the eye. This is called a *positive* eye-piece. In the negative eye-piece the rays pass through the field-lens just before coming to a focus, so that the image is formed just within that lens. The positive eye-piece is used when it is required to use a micrometer in the focal plane; but for mere looking the negative ocular is best. All telescopes are supplied with a number of eye-pieces, by changing which the magnifying power may be altered to suit the observer.

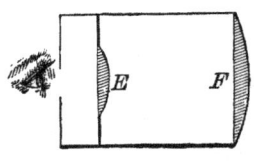

FIG. 33.—Section of eye-piece of a telescope.

The astronomical telescope used with these eye-pieces always shows objects upside down and right side left. This causes no inconvenience in celestial observations. But for viewing terrestrial objects the eye-piece must have two pairs of lenses, the first of which forms a new image of the object restored to its proper position, which image is viewed by the eye-piece formed of the second pair. This combination is called an erecting or terrestrial eye-piece.

§ 3. *The Mounting of the Telescope.*

If the earth did not revolve, so that each heavenly body would be seen hour after hour and day after day in nearly the same direction, the problem of using great telescopes would be much simplified. The objective and the eye-piece could be fixed so as to point at the object, and the observer could scrutinize it at his leisure. But actually, when we use a telescope, the diurnal revolution of the earth is apparently increased in proportion to the magnifying power of the instrument; and if the latter is fixed, and a high power is used, the object passes by with such rapidity that it is impossible to scrutinize it. Merely to point a telescope at an object needs many special contrivances, because, unless the pointing is accurate, the object cannot be found at all. With a telescope, and nothing more, an observer might spend half an hour in vain efforts to point it at Sirius so accurately that the image of the star should be brought into the field of view; and then, before he got one good look, it might flit away and be lost again. If this is the case with a bright star, how much harder must it be to point at the planet Neptune, an object invisible to the naked eye, which is not in the same direction two minutes in succession! It will readily be understood that, to make any astronomical use of a large telescope, two things are absolutely necessary: first, the means of pointing the telescope at any object, visible or invisible; and, second, the means of moving the telescope so that it shall follow the object in its diurnal motion, and thus keep its image in the field of view. The following are the mechanical contrivances by which these objects are effected:

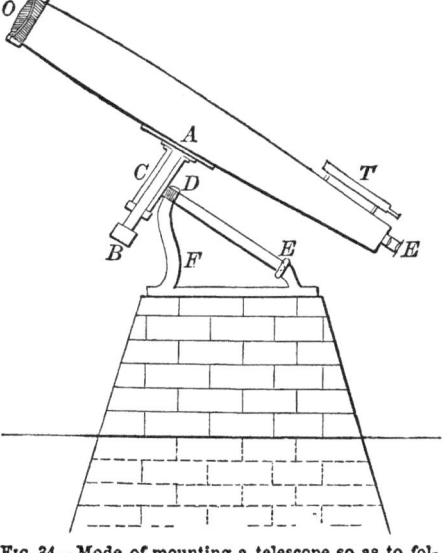

FIG. 34.—Mode of mounting a telescope so as to follow a star in its diurnal motion.

The object-glass is placed in one end of a tube, OE, the length of the tube being nearly equal to the focal length of the objective. The eye-piece is fitted into a projection at the lower end of the tube, E. The object of the tube is to keep the glasses in their proper relative positions, and to protect the eye of the observer from stray light.

The tube has an axis, AB, firmly fastened to it at A near its middle, which axis passes through a cylindrical case, C, into which it neatly fits, and in which it can turn. By turning the telescope on this axis, the end E can be brought towards the reader, and O from him, or *vice versa*. This axis is called the *declination axis*. The case, C, is firmly fastened to a second axis, DE, supported at D and E called the *polar axis*. This axis points to the pole of the heavens, and, by turning it, the whole telescope, with the part, AC, of the case, may be brought towards the observer, while the end B will recede from him; or *vice versa*. In order that the weight of the telescope may not make it turn on the polar axis, it is balanced by a weight at B, on the other end of the declination axis. This weight is commonly divided, a part being carried by the axis, and a part by the case, C. The polar axis is carried by a frame, F, well fastened on top of a pier of masonry.

Such is the general nature of the mechanism by which an astronomical telescope is mounted. The essential point is that there shall be two axes—one fixed, and pointing at the pole, and one at right angles to it, and turning with it. In the arrangement of these axes there are great differences in the telescopes of different makers; but Fig. 34 shows what is essential in the plan of mounting now very generally adopted.

In the figure the telescope is represented as east of the spectator, and as pointed at the pole, and therefore parallel to the polar axis. Suppose now that the telescope be turned on the declination axis, AB, through an arc of 90°, the eye-piece, E, being brought towards the spectator; the object end will then point towards the east horizon, and therefore towards the celestial equator, the eye end pointing directly towards the spectator. Then let the whole instrument be turned on the polar axis, the eye-piece being brought downwards. The telescope will then move along the celestial equator, or the path of a star, 90° from the pole. And at whatever distance from the pole we set it by turning it on the declination axis, if we turn it on the polar axis it will describe a circle having the pole at its centre; that is, the same circle which a star follows by its diurnal motion. So, to observe a star with the telescope, we have first to turn it on the declination axis to the polar distance of the star, and then on the polar axis till it points at the star. This pointing is effected by circles divided into degrees and minutes, not shown in the figure, by which the distance which the telescope points from the pole and from the meridian may be found at any time.

In order that the star, when once found, may be kept in the field of view, the telescope is furnished with a system of clockwork, by which the polar axis is slowly turned at the rate of one revolution a day. By starting this clock-work, the telescope is made to follow the star in its diurnal motion; or, to speak with greater astronomical precision, as the earth turns on its axis from west to east, the telescope turns from east to west with the same angular velocity, so that the direction in which it points in the heavens remains unaltered.

In order to facilitate the finding or recognition of an object, the telescope is furnished with a "finder," T, consisting of a small telescope of low power pointing in the same direction with the larger one. An object can be seen in the small telescope without the pointing being so accurate as is necessary in the case of the large one; and, when once seen, the telescope is moved until the object is in the middle of the field of view, when it is also in the field of view of the large one.

§ 4. *The Reflecting Telescope.*

Two radically different kinds of telescopes are made: the one just described, known as the refracting telescope, because dependent on the refraction of light through glass lenses; and the other, the reflecting telescope, so called because it acts by reflecting the light from a concave mirror. The name of the first inventor of this instrument is disputed; but Sir Isaac Newton was among the first to introduce it. It was designed by him to avoid the difficulty growing out of the chromatic aberration of the refracting telescopes of his time, which, it will be remembered, were not achromatic. If parallel rays of light from a distant object fall upon a concave mirror, as shown in Fig. 35, they will all be reflected back to a focus, *F*, halfway between the centre of curvature, *G*, and the surface of

Fig. 35.—Speculum bringing rays to a single focus by reflection.

the mirror. In order that the rays may be all reflected to absolutely the same focus, the section of the mirror must be a parabola, and the point where the rays meet will be the focus of the parabola. If the rays emanate from the various points of an object, an image of this object will be formed in and near the focus, as in the case of a lens. This image is to be viewed with a magnifying eye-piece like that of a refracting telescope. Such a mirror is called a *speculum*.

Here, however, a difficulty arises. The image is formed on the same side of the mirror on which the object lies; and in order that it may be seen directly, the eye of the observer and the eye-piece must be between *F* and *G*, directly in the rays of light emanating from the object. By placing the eye here, not only would a great deal of the light be cut off by the body of the observer, but the definition of the image would be greatly injured by the interposition of so large an object. Three plans have been devised for evading this difficulty, which are due, respectively, to Gregory, Newton, and Herschel.

The Herschelian Telescope.—In this form of telescope the mirror is slightly tipped, so that the image, instead of being formed in the centre of the tube, is formed near one side of it, as in Fig. 36. The observer can then view it without putting his head inside the tube, and, therefore, without cutting off any material portion of the light. In observation, he must stand at the upper, or outer, end of the tube, and look into it, his back being turned towards the object. From his looking

THE REFLECTING TELESCOPE.

directly into the mirror, it was also called the "front-view" telescope. The great disadvantage of this arrangement is that

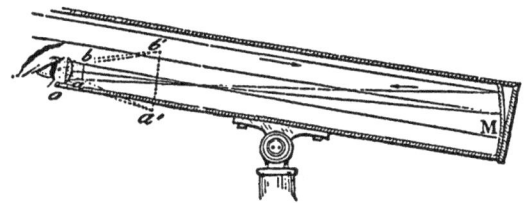

FIG. 36.—Herschelian telescope.

the rays cannot be brought to an exact focus when they are thrown so far to one side of the axis, and the injury to the definition is so great that the front-view plan is now entirely abandoned.

The Newtonian Telescope.—The plan proposed by Sir Isaac Newton was to place a small plane mirror just inside the focus, inclined to the telescope at an angle of 45°, so as to throw the rays to the side of the tube, where they come to a focus, and form the image. An opening is made in the side of the tube, just below where the image is formed in which the eyepiece is inserted. This mirror cuts off some of the light, but not enough to be a serious defect. An improvement which lessens this defect has been made by Professor Henry Draper.

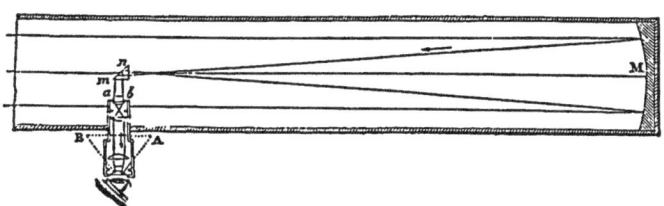

FIG. 37.—Horizontal section of a Newtonian telescope. This section shows how the luminous rays reflected from the parabolic mirror M meet a small rectangular prism $m\ n$, which replaces the inclined plane mirror used in the old form of Newtonian telescope. After undergoing a total reflection from $m\ n$, the rays form at $a\ b$ a very small image of the heavenly body.

The inclined mirror is replaced by a small rectangular prism, by reflection from which the image is formed very near the prism. A pair of lenses are then inserted in the course of the rays, by which a second image is formed at the opening in the side of the tube, and this second image is viewed by an ordinary eye-piece. The four lenses together form an erecting eye-piece.

The Gregorian Telescope.—This is a form proposed by James Gregory, who probably preceded Newton as an inventor of the reflecting telescope. Behind the focus, F, a small concave mirror, R, is placed, by which the light is reflected back again

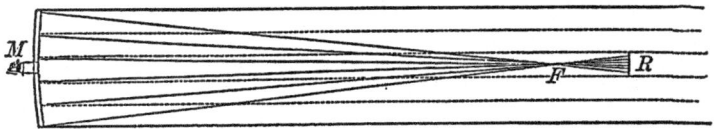

FIG. 38.—Section of the Gregorian telescope.

down the tube. The larger mirror, M, has an opening through its centre, and the small mirror, R, is so adjusted as to form a second image of the object in this opening. This image is then viewed by an eye-piece which is screwed into the opening.

The Cassegrainian Telescope—In principle the same with the Gregorian, differs from it only in that the small mirror, R, is convex, and is placed inside the focus, F, so that the rays are reflected from it before reaching the focus, and no image is formed until they reach the opening in the large mirror. This form has an advantage over the Gregorian in that the telescope may be made shorter, and the small mirror can be more easily shaped to the required figure. It has therefore entirely superseded the original Gregorian form.

Optically, these forms of telescope are inferior to the Newtonian. But the latter is subject to the inconvenience that the observer must be stationed at the upper end of the telescope, where he looks into an eye-piece screwed into the side of the tube. If the telescope is a small one, this inconvenience is not felt; but with large telescopes, twenty feet long or upwards, the case is entirely different. Means must then be provided by which the observer may be carried in the air at a height equal to the length of the instrument, and this requires considerable mechanism, the management of which is often very troublesome. On the other hand, the Cassegrainian telescope is pointed directly at the object to be viewed, like a refractor, and the observer stands at the lower end, and looks in at the opening through the large mirror. This is, therefore, the most convenient form of all in management. One drawback is, that there are two mirrors to be looked after, and, unless the figure of both is perfect, the image will be distorted. Another is the great size of the image, which forces the observer to use either a high magnifying power, or an eye-piece of corresponding size.* But these defects are of little importance compared with the great advantage of convenient use.

§ 5. *The Principal Great Reflecting Telescopes of Modern Times.*

The reflecting telescopes made by Newton and his contemporaries were very small indeed, none being more than a few inches in diameter. Though vastly more manageable than the immensely long refractors of Huyghens, they do not seem to have exceeded them in effectiveness. We might, therefore, have expected the achromatic telescope to supersede the reflector entirely, if it could be made of large size. But in the time of Dollond it was impossible to produce disks of flint-glass of sufficient uniformity for a telescope more than a very few inches in diameter. An achromatic of four inches aperture was then considered of extraordinary size, and good ones of more than two or three inches were rare. Consequently, for the purpose of seeing the most faint and difficult objects, the earlier achromatics were little, if any, better than the long telescopes of Huyghens and Cassini. As there were no such obstacles to the polishing of large mirrors, it was clear that it was to the reflecting telescope that recourse must be had for any great increase in optical power. Before the middle of the last century the reflectors were little larger than the refractors, and had not exceeded them in their optical performance. But a genius now arose who was to make a wonderful improvement in their construction.

---

* The Melbourne telescope has an eye-lens six inches in diameter.

William Herschel, in 1766, was a church-organist and teacher of music of very high repute in Bath, who spent what little leisure he had in the study of mathematics, astronomy, and optics. By accident a Gregorian reflector two feet long fell into his hands, and, turning it to the heavens, he was so enraptured with the views presented to him that he sent to London to see if he could not purchase one of greater power. The price named being far above his means, he resolved to make one for himself. After many experiments with metallic alloys, to learn which would reflect most light, and many efforts to find the best way of polishing his mirror, and giving it a parabolic form, he produced a five-foot Newtonian reflector, which revealed to him a number of interesting celestial phenomena, though, of course, nothing that was not already known. Determined to aim at nothing less than the largest telescope that could be made, he attempted vast numbers of mirrors of constantly increasing size. The large majority of the individual attempts were failures; but among the results of the successful attempts were telescopes of constantly increasing size, until he attained the hitherto unthought-of aperture of two feet, with a length of twenty feet. With one of these he discovered the planet Uranus. The fame of the musician-astronomer reaching the ears of King George III., that monarch gave him a pension of £200 per annum, to enable him to devote his life to a career of astronomical discovery. He now made the greatest stride of all by completing a reflector four feet in diameter and forty feet long, with which he discovered two new satellites of Saturn.

Herschel now found that he had attained the limit of manageable size. The observer had to be suspended perhaps thirty or forty feet in the air, in a room large enough to hold, not only himself, but all the means necessary for recording his observations; and this room had to follow the telescope as it moved, to keep a star in the field. To this was added the difficulty of keeping the mirror in proper figure, the mere change of temperature in the night operating injuriously in this respect. We need not, therefore, be surprised to learn

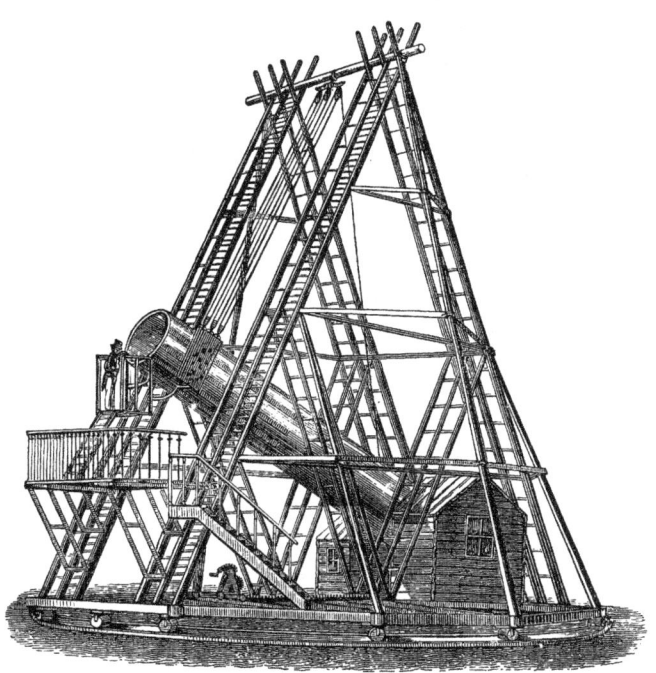

Fig. 39.—Herschel's great telescope.

that Herschel made very little use of this instrument, and preferred the twenty-foot even in scrutinizing the most difficult objects.*

---

* Herschel's great instrument is still preserved, but is not mounted for use; indeed, it is probable that the mirror lost all its lustre long years ago. In 1839, Sir John Herschel dismounted it, laid it in a horizontal position, and closed it up after a family celebration inside the tube, at which the following song was sung:

THE OLD TELESCOPE.

[*To be sung on New-year's-eve, 1839–'40, by Papa, Mamma, Madame Gerlach, and all the Little Bodies in the Tube thereof assembled.*]

In the old Telescope's tube we sit,
And the shades of the past around us flit;
His requiem sing we with shout and din,
While the old year goes out, and the new comes in.
    Chorus.—Merrily, merrily let us all sing,
        And make the old telescope rattle and ring!

The only immediate successor of Sir William Herschel in the construction of great telescopes was his son, Sir John Herschel. But the latter made none to equal the largest of his father's in size, and it is doubtful whether they exceeded them in optical power.

The first decided advance on the great telescope was the celebrated reflector of the Earl of Rosse,* at Parsonstown, Ire-

---

>Full fifty years did he laugh at the storm,
>And the blast could not shake his majestic form;
>Now prone he lies, where he once stood high,
>And searched the deep heaven with his broad, bright eye.
>   *Chorus.*—Merrily, merrily, etc., etc.
>
>There are wonders no living sight has seen,
>Which within this hollow have pictured been;
>Which mortal record can never recall,
>And are known to Him only who made them all.
>   *Chorus.*—Merrily, merrily, etc., etc.
>
>Here watched our father the wintry night,
>And his gaze has been fed with preadamite light.
>His labors were lightened by sisterly love,
>And, united, they strained their vision above.
>   *Chorus.*—Merrily, merrily, etc., etc.
>
>He has stretched him quietly down, at length,
>To bask in the starlight his giant strength;
>And Time shall here a tough morsel find
>For his steel-devouring teeth to grind.
>   *Chorus.*—Merrily, merrily, etc., etc.
>
>He will grind it at last, as grind it he must,
>And its brass and its iron shall be clay and rust;
>But scathless ages shall roll away,
>And nurture its frame, and its form's decay.
>   *Chorus.*—Merrily, merrily, etc., etc.
>
>A new year dawns, and the old year's past;
>God send it, a happy one like the last
>(A little more sun and a little less rain
>To save us from cough and rheumatic pain).
>   *Chorus.*—Merrily, merrily, etc., etc.
>
>God grant that its end this group may find
>In love and in harmony fondly joined!
>And that some of us, fifty years hence, once more
>May make the old Telescope's echoes roar.
>   *Chorus.*—Merrily, merrily, etc., etc.

---

* William Parsons, third Earl of Rosse, the original constructor of this telescope, died in 1867. The work of the instrument is continued by his son, the present earl.

Fig. 40.—Lord Rosse's great telescope.

land. The speculum of this telescope is six feet in diameter, and about fifty-four feet focal length, and was cast in 1842. One of the great improvements made by the Earl of Rosse was the introduction of steam machinery for grinding and polishing the great mirror, an instrumentality of which Herschel could not avail himself. The mounting of this telescope is decidedly different from that adopted by Herschel. The telescope is placed between two walls of masonry, which only allow it to move about 10° on each side of the meridian, and it turns on a pivot at the lower end of the tube. It is moved north and south in the meridian by an ingenious combination of chains, and may thus be set at the polar distance of any star which it is required to observe. It is then moved slowly towards the west, so as to follow the star, by a long screw driven by an immense piece of clock-work. It is commonly used as a Newtonian, the observer looking into the side of the tube near the upper end. To enable him to reach the mouth of the tube, various systems of movable platforms and staging are employed. One of the platforms is suspended south of the piers; it extends east and west by the distance between the walls, and may be raised by machinery so as to be directly under the mouth of the telescope so long as the altitude of the latter is less than 45°. When the altitude is greater than this, the observer ascends a stairway to the top of one of the walls, where he mounts one of several sliding stages, by which he can be carried to the mouth of the telescope, in any position of the latter. This instrument has been employed principally in making drawings of lunar scenery and of the planets and nebulæ. Its great light-gathering power peculiarly fits it for the latter object.

Other Reflecting Telescopes.—Although no other reflector approaching the great one of the Earl of Rosse in size has ever been made, some others are worthy of notice, on account of their perfection of figure and the importance of the discoveries made with them. Among these the first place is due to the great reflectors of Mr. William Lassell, of England. This gentleman made a reflector of two feet aperture about the same time that Rosse constructed his immense six-foot. The perfection of figure of the mirror was evinced by the discovery of two satellites of Uranus, which had been previously unknown and unseen, unless, as is possible, Herschel and Struve caught glimpses of them on a few occasions. He afterwards made one of four feet aperture, which, in 1863, he took to the island of Malta, where he made a series of observations on satellites and nebulæ.

Fig. 41.—Mr. Lassell's great four-foot reflector, as mounted at Malta.

In 1870, a reflecting telescope four feet in diameter, on the Cassegrainian plan, was made by Thomas Grubb & Son, of Dublin, for the Observatory of Melbourne, Australia. This instrument is remarkable, not only for its perfection of figure, but as being probably the most easily managed large reflector ever made.

Fig. 42.—The new Paris reflector.

The only American who has ever successfully undertaken the construction of large reflecting telescopes is Professor Henry Draper, of New York, who has one of twenty-eight inches aperture, the work of his own hands. This instrument was mounted about 1872 in the owner's private observatory at Hastings, on the Hudson. The mirror is not of speculum metal, but of silvered glass, and is almost perfect in figure. This telescope has been principally employed in making photographs of celestial objects, and can be used either as a Newtonian or a Cassegrainian.

An attempt has recently been made at the Paris Observatory to construct a reflecting telescope with a mirror of silvered glass, as large as the great specula of Lassell and the Melbourne Observatory. The diameter of the glass is 120 centimetres, a fraction of an inch short of four English feet. It was figured, polished, and silvered at the Paris Observatory by M. Martin, using the methods devised by Foucault. It was mounted in 1875; but, unfortunately, the proper measures were not taken to prevent the glass from bending under its own weight, and thus destroying the perfection of the parabolic figure which M. Martin had succeeded in obtaining. It was therefore taken from its tube to have this defect of mounting remedied. The machinery for supporting and moving this telescope being in some respects peculiar, we present a view of it in Fig. 42, on page 134.

§ 6. *Great Refracting Telescopes.*

We have already remarked that, in the early days of the achromatic telescope, its progress was hindered by the difficulty of making large disks of flint-glass. About the beginning of the present century, Guinand, a Swiss mechanic, after a long series of experiments, discovered a method by which he could produce disks of flint-glass of a size before unheard of. The celebrated Fraunhofer was then commencing business as an optician in Munich, and hearing of Guinand's success induced him to come to Munich and commence the manufacture of optical glass. Fraunhofer was a physicist of a

Fig. 43.—The great Melbourne reflector. T, the tube containing the great mirror near its lower end. Y, the small mirror throwing the light back to the eye-piece, y. O N, the polar axis. U, the counterpoise at the end of the declination axis. Z, the clock-work which moves the telescope by the jointed rods z e e E, and the clamp F.

high order, and made a more careful and exhaustive study of the optical qualities of glass, and the conditions for making the best telescope, than any one before him had ever attempted. With the aid of the large disks furnished by Guinand, he was able to carry the aperture of his telescopes up to ten inches. Dying in 1826, his successors, Merz and Mahler, of Munich, made two telescopes of fifteen inches aperture, which were then considered most extraordinary. One of these belongs to the Pulkowa Observatory, in Russia; and the other was purchased by a subscription of citizens of Boston for the observatory of Harvard University.

No rival of the house of Fraunhofer in the construction of great refractors arose until he had been dead thirty years, and then it arose where least expected. In 1846, Mr. Alvan Clark was a citizen of Cambridgeport, Massachusetts, unknown to fame, who made a modest livelihood by pursuing the self-taught art of portrait-painting, and beguiled his leisure by the construction of small telescopes. Though without the advantage of a mathematical education, he had a perfect knowledge of optical principles to just the extent necessary to enable him to make and judge a telescope. Having been led by accident to attempt the grinding of lenses, he soon produced objectives equal in quality to any ever made, and, if he had been a citizen of any other civilized country, would have found no difficulty in establishing a reputation. But he had to struggle ten years with that neglect and incredulity which is the common lot of native genius in this country; and, extraordinary as it may seem, it was by a foreigner that his name and powers were first brought to the notice of the astronomical world. Rev. W. R. Dawes, one of the leading amateur astronomers of England, and an active member of the Royal Astronomical Society, purchased an object-glass from Mr. Clark in 1853. He found it so excellent that in the course of the next two or three years he ordered several others, and, finally, an entire telescope. He also made several communications to the Astronomical Society, giving lists of difficult double stars detected by Mr. Clark with telescopes of his own construction, and showing that Mr. Clark's objectives were almost perfect in definition.

The result of this was that the American artist began to be appreciated in his own country; and in 1860 he received an order from the University of Mississippi, of which Dr. F. A. P. Barnard* was then president, for a refractor of eighteen

---

* Now President of Columbia College, New York City.

inches aperture, which was three inches greater than the largest that had then been made. Before the glass was finished, it was made famous by the discovery of the companion of Sirius, a success for which the Lalande medal was awarded by the French Academy of Sciences. While this telescope was in progress, the civil war broke out, and prevented the party originally ordering it from taking it; but it was soon sold to the Astronomical Society of Chicago, in which city it was mounted in 1863. The definition of this telescope is very fine; but the defects of the dome in which it is mounted, and the want of means to support an astronomer, have greatly interfered with its efficiency.

This instrument did not long retain its supremacy. The firm of Thomas Cooke & Sons, of York, England, in 1870, mounted a refractor of twenty-five inches clear aperture for R. S. Newall, Esq., of Gateshead, England, of which the definition is very good. This instrument was intended by its owner to be transported to some finer climate than that of England; but this project has not been put into execution. In the summer of 1874 it was used by Mr. Lockyer, in a study of Coggia's comet.

During the time that these immense telescopes were being made on every hand, and after it was proved that telescopes of more than two feet aperture could be made, the National Observatory of the United States had nothing better than an old Munich refractor of nine and a half inches, such as Fraunhofer used to make early in the century. The attention of Congress was so forcibly called to this deficiency, and to the abilities of the firm of Alvan Clark & Sons to remedy it, that, in 1870, a bill was passed authorizing the superintendent of the observatory to contract for a telescope of the largest size of American manufacture. The aperture agreed on was twenty-six inches, exceeding that of Mr. Newall's telescope by only one inch. It proved extremely difficult to obtain disks of rough glass even of this size, and more than a year elapsed after Messrs. Chance & Co. received the order from Mr. Clark before they were able to complete good disks of the required size. The glass arrived in December, 1871, and work was commenced in January following. The labor of polishing the glasses was completed in October, 1872; the whole instrument was completed in a year more, and was finally mounted and ready for observation in November, 1873. The figure of this glass is almost perfect, its principal defect arising from the secondary aberration which is inseparable from a large refractor. It has been principally employed in observing the satellites of Saturn, Uranus, and Neptune, with the view of determining the masses of these planets.

§ 7. *The Magnifying Powers of the Two Classes of Telescopes.*

Questions which now very naturally arise are, Which of the two classes of telescopes we have described is the more powerful, the reflector or the refractor? and is there any limit to the magnifying power of either? To these questions it is difficult to return a decided answer, because each class has its peculiar advantages, and in each class many difficulties lie in the way of obtaining the highest magnifying power. The fact is, that very exaggerated ideas of the magnifying power of great telescopes are entertained by the public. It will, therefore, be instructive to state what the circumstances are which prevent these ideas from being realized, and what the conditions are on which the seeing power of telescopes depends.

We note, first, that when we look at a luminous point—a star, for instance—without a telescope, we see it by the aid of the cone of light which enters the pupil of the eye. The diameter of the pupil being about one-fifth of an inch, as much light from the star as falls on a circle of this diameter is brought to a focus on the retina, and unless this quantity of light is sufficient to be perceptible, the star will not be seen. Now, we may liken the telescope to a "Cyclopean eye," of which the object-glass is the pupil, because, by its aid, all the light which falls on the object-glass is brought to a focus on the retina, provided that a sufficiently small eye-piece is used. Of course, we must except that portion of the light which is lost in passing through the glasses. Since the quantity of light which falls on a surface is proportional to the extent of the surface, and therefore to the square of its diameter, it follows that, because a telescope of one-inch clear aperture has five times the diameter of the pupil, it will admit 25 times the light; a six-inch will admit 900 times the light which the pupil will; and so with any other aperture. A star viewed with the telescope will, therefore, appear brighter than to the naked eye in proportion to the square of the aperture of the instrument. But the star will not be magnified like a planet, because a point is only a point, no matter how often we multiply it. It is true that a bright star in the telescope sometimes appears to have a perceptible disk; but this is owing to various imperfections of the image, having their origin in the air, the instrument, and the eye, all of which have the effect of slightly scattering a portion of the light which comes from the star. Hence, with perfect vision the apparent brilliancy of a star will be proportional to the square of the aperture of the telescope. It is said that Sir William Herschel, at a time when by accident his telescope was so pointed that Sirius was about to enter its field of view, was first apprised of what was coming by the appearance of a dawn like the morning. The light increased rapidly, until the star itself appeared with a dazzling splendor which reminded him of the rising sun. Indeed, in any good telescope of two feet aperture or upwards, Sirius is an almost dazzling object to an eye which has rested for some time in darkness.

But in order that all the light which falls on the object-glass, or mirror, of a telescope may enter the pupil of the eye, it is necessary that the magnifying power be at least equal to the ratio which the aperture of the telescope bears to that of the pupil. The latter is generally about one-fifth of an inch. We must, therefore, employ a magnifying power of at least five for every inch of aperture, or we will not get the full advantage of our object-glass. The reason of this will be apparent by studying Fig. 29, p. 109, from which it will be seen that a pencil of parallel rays falling on the object-glass, and passing through the eye-piece, will be reduced in diameter in the ratio of the focal distance of the objective to that of the eye-piece, which is the same as the magnifying power. For instance, if to a twenty-four-inch telescope we attached an eye-piece so large that the magnifying power was only 48, and pointed it at a bright star, the "emergent pencil" of rays from the eye-piece would be half an inch in diameter, and the whole of them could not possibly enter the pupil. By increasing the magnifying power, we would increase the apparent brilliancy of the star, until we reached the power 120, after which no further increase of brilliancy would be possible.

All this supposes that we are viewing a star or other luminous point. If the object has a sensible surface, like the moon, or a large nebula, and we consider its apparent superficial brilliancy, the case will be in part reversed. The object will then appear equally illuminated, with all powers below five for each inch of aperture, but will begin to grow darker when we pass above that limit. The reason of this is, that as we increase the magnifying power the light is spread over a larger surface of the retina, and is thus enfeebled. So long as our magnifying power is below the limit, the increased quantity of light which enters the pupil by an increase of magnifying power just compensates for the greater surface over which it is spread, so that the brilliancy is constant. Above the limit of five to the inch, the surface over which the light is spread, or the apparent magnitude of the object, still increases with the magnifying power, but there is no increase of light; hence, the object looks fainter. What may at first sight seem paradoxical is, that the degree of illumination to which we now refer can never be increased by the use of the telescope, but, at the best, will be the same as to the naked eye. Indeed, as some light is necessarily lost in passing through any telescope, the illumination is always less with the telescope. With the best reflectors of speculum metal, the illumination will be reduced to one-half, or less, if the polish is not perfect; and with refractors it will be reduced to seven or eight tenths. As examples of these conclusions, the sky can never be made to appear as bright through a telescope as to the naked eye; the moon or a large nebula will appear more brightly illuminated through a refracting telescope than through a reflector. If the object is a very brilliant one, like the sun or Venus, the loss of brilliancy by magnifying, which we have described, will not cause any inconvenience; but the outer planets and many of the nebulæ are so faintly illuminated that a magnifying power many times exceeding the limit cannot be used with advantage.

Still another cause which places a limit to the power of telescopes is diffraction. When the "emergent pencil" is reduced below $\frac{1}{50}$ of an inch in diameter—that is, when the magnifying power is greater than 50 for every inch of aperture of the object-glass—the outlines of every object observed become confused and indistinct, no matter how bright the illumination or how perfect the glass may be. The effect is the same as if we looked through a small pin-hole in a card, an experiment which anyone may try. This effect is owing to the diffraction of the light at the edge of the object-glass or mirror, and it increases so rapidly with the magnifying power that when we carry the latter above 100 to the inch, the increase of indistinctness neutralizes the increase of power. If, then, we multiply the aperture of the telescope in inches by 100, we shall have a limit beyond which there is no use in magnifying. Indeed, it is doubtful if any real advantage is gained beyond 60 to the inch. In a telescope of two feet (24 inches) aperture this limit would be 2400. Such a limit cannot be set with entire exactness; but, even under the most favorable circumstances, the advantage in attempting to surpass a power of 70 to the inch will be very slight.

The foregoing remarks apply to the most perfect telescopes, used under the most favorable circumstances. But the best telescope has imperfections which would nearly always prevent the use of the highest magnifying powers in astronomical observations. In the refracting telescope the principal defect arises from the secondary aberration already explained, which, arising from an inherent quality of the glass itself, cannot be obviated by perfection of workmanship. In the case of the reflector, the corresponding difficulty is to keep the mirror in perfect figure in every position. As the telescope is moved about, the mirror is liable to bend, through its own weight and elasticity, to such an extent as greatly to injure or destroy the image in the focus; and, though this liability is greatly diminished by the plan now adopted, of supporting the mirror on a system of levers or on an air-cushion, it is generally troublesome, owing to the difficulty of keeping the apparatus in order.

If we compare the refracting and reflecting telescopes which have hitherto been made, it is easy to make a summary of their relative advantages. If properly made and attended to, the refractor is easy to manage, convenient in use, and always in order for working with its full power. If its greatest defect, the secondary spectrum, cannot be diminished by skill, neither can it be increased by the want of skill on the part of the observer. So important is this certainty of operation, that far the greater part of the astronomical observations of the present century have been made with refractors, which have always proved themselves the best working instruments. Still, the defects arising from the secondary spectrum are inherent in the latter, and increase with the aperture of the glass to such an extent that no advantage can ever be gained by carrying the diameter of the lenses beyond a limit which may be somewhere between 30 and 36 inches. On the other hand, when we consider mere seeing-power, calculation at least gives the preference to the reflector. It is easy to compute that Lord Rosse's "Leviathan," and the four-foot reflectors of Mr. Lassell and of the Paris and Melbourne observatories, must collect from two to four times the light of the great Washington telescope. But when, instead of calculation, we inquire what difficult objects have actually been seen with the two classes of instruments, the result seems to indicate that the greatest refractor is equal in optical power to the great reflectors. No known object seen with the latter is too faint to be seen with the former. Why this discrepancy between the calculated powers of the great reflectors and their actual performance? The only causes we can find for it are imperfections in the figure and polish of the great mirrors. The great refractors are substantially perfect in their workmanship; the reflectors do not appear to be perfect, though what the imperfections may be, it is impossible to say with entire certainty. Whether the great telescope of the future shall belong to the one class or the other must depend upon whether the imperfections of the reflecting mirror can be completely overcome. Mr. Grubb, the maker of the great Melbourne telescope, thinks he has completely succeeded in this, so as to insure a mirror of six, seven, or even eight feet in diameter which shall be as perfect as an object-glass. If he is right — and there is no mechanician whose opinion is entitled to greater confidence — then he has solved the problem in favor of the reflector, so far as optical power is concerned. But so large a telescope will be so difficult to manipulate, that we must still look to the refractor as the working instrument of the future as well as of the past; though, for the discovery and examination of very faint objects, it may be found that the advantage will all be on the side of the future great reflector.

The great foe to astronomical observation is one which people seldom take into account, namely, the atmosphere. When we look at a distant object along the surface of the ground on a hot summer day, we notice a certain waviness of outline, accompanied by a slight trembling. If we look with a telescope, we shall find this waving and trembling magnified as much as the object is, so that we can see little better with the most powerful telescope than with the naked eye. The cause of this appearance is the mixing of the hot air near the ground with the cooler air above, which causes an irregular and constantly changing refraction, and the result is that astronomical observations requiring high magnifying power can very rarely be advantageously made in the daytime. By night the air is not so much disturbed, yet there are always currents of air of slightly different temperatures, the crossing and mixing of which produce the same effects in a small degree. To such currents is due the twinkling of the stars; and we may lay it down as a rule, that when a star twinkles the finest observation of it cannot be made with a telescope of high power. Instead of presenting the appearance of a bright, well-defined point, it will look like a blaze of light flaring about in every direction, or like a pot of molten boiling metal; and the higher the magnifying power, the more it will flare and boil. The amount of this atmospheric disturbance varies greatly from night to night, but it is never entirely absent. If no continuous disturbance of the image could be seen with a power of 400, most astronomers would regard the night as a very good one; and nights on which a power of more than 1000 can be advantageously employed are quite rare, at least in this climate.

It has sometimes been said that Sir William Herschel employed a power as high as 6000 with one of his great telescopes, and, on the strength of this, that the moon may have been brought within an apparent distance of forty miles. If such a power was used on the moon, we must suppose, not merely that the moon was seen as if at the distance of forty miles, even if Herschel used his largest telescope — that of four feet aperture—but that the vision would be the same as if he had looked through a pin-hole $\frac{1}{125}$ of an inch in diameter, and through several yards of running water, or many miles of air. It is doubtful whether the moon has ever been seen with any telescope so well as it could be seen with the naked eye at a distance of 500 miles. If such has been the case, we may be sure that the magnifying power did not exceed 1000.

If seeing depended entirely on magnifying power, we could not hope to gain much by further improvement of the telescope, unless we should mount our instrument in some place where there is less atmospheric disturbance than in the regions where observatories have hitherto been built. It is supposed that, on the mountains or table-lands in the western and south-western regions of North America, the atmosphere is clear and steady in an extraordinary degree; and if this supposition is entirely correct, a great gain to astronomy might result from establishing an observatory in that region.

CHAPTER II.

APPLICATION OF THE TELESCOPE TO CELESTIAL MEASUREMENTS.

§ 1. *Circles of the Celestial Sphere, and their Relations to Positions on the Earth.*

IN the opening chapter of this work it was shown that all the heavenly bodies seem to lie and move on the surface of a sphere, in the interior of which the earth and the observer are placed. The operations of Practical Astronomy consist largely in determining the apparent positions of the heavenly bodies on this sphere. These positions are defined in a way analogous to that in which the position of a city or a ship is defined on the earth, namely, by a system of celestial latitudes and longitudes. That measure which, in the heavens, corresponds most nearly to terrestrial longitude is called *Right Ascension*, and that which corresponds to terrestrial latitude is called *Declination*.

In Fig. 45 let the globe be the celestial sphere, represented as if viewed from the outside by an observer situated towards the east, though we necessarily see the actual sphere from the centre. P is the north pole, AB the horizon, Q the south pole (invisible in northern latitudes because below the horizon), EF the equator, Z the zenith. The meridian lines radiate from the north pole in every direction, cross the equator at right angles, and meet again at the south pole, just like meridians on the earth. The meridian from which right ascensions are counted, corresponding in this respect to the meridian of Greenwich on the surface of the earth, is that which passes through the vernal equinox, or point of crossing of the equator and ecliptic. It is called the first meridian. Three bright stars near which this meridian now passes may be seen during the autumn: they are α Andromedæ and γ Pegasi, on Maps II. and V., and β Cassiopeiæ, on Map I. The right ascension of any star on this meridian is zero, and the right ascension of any other star is measured by the angle which the meridian passing through it makes with the first meridian, this angle being always counted towards the east. For reasons which will soon be explained, right ascension is generally reckoned, not in degrees, but in hours, minutes, and seconds of time.

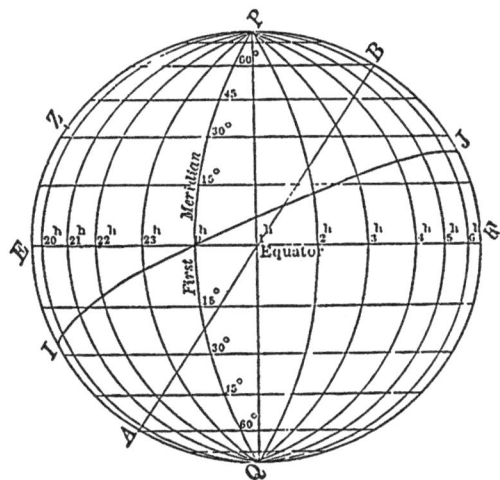

Fig. 44.—Circles of the celestial sphere.

IJ is the ecliptic, crossing the equator at its point of intersection with the first meridian, and making an angle of $23\frac{1}{2}°$ with it. The declination of a star is its distance from the celestial equator, whether north or south, exactly as latitude on the earth is distance from the earth's equator. Thus, when the right ascension and declination of a heavenly body are given, the astronomer knows its position in the celestial sphere, just as we know the position of a city on the earth when its longitude and latitude are given.

It must be observed that the declinations of the heavenly

bodies are, in a certain sense, referred to the earth. In astronomy the equator is regarded as a plane passing through the centre of the earth, at right angles to its axis, and dividing it into two hemispheres. The line where this plane intersects the surface of the earth is our terrestrial, or geographical, equator. If an observer standing on the geographical equator imagines this plane running east and west, and cutting into and through the earth, where he stands he will have the astronomical equator, which differs from the geographical equator only in being the plane in which the latter is situated. Now imagine this plane continued in every direction without limit till it cuts the infinite celestial sphere as in Fig. 17, page 62. The circle in which it intersects this sphere will be the celestial equator. It will pass directly over the head of the observer at the equator.

There is a general correspondence between latitude on the earth and declination in the heavens, which may be seen by referring to the same figure. Here the reader must conceive of the earth as a globe, *ep*, situated in the centre of the celestial sphere, *EPQS*, which is infinitely larger than the earth. The plane represented by *EQ* is the astronomical equator, dividing both the earth and the imaginary celestial sphere into two equal hemispheres. Suppose, now, that the observer, instead of standing under the equator, is standing under some other parallel, say that of 45° N. (Being in this latitude means that the plumb-line where he stands makes an angle of 45° with the plane of the equator.) The point over his head will then be in 45° celestial declination. If we imagine a pencil of infinite length rising vertically where the observer stands so that its point shall meet the celestial sphere in his zenith, and if, as the earth performs its diurnal revolution on its axis, we imagine this pencil to leave its mark on the celestial sphere, this mark will be the parallel of 45° N. declination, or a circle everywhere equally distant from the equator and from the pole. The same observer will see the celestial pole at an elevation equal to his latitude, that is, at the angle 45°. We have now the following rules for determining the latitude of a place:

CIRCLES OF THE CELESTIAL SPHERE. 149

1. *The latitude is equal to the declination of the observer's zenith.*
2. *It is also equal to the altitude of the pole above his horizon.*

Hence, if the astronomer at any unknown station wishes to determine his latitude, he has only to find what parallel of declination passes through his zenith, the latter being marked by the direction of the plumb-line, or by the perpendicular to the surface of still water or quicksilver. If he finds a star passing exactly in his zenith, and knows its declination, he has his latitude at once, because it is the same as the star's declination. Practically, however, an observer will never find a known star exactly in his zenith; he must therefore find at what angular distance from the zenith a known star passes his meridian, and by adding or subtracting this distance from the star's declination he has his latitude. If he does not know the declination of any star, he measures the altitudes above the horizon at which any star near the pole passes the meridian, both above the pole and under the pole. The mean of the two gives the latitude.

Let us now consider the more complex problem of determining longitudes. If the earth did not revolve, the observer's longitude would correspond to the right ascension of his zenith in the same fixed manner that his latitude corresponds to its declination. But, owing to the diurnal motion, there is no such fixed correspondence. It is therefore necessary to have some means of representing the constantly varying relation.

Wherever on the earth's surface an observer may stand, his meridian, both terrestrial and celestial, is represented astronomically by an imaginary plane similar to the plane of the equator. This plane is vertical to the observer, and passes through the poles. It divides the earth into two hemispheres, and is perpendicular to the equator. In Fig. 17, the celestial and terrestrial spheres are supposed to be cut through by this plane; it cuts the earth when the observer stands in a line running north and south from pole to pole, and thus forms a terrestrial meridian. The same plane intersects the celestial sphere in a great circle, which, rising above the observer's horizon in the north, passes through the pole and the zenith, and disappears at the south horizon. Two observers north and south of each other have the same meridian; but in different longitudes they have different meridians, which, however, all pass through each pole.

In consequence of the earth's diurnal motion, the meridian of every place is constantly moving among the stars in such a way as to make a complete revolution in 23 hours 56 minutes 4.09 seconds. The reader will find it more easy to conceive of the celestial sphere as revolving from east to west, the terrestrial meridian remaining at rest; the effect being geometrically the same whether we conceive of the true or the apparent motion. There are, then, two sets of meridians on the celestial sphere. One set (that represented in Fig. 45) is fixed among the stars, and is in constant apparent motion from east to west with the stars, while the other set is fixed by the earth, and is apparently at rest.

As differences of latitude are measured by angles in the heavens, so differences of terrestrial longitude are measured by the time it takes a celestial meridian to pass from one terrestrial meridian to another; while differences of right ascension are measured by the time it takes a terrestrial meridian to move from one celestial meridian to another. Ordinary solar time would, however, be inconvenient for this measure, because a revolution does not take place in an exact number of hours. A different measure, known as *sidereal time*, is therefore introduced. The time required for one revolution of the celestial meridian is divided into 24 hours, and these hours are subdivided into minutes and seconds. *Sidereal noon* at any place is the moment at which the vernal equinox passes the meridian of that place, and sidereal time is counted round from 0 hour to 24 hours, when the equinox will have returned to the meridian, and the count is commenced over again. Since right ascensions in the heavens are counted from the equinox, when it is sidereal noon, or 0 hour, all celestial objects on the meridian of the place are in 0° of right ascension. At 1 hour sidereal time, the meridians have moved 15°, and objects now on the meridian are in 15° of right ascension. Throughout its whole diurnal course the right ascension of the meridian constantly increases at the rate of 15° per hour, so that the right ascension is always found by multiplying the sidereal time by 15. To avoid this constant multiplication, it is customary in astronomy to express both right ascensions and terrestrial longitudes by hours. Thus the Pleiades are said to be in 3 hours 40 minutes right ascension, meaning that they are on the meridian of any place at 3 hours 40 minutes sidereal time. The longitude of the Washington Observatory from Greenwich is 77° 3'; but in astronomical language the longitude is said to be 5 hours 8 minutes 12 seconds, meaning that it takes 5 hours 8 minutes 12 seconds for any celestial meridian to pass from the meridian of Greenwich to that of Washington. In consequence, when it is 0 hour, sidereal time at Washington, it is 5 hours 8 minutes 12 seconds sidereal time at Greenwich.

About March 22d of every year, sidereal 0 hour occurs very nearly at noon. On each successive day it occurs about 3 minutes 56 seconds earlier, which in the course of a year brings it back to noon again. Since the sidereal time gives the position of the celestial sphere relatively to the meridian of any place, it is convenient to know it in order to find what stars are on the meridian. The following table shows the sidereal time of mean, or ordinary civil, noon at the beginning of each month:

	Hrs. Min.		Hrs. Min.
January	18 45	July	6 38
February	20 47	August	8 40
March	22 37	September	10 43
April	0 40	October	12 41
May	2 38	November	14 43
June	4 40	December	16 42

The sidereal time at any hour of the year may be found from the preceding table by the following process within a very few minutes: To the number of the preceding table corresponding to the month add 4 minutes for each day of the month, and the hour past noon. The sum of these numbers, subtracting 24 hours if the sum exceeds that quantity, will give the sidereal time. As an example, let it be required to find the sidereal time corresponding to November 13th at 3 A.M. This is 15 hours past noon. So we have

	Hrs.	Min.
November, from table.......................................	14	43
13 days × 4..	0	52
Past noon...	15	0
Sum...	30	35
Subtract..	24	0
Sidereal time required...........................	6	35

The sidereal time obtained in this way will seldom or never be more than five minutes in error during the remainder of this century. In every observatory the principal clock runs by sidereal time, so that by looking at its face the astronomer knows what stars are on or near the meridian. Having the sidereal time, the stars which are on the meridian may be found by reference to the star maps, where the right ascensions are shown on the borders of the maps.

§ 2. *The Meridian Circle, and its Use.*

As a complete description of the various sorts of instruments used in astronomical measurements, and of the modes of using them, would interest but a small class of readers, we shall confine ourselves for the present to one which may be called the fundamental instrument of modern astronomy, the application of which has direct and immediate reference to the circles of the celestial sphere described in the preceding section. This one is termed the *Meridian Circle*, or *Transit Circle*. Its essential parts are a moderate-sized telescope balanced on an axis passing through its centre, with a system of fine lines in the eye-piece; one or two circles fastened on the axis, revolving with the telescope, and having degrees and subdivisions cut on their outer edges; and a set of microscope micrometers for measuring between the lines so cut. It is absolutely necessary that every part of the instrument shall be of the most perfect workmanship, and that the masonry piers on which it is mounted shall be as stable as it is possible to make them.

There are many differences of detail in the construction and mounting of different meridian circles, but they all turn on an east and west horizontal axis, and therefore the telescope moves only in the plane of the meridian. Fig. 45 shows the

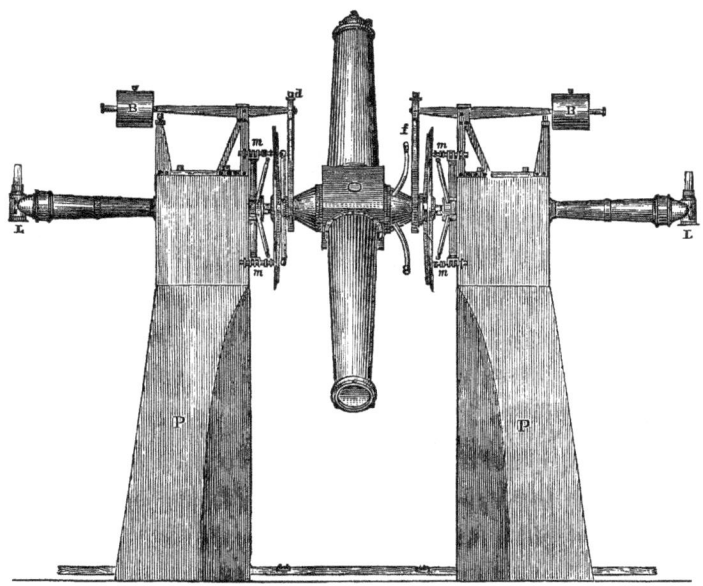

FIG. 45.—The Washington transit circle.

construction of the great circle in the Naval Observatory, Washington. The marble piers, *PP*, are supported on a mass of masonry under the floor, the bottom of which is twelve feet below the surface of the ground. The middle of the telescope is formed of a large cube, about fifteen inches on each side. From the east and west side of this cube extend the trunnions, which are so large next the cube as to be nearly conical in shape. The outer ends terminate in finely ground steel pivots two and a half inches in diameter, which rest on brass V's firmly fixed to heavy castings set into the piers with hydraulic cement. In order that the delicate pivots may not be worn by the whole weight of the instrument resting on them, the counterpoises, *BB*, support all the weight except 30 or 40 pounds. Near the ends of the axis are the circles, seen edgewise, which are firmly screwed on the trunnions, and therefore turn with the instrument. Each pier carries four arms, and each of these arms carries a microscope, marked *m*, having in its focus the face of the circle on which the lines are cut. These lines divide the circle into 360°, and each degree into thirty spaces of two minutes each, so that there are 10,800 lines cut on the circle. They are cut in a silver band, and are so fine as to be invisible to the naked eye unless the light is thrown upon them in a particular way. On each side of the instrument, in a line with the axis, is a lamp which throws light into the telescope so as to illuminate the field of view. Reflecting prisms inside of the pier throw some of the light upon those points of the circle which are viewed by the microscopes, so as to illuminate the fine divisions on the circle. Being thus limited in its movements, an object can be seen with the telescope only when on, or very near, the meridian. The sole use of the instrument is to observe the exact times at which stars cross the meridian, and their altitudes above the horizon, or distances from the zenith, at the time of crossing. To give precision to these observations, the eye-piece of the instrument is supplied with a system of fine black lines, usually made of spider's web, as shown in Fig. 46. These lines are set in the focus, so that the image of a star crossing the meridian passes over them. The middle vertical spider line marks the meridian; and to find the time of meridian transit of a star it is only necessary to note the moment of passage of its image over this line. But, to give greater precision and certainty to his

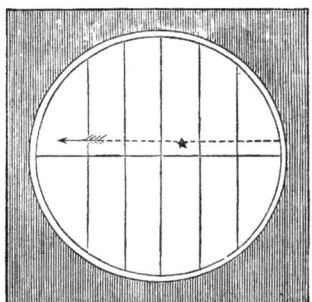

Fig. 46.—Spider lines in field of view of a meridian circle.

THE MERIDIAN CIRCLE, AND ITS USE. 155

observation, the astronomer generally notes the moments of transit over five or more lines, and takes the average of them all.

Formerly the astronomer had to find the times of transit by listening to the beat of his sidereal clock, counting the seconds, and estimating the tenths of a second at which the transit over a line took place. If, for instance, he should find that the star had not reached the line when the tick of twenty-three seconds was heard, but crossed before the twenty-fourth second was ticked, he would know that the time was twenty-three seconds and some fraction, and would have to estimate what that fraction was. A skilful observer will generally make this estimate within a tenth of a second, and will only on rare occasions be in error by as much as two tenths.

Shortly after the introduction of the electric-telegraph, the American astronomers of that day introduced a much easier method of determining the time of transit of a star, by means of the *electro-chronograph*. As now made, this instrument consists of a revolving cylinder, having a sheet of paper wrapped around it, and making one revolution per minute. A pen or other marker is connected with a telegraphic apparatus in such a way that whenever a signal is sent to the pen it makes a mark on the moving paper. This pen moves lengthwise of the cylinder at the rate of about an inch in ten minutes, so that, in consequence of the turning of the cylinder on its axis, the marks of the pen will be along a spiral, the folds of which are one-tenth of an inch apart. The galvanic circuit which works the pen is connected with the sidereal clock, so that the latter causes the pen to make a signal every second. The same pen may be worked by a telegraphic key in the hand of the observer. The latter, looking into his telescope, and watching the approach of the image of the star to each wire, makes a signal at the moment at which the star crosses. This signal is recorded on the chronograph in its proper place among the clock signals, from which it may be distinguished by its greater strength. The record is permanent, and the sheet may be taken off and read at leisure, the exact tenth of a second at which each signal was made being seen by its position among the clock signals. The great advantages of this method are, that great skill and practice are not required to make good observations, and that the observer need not see either the clock or his book, and can make a great number of observations in the course of the evening which may be read off at leisure. In the case of the most skilful observers there is no great gain in accuracy, for the reason that they can estimate the fraction of a second by the eye and ear with nearly the same accuracy that they can give the signal.

The zenith distance of the star, from which its declination is determined, is observed by having in the reticule a horizontal spider line which is made to bisect the image of the star as it passes the meridian line. The observer then goes to the microscopes, ascertains what lines cut on the circle are under them, and what number of seconds the nearest line is from the proper point in the field of the microscope. The mean of the results from the four microscopes is called the *circle-reading*, and can be determined within two or three tenths of a second of arc, or even nearer, if all the apparatus is in the best order. The minuteness of this angle may be judged by the circumstance that the smallest round object a keen eye can see subtends an angle of about forty seconds.

We have described only the leading operations necessary in determinations with a meridian circle. To complete the determination of the position of a star as accurately as a practised observer can bisect it with the spider line is a much more complicated matter, owing to the unavoidable errors and imperfections of his instrument. It is impossible to set the latter in the meridian with mathematical precision, and, if it were done, it would not remain so a single day. When the astronomer comes to tenths of seconds, he has difficulties to contend with at every step. The effects of changes of temperature and motions of the solid earth on the foundations of his instrument are such as to keep it constantly changing; his clock is so far from going right that he never attempts to set it perfectly right, but only determines its error from his observations. Every observation must, therefore, be corrected for a number of instrumental errors before the result is accurate, an operation many times more laborious than merely making the observation.

§ 3. *Determination of Terrestrial Longitudes.*

The telegraphic mode of recording observations, described in the last section, affords a method of determining differences of longitude between places connected by telegraph of extraordinary elegance and perfection. We have already shown that the difference of longitude between two points is measured by the time it takes a star to move from the meridian of the easternmost point to that of the westernmost point. We have also explained in the last section how an observer with a meridian circle determines and records the passage of a star over his meridian within a tenth of a second. Since the zenith distance of the star is not required in this observation, the circles and microscopes may be dispensed with, and the instrument is then much simpler in construction, and is termed a *Transit Instrument.* When the observer makes a telegraphic record of the moment of transit of a star by striking a key in the manner described, it is evident that the electro-chronograph on which his taps are recorded may be at any distance to which the electric current can carry his signal. It may, therefore, be in a distant city. There is no difficulty in a Washington observer recording his observations in Cincinnati.

On this system, the mode of operation is about as follows: the Washington and Cincinnati stations each has its transit instrument, its observer, and its chronograph; but the chronographs are connected by telegraph, so that any signal made by either observer is recorded on both chronographs. As the Washington observer sees a star previously agreed on pass over the lines in the focus of his instrument, he makes signals with his telegraphic key, which are recorded both on his own chronograph and on that of Cincinnati. When the star reaches the meridian of the latter city, the observer there signals the transit of the star in like manner, and the moment of passage over each line in the focus of his instrument is recorded, both in Cincinnati and Washington. The elapsed time is then found by measuring off the chronograph sheets.

The reason for having all the observations recorded on both chronographs is that the results may be corrected for the time it takes the electric current to pass between the two cities, which is quite perceptible at great distances. In consequence of this "wave-time," the Washington observation will be recorded a little too late at Cincinnati, so that the difference of longitude on the Cincinnati chronograph will be too small. The Cincinnati observation, which comes last, being recorded a little too late at Washington, the difference of time on the Washington chronograph will be a little too great. The mean of the results on the two chronographs will be the correct longitude, while their difference will be twice the time it takes the electric current to pass between the two cities. The results thus obtained for the velocity of electricity are by no means accordant, but the larger number do not differ very greatly from 8000 miles per second.

A celestial meridian moves over the earth's surface at the rate of fifteen degrees an hour, or a minute of arc in four seconds of time. More precisely, this is the rate of rotation of the earth. The length of a minute of arc in longitude depends on the latitude. It is about 6000 feet, or a mile and a sixth at the equator, but diminishes whether we go north or south, owing to the approach of the meridians on the globular earth, as can be seen on a globe. In the latitude of our Middle States it is about 4600 feet, so that the surface of the earth there moves over 1150 feet a second. At the latitude of Greenwich it is 3800 feet, so that the motion is 950 feet per second. Two skilful astronomers, by making a great number of observations, can determine the time it takes the stars to pass from one meridian to another within one or two hundredths of a second of time, and can therefore make sure of the difference of longitude between two distant cities within six or eight yards.

Of late the telegraphic method of determining longitudes

has been applied in a way a little different, though resting on the same principles. Instead of recording the transits of stars on both chronographs, each observer determines the error of his clock by transits of stars of which the right ascension has been carefully determined. Each clock is then connected with both chronographs by means of the telegraphic lines, and made to record its beats for the space of a few minutes only. Thus the difference between the sidereal times at the two stations for the same moment of absolute time can be found, and this difference is the difference of longitude in time. A few years ago, when the difference of longitude between points on the Atlantic and Pacific coasts was determined by the Coast Survey, a clock in Cambridge was made to record its beats on a chronograph in San Francisco, and *vice versa*. In 1866, as soon as the Atlantic cable had been successfully laid, Dr. B. A. Gould went to Europe, under the auspices of the Coast Survey, to determine the difference of longitude between Europe and America. Owing to the astronomical importance of this determination, it has since been twice repeated, once under the direction of Mr. Dean, and, lastly, under that of Mr. Hilgard, both of the Survey. These three campaigns gave the following separate results for the difference of longitude between the Royal Observatory, Greenwich, and the Naval Observatory, Washington:

	Hrs.	Min.	Sec.
Dr. Gould, 1867	5	8	12.11
Mr. Dean, 1870	5	8	12.16
Mr. Hilgard, 1872	5	8	12.09

The extreme difference, it will be seen, is less than a tenth of a second, and would probably have been smaller but for the numerous difficulties attendant on a determination through a long ocean cable, which are much greater than through a land line.

The use of the telegraph for the determination of longitude is necessarily limited, and other methods must therefore generally be used. The general problem of determining a longitude, whether that of a ship upon the ocean or of a station upon the land, depends on two requirements: (1) a knowledge of the local time at the station, and (2) a knowledge of the corresponding time at Greenwich, Washington, or some other standard meridian. The difference of these two represents the longitude.

The first determination, that of the local time, is not a difficult problem when the utmost accuracy is not required. We have already shown how it is determined with a transit instrument. But this instrument cannot be used at all at sea, and is somewhat heavy to carry and troublesome to set up on the land. For ships and travellers it is, therefore, much more convenient to use a sextant, by which the altitude of the sun or of a star above the horizon can be measured with very little time or trouble. To obtain the time, the observation is made, not when the object is on the meridian, but when it is as nearly as practicable east or west. Having found the altitude, the calculation of a spherical triangle from the data given in the *Nautical Almanac* at once gives the local time, or the error of the chronometer on local time.

The difficult problem is to determine the Greenwich time. So necessary to navigation is some method of doing this, that the British Government long had a standing offer of a reward of £10,000 to any one who would find a successful method of determining the longitude at sea. When the office of Astronomer Royal was established, which was in 1675, the duty of the incumbent was declared to be "to apply himself with the most exact care and diligence to the rectifying the Tables of the Motions of the Heavens, and the places of the Fixed Stars, in order to find out the so much desired Longitude at Sea for the perfecting the Art of Navigation." The reward above referred to was ultimately divided between an astronomer, Tobias Mayer, who made a great improvement in the tables of the moon, and a watch-maker who improved the marine chronometer.

The moon, making her monthly circuit of the heavens, may be considered a sort of standard clock from which the astronomer can learn the Greenwich time, in whatever part of the

DETERMINATION OF TERRESTRIAL LONGITUDES. 161

world he may find himself. This he does by observing her positions among the stars. The *Nautical Almanac* gives the predicted distance of the moon from certain other bodies—sun, planets or bright stars—for every three hours of Greenwich time; and if the astronomer or navigator measures this distance with a sextant, he has the means of finding at what Greenwich time the distance was equal to that measured. Unfortunately, however, this operation is much like that of determining the time from a clock which has nothing but an hour-hand. The moon moves among the stars only about 13° in a day, and her own diameter in an hour. If the observer wants his Greenwich time within half a minute, he must determine the position of the moon within the hundred and twentieth of her diameter. This is about as near as an ordinary observer at sea can come with a sextant; and yet the error would be $7\frac{1}{2}$ miles of longitude. Even this degree of exactness can be obtained only by having the moon's place relatively to the stars predicted with great accuracy; and here we meet with one of the most complex problems of astronomy, the efforts to solve which have already been mentioned.

In addition to the uncertainty of which we have spoken, this method is open to the objection of being difficult, owing to the long calculation necessary to free the measured distance from the effects of the refraction of both bodies by the atmosphere, and of the parallax of the moon. On ordinary voyages navigators prefer to trust to their chronometers. The error of the chronometer on Greenwich time and its daily rate are determined at ports of which the longitude is known, and the navigator can then calculate this error on the supposition that the chronometer gains or loses the same amount every day. On voyages between Europe and America a good chronometer will not generally deviate more than ten or fifteen seconds from its calculated rate, so that it answers all the purposes of navigation.

Still another observation by which Greenwich time may be obtained to a minute in any part of the world is that of the eclipses of Jupiter's first satellite. The Greenwich or Washington times at which the eclipses are to occur are given in the *Nautical Almanac*, so that if the traveller can succeed in observing one, he has his Greenwich time at once, without any calculation whatever. But the error of his observation may be half a minute, or even an entire minute, so that this method is not at all accurate.

Where an astronomer can fit up a portable observatory, the observation of the moon affords him a much more accurate longitude than it does the navigator, because he can use better instruments. If he has a transit instrument, he determines from observation the right ascension of the moon's limb as she passes his meridian, and then, referring to the *Nautical Almanac*, he finds at what Greenwich time the limb had this right ascension. A single transit would, if the moon's place were correctly predicted, give a longitude correct within six or eight seconds of time. It is found, however, that, owing to the errors of the moon's tables, it is necessary for the astronomer to wait for corresponding observations of the moon at some standard observatory before he can be sure of this degree of accuracy.

§ 4. *Mean, or Clock, Time.*

We have hitherto described only sidereal time, which corresponds to the diurnal revolution of the starry sphere, or, more exactly yet, of the vernal equinox. Such a measure of time would not answer the purposes of civil life, and even in astronomy its use is generally confined to the determination of right ascensions. Solar time, regulated by the diurnal motion of the sun, is almost universally used in astronomical observations as well as in civil life. Formerly, solar time was made to conform absolutely to the motion of the sun; that is, it was noon when the sun was on the meridian, and the hours were those that would be given by a sundial. If the interval between two consecutive transits of the sun were always the same, this measure would have been adhered to. But there are two sources of variation in the motion of the sun in right ascension, the effect of which is to make these intervals unequal:

1. The eccentricity of the earth's orbit. In consequence of this, as already explained, the angular motion of the earth round the sun is more rapid in December, when the earth is nearest the sun, than in June, when it is farthest. The average, or mean, motion is such that the sun is 3 minutes 56 seconds longer in returning to the meridian than a star is. But, owing to the eccentricity, this motion is actually one-thirtieth greater in December, and the same amount less in June; so that it varies from 3 minutes 48 seconds to 4 minutes 4 seconds.

2. The principal source of the inequality referred to is the obliquity of the ecliptic. When the sun is near the equinoxes, his motion among the stars is oblique to the direction of the diurnal motion; while the latter motion is directly to the west, the former is $23\frac{1}{2}°$ north or south of east. If, then, sun and star cross the meridian together one day near the equinox, he will not be 3 minutes 56 seconds later than the star in crossing the next day, but about one-twelfth less, or 20 seconds. Therefore, at the times of the equinoxes, the solar days are about 20 seconds shorter than the average. At the solstices, the opposite effect is produced. The sun, being $23\frac{1}{2}°$ nearer the pole than before, the diurnal motion is slower, and it takes the sun 20 seconds longer than the regular interval of 3 minutes 56 seconds for that motion to carry the sun over the space which separates him from the star which culminated with him the day before. The days are then 20 seconds longer than the average, from this cause.

So long as clocks could not be made to keep time within 20 seconds a day, these variations in the course of the sun were not found to cause any serious inconvenience. But when clocks began to keep time better than the sun, it became necessary either to keep putting them ahead when the sun went too fast, and behind when he went too slow, or to give up the attempt to make them correspond. The latter course is now universally adopted, where accurate time is required; the standard sun for time being, not the real sun, but a "mean sun," which is sometimes ahead of the real one, and sometimes behind it. The irregular time depending on the motion of the true sun, or that given by a sundial, is called *Apparent Time*, while that given by the mean sun, or by a clock going at a uniform rate, is called *Mean Time*. The two measures coincide four times in a year; during two intermediate seasons the mean time is ahead, and during two it is behind. The following are the dates of coincidence, and of maximum deviation, which vary but slightly from year to year:

February 10th............................ True sun 15 minutes slow.
April 15th................................ "   " correct.
May 14th................................. "   " 4 minutes fast.
June 14th................................ "   " correct.
July 25th................................ "   " 6 minutes slow.
August 31st.............................. "   " correct.
November 2d .... "   " 16 minutes fast.
December 24th........................... "   " correct.

When the sun is slow, it passes the meridian after mean noon, and the clock is faster than the sundial, and *vice versa*. These wide deviations are the result of the gradual accumulations of the deviations of a few seconds from day to day, the cause of which has just been explained. Thus, during the interval between November 2d and February 12th, the sun is constantly falling behind the clock at an average rate of 18 or 19 seconds a day, which, continued through 100 days, brings it from 16 minutes fast to 15 minutes slow.

This difference between the real and the mean sun is called the *Equation of Time*. One of its effects, which is frequently misunderstood, is that the interval from sunrise until noon, as given in the almanacs, is not the same as that between noon and sunset. This often leads to the inquiry whether the forenoons can be longer or shorter than the afternoons. If by "noon" we meant the passage of the real sun across the meridian, they could not; but the noon of our clocks being sometimes 15 minutes before or after noon by the sun, the former may be half an hour nearer to sunrise than to sunset, or *vice versa*.

CHAPTER III.

MEASURING DISTANCES IN THE HEAVENS.

§ 1. *Parallax in General.*

THE determination of the distances of the heavenly bodies from us is a much more complex problem than merely determining their apparent positions on the celestial sphere. The latter depend entirely on the direction of the bodies from the observer; and two bodies which lie in the same direction will seem to occupy the same position, no matter how much farther one may be than the other. Notwithstanding the enormous differences between the distances of different heavenly bodies, there is no way of telling even which is farthest and which nearest by mere inspection, much less can the absolute distance be determined in this way.

The distances of the heavenly bodies are generally determined from their *Parallax*. Parallax may be defined, in the most general way, as the *difference between the directions of a body as seen from two different points*. Other conditions being equal, the more distant the body, the less this difference, or the less the parallax. To show, in the most elementary way, how difference of direction depends on distance, suppose an observer at O to see two lights, A and B, at night. He cannot tell by mere inspection which is the more distant. But suppose he walks over to the point P. Both lights will then seem to change their direction, moving in the direction opposite to that in which he goes. But the light A will change more than the light B, for, being to the right of B when the

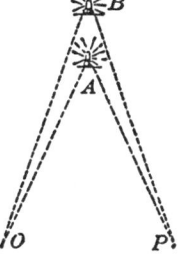

FIG. 47.—Diagram illustrating parallax.

observer was at *O*, it is now to the left of it. The observer can then say with entire certainty that *A* is nearer than *B*.

As a steamship crosses the ocean, near objects at rest change their direction rapidly, and soon flit by, while more distant ones change very slowly. The stars are not seen to change at all. If, however, the moon did not move, the passenger would see her to have changed her apparent position about one and a half times her diameter in consequence of the journey. If, when the moon is near the meridian, an observer could in a moment jump from New York to Liverpool, keeping his eye fixed upon her, he would see her apparently jump in the opposite direction about this amount.

Astronomically, the direction of an object from an observer is determined by its position on the celestial sphere; that is, by its right ascension and declination. In consequence of parallax, the declination of a body is not the same when seen from different parts of the earth. As the moon passes the meridian of the Cape of Good Hope, her measured declination may be a degree or more farther north than it is when she passes the meridian of Greenwich. The determination of the parallax of the moon was one of the objects of the British Government in establishing an observatory at the Cape, and so well has this object been attained that the best determinations of the parallax have been made by comparing the Greenwich and Cape observations of the moon's declination.

The determination of the distance of a celestial object from the parallax depends on the solution of a triangle. If, in Fig. 48, we suppose the circle to represent the earth, and imagine an observer at *A* to view a celestial object, *M*, he will see it projected on the infinite celestial sphere in the direction *AM* continued. Another observer at *A'* will see it in the direction *A'M*. The difference of these directions is the angle at *M*. Knowing all the angles of the quadrilateral

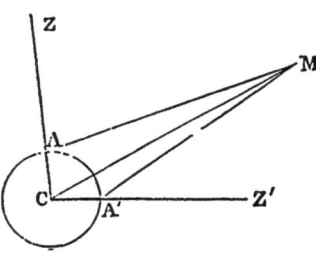

Fig. 48.—Diagram illustrating parallax.

$ACA'M$, and the length of the earth's radius, CA, the distance of the object from the three points, A, A', and C, can be found by solving a simple problem of trigonometry.

The term *parallax* is frequently used in a more limited sense than that in which we have just defined and elucidated it. Instead of the difference of directions of a celestial body seen from any two points, the astronomer generally means the difference between the direction of the body as it would appear from the centre of the earth, and the direction seen by an observer at the surface. Thus, in Fig. 49, an observer at the centre of the earth, C, would see the object M' in the direction CM', while one on the surface at P will see it in the direction PM'. The difference of these directions is the angle $PM'C$. If the observer

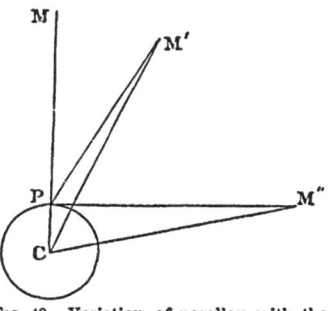

Fig. 49.—Variation of parallax with the altitude.

should be at the point where the line $M'C$ intersects the surface of the earth, there would be no parallax: in this case, the object would be in his geocentric zenith. If, on the other hand, the observer has the object in his horizon, so that the line PM'' is tangent to the surface of the earth, the angle $CM''P$ is called the horizontal parallax. *The horizontal parallax is equal to the angle which the radius of the earth subtends as seen from the object.* When we say that the horizontal parallax of the moon is 57″, and that of the sun 8″.85, it is the same thing as saying that the diameter of the earth subtends twice those angles as seen from the moon and sun respectively.

Owing to the ellipticity of the earth, all its diameters will not subtend the same angle; the polar diameter being the shortest of all, and the equatorial the longest. The equatorial diameter is, therefore, adopted by astronomers as the standard for parallax. The corresponding parallax, that is, the equatorial radius of the earth as seen from a celestial body, is called the *Equatorial Horizontal Parallax* of that body.

To measure directly the distance of the moon or any other heavenly body, the line *PC* must be replaced by the line joining the positions of the two observers, called the base-line. Knowing the length and direction of this base-line, and the difference of directions, or parallax, the distance is at once obtained. If the absolute length of the base-line should not be known, the astronomer could still determine the proportion of the distance of the object to the base-line, leaving the final determination of the absolute distances to be made when the base-line could be measured.

It is not always necessary for two observers actually to station themselves in two distant parts of the earth to determine a parallax. If the observer himself could move along the base-line, and keep up a series of observations on the object, to see how it seemed to move in the opposite direction, he would still be able to determine its distance. Now, every observer is actually carried along by two such motions, because he is on the moving earth. He is carried round the sun every year, and round the axis of the earth every day. We have already shown how, in consequence of the first motion, all the planets seem to describe a series of epicycles. This apparent motion is an effect of parallax, and by means of it the proportions of the solar system can be determined with extreme accuracy. The base-line is the diameter of the earth's orbit. But the parallax in question does not help us to determine this base-line. To find it, we must first know the distance of the earth from the sun, and here we have no base-line but the diameter of the earth itself. Nor can the annual motion of the earth round the sun enable us to determine the distance of the moon, because the latter is carried round by the same motion.

The result of the daily revolution of the observer round the earth's axis is, that the apparent movement of the planet along its course is not perfectly uniform: when the observer is east, the planet is a little to the west, and *vice versa*. By observing the small inequalities in the motion of the planet corresponding to the rotation of the earth on its axis, we have the means of observing its distance with the earth's diameter as a base-

PARALLAX IN GENERAL. 169

line, and this diameter is well known. Unfortunately, however, the earth is so small compared with the distances of the planets, that the parallax in question almost eludes measurement, except in the case of those planets which are nearest the earth, and even then it is so minute that its accurate determination is one of the most difficult problems of modern astronomy.

The principal difficulty in determining a parallax from the revolution of the observer around the earth's axis is that the observations are not to be made in the meridian, but when the planet is near the horizon in the east and west. Hence the most accurate and convenient instrument of all, the meridian circle, cannot be used, and recourse must be had to methods of observation subject to many sources of error.

In measuring very minute parallaxes, it may be doubtful whether the position of the body on the celestial sphere can be determined with the necessary accuracy. In this case resort is sometimes had to relative parallax. By this is meant the difference between the parallaxes of two bodies lying nearly in the same direction. The most notable example of this is afforded by a transit of Venus over the face of the sun. To determine the absolute direction of Venus when nearest the earth with the accuracy required in measurements of parallax has not hitherto been found practicable, because the observation must be made in the daytime, when the atmosphere is much disturbed by the rays of the sun, and also because only a small part of the planet can then be seen. But if the planet is actually between us and the sun, so as to be seen projected on the sun's face, the apparent distance of the planet from the centre or from the limb of the sun may be found with considerable accuracy. Moreover, this distance will be different as seen from different parts of the earth's surface at the same moment, owing to the effect of parallax; that is, different observers will see Venus projected on different parts of the sun's face. But the change thus observed will be only that due to the difference of the parallaxes of the two bodies; while both change their directions, that nearest the observer changes the more, and thus seems to move past the other, exactly as in the diagram of the lights.

It may be asked how the parallax of the sun can be found from observations of the transit of Venus, if such observations show only the difference between the parallax of Venus and that of the sun. We reply that the ratio of the parallaxes of the two bodies is known with great precision from the proportions of the system. We have already shown that these proportions are known with great accuracy from the third law of Kepler, and from the annual parallax produced by the revolution of the earth round the sun. It is thus known that at the time of the transit of Venus, in 1874, the sun was nearly four times the distance of Venus, or, more exactly, that he was 3.783 times as far as that planet. Consequently, the parallax of Venus was then 3.783 times that of the sun. The difference of the parallaxes, that is, the relative parallax, must then have been 2.783 times the sun's parallax. Consequently, we have only to divide the relative parallax found from the observations by 2.783 to have the parallax of the sun itself.

Still another parallax, seldom applied except to the fixed stars, is the *Annual Parallax*. This is the parallax already explained as due to the annual revolution of the earth in its orbit. It is equal to the angle subtended by the line joining the earth and sun, as seen from the star or other body. When we say that the annual parallax of a star is one second of arc, it is the same thing as saying that at the star the line joining the earth and sun would subtend an apparent angle of one second, or that the diameter of the earth's orbit would appear under an angle of two seconds.

It will be seen that the measurement of the heavens involves two separate operations. The one consists in the determination of the distance between the earth and the sun, which is made to depend on the solar parallax, or the angle which the semidiameter of the earth subtends as seen from the sun, and which is the unit of distance in celestial measurements. The other consists in the determination of the distances of the stars and planets in terms of this unit, which gives what we may call the proportions of the universe. Knowing this proportion, we can determine all the distances of the universe when the length of our unit or the distance of the sun is known, but not before. The determination of this distance is, therefore, one of the capital problems of astronomy, as well as one of the most difficult, to the solution of which both ancient and modern astronomers have devoted many efforts.

§ 2. *Measures of the Distance of the Sun.*

We have already shown, in describing the phases of the moon, how Aristarchus attempted to determine the distance of the sun by measuring the angle between the sun and the moon, when the latter appeared half illuminated. From this measure, the sun was supposed to be twenty times as far as the moon; a result which arose solely from the accidental errors of the observations.

Another method of attacking the problem was applied by Ptolemy, but is probably due to Hipparchus. It rests on a very ingenious geometrical construction founded on the principle that the more distant the sun, the narrower will be the shadow of the earth at the distance of the moon. The actual diameter was determined from an ingenious combination of two partial eclipses of the moon, in one of which half of the moon was south of the limit of the shadow, while in the other three-fourths of her diameter was north of the limit; that is, one fourth of the moon's disk was eclipsed. It was thus found that the moon's apparent diameter was $31\frac{1}{3}'$, and the apparent diameter of the shadow $40\frac{2}{3}'$. The former number was certainly remarkably near the truth. From this it was concluded that the sun's parallax was $3'\,11''$, and his distance 1210 radii of the earth. This result was an entire mistake, arising from the uncertainty of any measure of so small an angle. Really, the parallax is so minute as to elude all measurement with any instrument in which the vision is not assisted by the use of a telescope. Yet this result continued to figure in astronomy through the fourteen centuries during which the "*Almagest*" of Ptolemy was the supreme authority, without, apparently, any astronomer being bold enough to seriously undertake its revision.

Kepler and his contemporaries saw clearly that this distance must be far too small; but all their estimates fell short of the truth. Wendell came nearest the truth, as he claimed that the parallax could not exceed 15″. But the best estimate of the seventeenth century was made by Huyghens,* the reason why it was the best being that it was not founded on any attempt to measure the parallax itself, which was then really incapable of measurement, but on the probable magnitude of the earth as a planet. The parallax of the sun is, as already explained, the apparent semidiameter of the earth as seen from the sun. If, then, we can find what size the earth would appear if seen from the sun, the problem would at once be solved. The apparent magnitudes of the planets, as seen from the earth, are found by direct measurement with the telescope. The proportions of the solar system being known, as already explained, it is very easy to determine the magnitudes of all the planets as seen from the sun, the earth alone excepted. The idea of Huyghens was that the earth, being a planet, its magnitude would probably be somewhere near that of the average of the two planets on each side of it, namely, Venus and Mars. So, taking the mean of the diameters of Venus and Mars, and supposing this to represent the diameter of the earth, he found the angle which the semidiameter of the supposed earth would subtend from the sun, which would be the solar parallax.

Although this method may look like a happy mode of guessing, it was much more reliable than any which had before been applied, for the reason that, in supposing the magnitude of the earth to be between those of Venus and Mars, he was likely to be nearer the truth than any measure of an angle entirely invisible to the naked eye would be. And, by a lucky accident, Huyghens's estimate was nearer the truth than any determinations made previous to the transit of Ve-

---

* At the close of his "Systema Saturnium."

nus in 1769, his result for the distance of the sun being 25,086 semidiameters of the earth, or 99 millions of miles. If he had used the correct diameters of Venus and Mars, he would have been farther from the truth, because the earth is considerably larger than the mean of Venus and Mars—in fact, rather larger than Venus herself. But the imperfect telescopes used by Huyghens showed the planets larger than they really were, so that when he took the mean diameter of these planets as they appeared in his telescopes, he just hit the diameter of the earth, and reached the true solution of the problem.

We now come to the modern methods of measuring the parallax of the sun. These consist, not in measuring this parallax directly, because this cannot even now be done with any accuracy, but in measuring the parallax of one of the planets Venus and Mars when nearest the earth. These planets passing from time to time much nearer to us than the sun does, have then a much larger parallax, and one which can easily be measured. Having the parallax of the planet, that of the sun is determined from the known proportion between their respective distances.

The first application of this method was made by the French astronomers to the planet Mars. In 1671 they sent an expedition to the colony of Cayenne, in South America, which made observations of the position of Mars during the opposition of 1672, while corresponding observations were made at the Paris Observatory. The difference of the two apparent positions, reduced to the same moment, gave the parallax of Mars. From a discussion of these observations, Cassini concluded the parallax of the sun to be $9''.5$, corresponding to a distance of the sun equal to 21,600 semidiameters of the earth. This distance was as much too small as Huyghens's was too great, so that, as we now know, no real improvement was made. Still, the data were much more certain than those on which the estimate of Huyghens was made, and for a hundred years it was generally considered that the sun's parallax was about $10''$, and his distance between 80 and 90 millions of miles.

The method by observations of Mars is still, in some of its

forms, among the most valuable which have been applied to the determination of the solar parallax. About once in sixteen years Mars approaches almost as near the earth as Venus does at the times of her transits, the favorable times being those when Mars at opposition is near his perihelion. His distance outside the earth's orbit is then only 0.373 of the astronomical unit, or $34\frac{1}{2}$ millions of miles, while at his aphelion the distance is nearly twice as great. At the nearest oppositions, his parallax is over $23''$, an angle which can be measured with some accuracy. The plan of observation has generally been to send an observing party to the southern hemisphere in advance, for the purpose of making observations of the position of Mars on the celestial sphere, or of its distance from certain selected stars, from night to night, while corresponding observations are made at the fixed observatories of the northern hemisphere. The displacement of the planet due to parallax is then found by comparing the results of these observations.

The last expedition of this sort was that of Captain James M. Gilliss, late of the United States Navy, who went out to Chili under the auspices of the American Government in 1849, and remained till 1852, for the purpose of observing both Venus and Mars during the periods when the parallax was greatest. Several circumstances conspired to prevent this enterprise from producing results corresponding to its merits. The opposition of Mars was a very unfavorable one; observations of Venus could not be made with the necessary accuracy, and there was a lack of sufficient coöperation on the part of northern observers. The astronomical results of the expedition were, nevertheless, important, Captain Gilliss having prepared an immense catalogue of the stars of the southern hemisphere, while his instruments became the property of the Government of Chili, which employed them in fitting up a national observatory. Several observatories have since been founded in the southern hemisphere, so that there is no longer any need of sending out expeditions to observe the planet Mars for the purpose in question.

§ 3. Solar Parallax from Transits of Venus.

The most celebrated method of determining the solar parallax has been by transits of Venus over the face of the sun, by which the difference between the parallax of the planet and that of the sun can be found, as explained in § 1. We know from our astronomical tables that this phenomenon has recurred in a certain regular cycle four times every 243 years for many centuries past. This cycle is made up of four intervals, the lengths of which are, in regular order, $105\frac{1}{2}$ years, 8 years, $121\frac{1}{2}$ years, 8 years, after which the intervals repeat themselves. The dates of occurrence for eight centuries are as follows:

1518	June 2d.	1882	December 6th.
1526	June 1st.	2004	June 8th.
1631	December 7th.	2012	June 6th.
1639	December 4th.	2117	December 11th.
1761	June 5th.	2125	December 8th.
1769	June 3d.	2247	June 11th.
1874	December 9th.	2255	June 9th.

It has been only in comparatively recent times that this phenomenon could be predicted and observed. In the years 1518 and 1526 the idea of looking for such a thing does not seem to have occurred to any one. The following century gave birth to Kepler, who so far improved the planetary tables as to predict that a transit would occur on December 6th, 1631. But it did not commence until after sunset in Europe, and was over before sunrise next morning, so that it passed entirely unobserved. Unfortunately, the tables were so far from accurate that they failed to indicate the transit which occurred eight years later, and led Kepler to announce that the phenomenon would not recur till 1761. The transit of 1639 would, therefore, like all former ones, have passed entirely unobserved, had it not been for the talent and enthusiasm of a young Englishman. Jeremiah Horrox was then a young curate of eighteen, residing in the North of England, who, even at that early age, was a master of the astronomy of his times. Comparing different tables with his own observations of Venus, he found that a transit might be expected to occur on December 4th, and prepared to observe it, after the fashion then in vogue, by letting the image of the sun passing through his telescope fall on a screen behind it. Unfortunately, the day was Sunday, and his clerical duties prevented his seeing the ingress of the planet upon the solar disk—a circumstance which science has mourned for a century past, and will have reason to mourn for a century to come. When he returned from church, he was overjoyed to see the planet upon the face of the sun, but, after following it half an hour, the approach of sunset compelled him to suspend his observations.

During the interval between this and the next transit, which occurred in 1761, exact astronomy made very rapid progress, through the discovery of the law of gravitation and the application of the telescope to celestial measurements. A great additional interest was lent to the phenomenon by Halley's discovery that observations of it made from distant points of the earth could be used to determine the distance of the sun.

The principles by which the parallaxes, and therefore the distances, of Venus and the sun are determined by Halley's method are quite simple. In consequence of the parallax of Venus, two observers at distant points of the earth's surface, watching her course over the solar disk, will see her describe slightly different paths, as shown in Fig. 50. It is by the distance between these paths that the parallax has hitherto been determined.

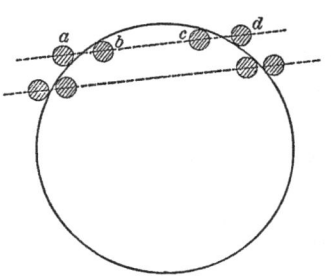

FIG. 50.—Apparent paths of Venus across the sun, as seen from different stations during the transit of 1874. The upper path is that seen from a southern station; the lower is that seen from a northern station, but the distance between the paths is exaggerated.

The essential principle of Halley's method consists in the mode of determining the distance between these apparent paths. An inspection of the figure will show that the path farthest from the sun's centre is shorter than the other, so that Venus will pass over the sun more quickly when watched from a southern station than when watched from a northern one. Halley therefore proposed that the different observers should, with a telescope and a chronometer, note the time it took Venus to pass over the disk, and the difference between these times, as seen from different stations, would give the means of determining the difference between the parallaxes of Venus and the sun. The ratio between the distances of the planet and the sun is known with great exactness by Kepler's third law, from which, knowing the differences of parallaxes, the distance of each body can be determined.

By this plan of Halley the observer must note with great exactness the times both of beginning and end of the transit. There are two phases which may be observed at the beginning and two at the end, making four in all.

The first is that when the planet first touches the edge of the solar disk, and begins to make a notch in it, as at *a*, Fig. 50. This is called *first external contact*.

The second is that when the planet has just entered entirely upon the sun, as at *b*. This is called *first internal contact*.

The third contact is that in which the planet, after crossing the sun, first reaches the edge of the disk, and begins to go off, as at *c*. This is called *second internal contact*.

The fourth contact is that in which the planet finally disappears from the face of the sun, as at *d*. This is called *second external contact*.

Now, it was the opinion of Halley, and a very plausible one, too, that the internal contacts could be observed with far greater accuracy than the external ones. He founded this opinion on his own experience in observing a transit of the planet Mercury at St. Helena in 1677. It will be seen by inspecting Fig. 51, which represents the position of the planet just before first internal contact, that as the planet moves forward on the solar disk the sharp horns of light on each side of it approach each other, and that the moment of internal contact is marked by these horns meeting each other, and forming a thread of light all the way across the dark space, as in Fig. 52. This thread of light is indeed simply the extreme edge of the sun's disk coming into view behind the planet. In observing the transit of Mercury, Halley felt sure that he could fix the moment at which the horns met, and the edge of the sun's disk appeared unbroken, within a single second; and he hence concluded that observers of the transit of Venus could observe the time required for Venus to pass across the sun within one or two seconds. These times would differ in different parts of the earth by fifteen or twenty minutes, in consequence of parallax. Hence it followed, that if Halley's estimate of the degree of accuracy attainable were correct, the parallax of Venus and the sun would be determined by the proposed system of observations within the six hundredth of its whole amount.

Fig. 51.—Venus approaching internal contact on the face of the sun. The planet is supposed to be moving upward.

Fig. 52.—Internal contact of the limb of Venus with that of the sun.

When the long-expected 5th of June, 1761, at length approached, which was a generation after Halley's death, expeditions were sent to distant parts of the world by the principal European nations to make the required observations. The French sent out from among their astronomers, Le Gentil to Pondicherry; Pingré to Rodriguez Island, in the neighborhood of the Mauritius; and the Abbé Chappe to Tobolsk, in Siberia. The war with England, unfortunately, prevented the first two from reaching their stations in time, but Chappe was successful. From England, Mason—he of the celebrated Mason and Dixon's Line—was sent to Sumatra; but he, too, was stopped by the war: Maskelyne, the Astronomer Royal, was sent to St. Helena. Denmark, Sweden, and Russia also sent out expeditions to various points in Europe and Asia.

With those observers who were favored by fine weather, the entry of the dark body of Venus upon the limb of the sun was seen very well until the critical moment of internal contact approached. Then they were perplexed to find that the planet, instead of preserving its circular form, appeared to assume the shape of a pear or a balloon, the elongated portion being connected with the limb of the sun. We give two figures, 52 and 53, the first showing how the planet ought to have looked, the last how it really did look. Now, we can readily see that the observer, looking at such an appearance as in Fig. 53, would be unable to say whether internal contact had or had not taken place. The round part of the planet is entirely within the sun, so that if he judged from this alone, he would say that internal contact is passed. But the horns are still separated by this dark elongation, or "black drop," as it is general-

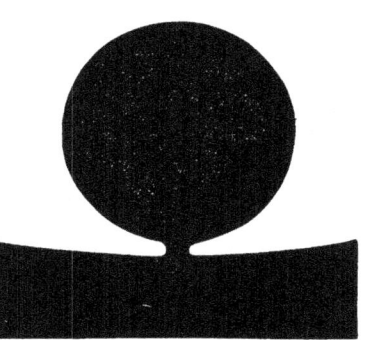

FIG. 53.—The black drop, or ligament.

ly called, so that, judging from this, internal contact has not taken place. The result was an uncertainty sometimes amounting to nearly a minute in observations which were expected to be correct within a single second.

When the parties returned home, and their observations were computed by various astronomers, the resulting values of the solar parallax were found to range from 8″.5, found by Short of England, to 10″.5, found by Pingré, of France, so that there was nearly as much uncertainty as ever in the value of the element sought. Nothing daunted, however, preparations yet more extensive were made to observe the transit of 1769. Among the observers was one whose patience and whose fortune must excite our warmest sympathies. We have said that Le Gentil, sent out by the French Academy to observe the transit of 1761 in the East Indies, was prevented from reaching his station by the war with England. Finding the first port he attempted to reach in the possession of the English, his commander attempted to make another, and, meeting with unfavorable winds, was still at sea on the day of the transit. He thereupon formed the resolution of remaining, with his instruments, to observe the transit of 1769. He was enabled to support himself by some successful mercantile adventures, and he also industriously devoted himself to scientific observations and inquiries. The long-looked-for morning of June 4th, 1769, found him thoroughly prepared to make the observations for which he had waited eight long years. The sun shone out in a cloudless sky, as it had shone for a number of days previously. But just as it was time for the transit to begin, a sudden storm arose, and the sky became covered with clouds. When they cleared away the transit was over. It was two weeks before the ill-fated astronomer could hold the pen which was to tell his friends in Paris the story of his disappointment.

In this transit the ingress of Venus on the limb of the sun occurred just before the sun was setting in Western Europe, which allowed numbers of observations of the first two phases to be made in England and France. The commencement was also visible in this country—which was then these colonies—under very favorable circumstances, and it was well observed by the few astronomers we then had. The leader among these was the talented and enthusiastic Rittenhouse, who was already well known for his industry as an observer. The observations were organized under the auspices of the American Philosophical Society, then in the vigor of its youth, and parties of observers were stationed at Norristown, Philadelphia, and Cape Henlopen. These observations have every appearance of being among the most accurate made on the transit;

but they have not received the consideration to which they are entitled, partly, we suppose, because the altitude of the sun was too great to admit of their being of much value for the determination of parallax, and partly because they were not very accordant with the European observations.

The phenomena of the distortion of the planet and the "black drop," already described, were noticed in this, as in the preceding transit. It is strongly indicative of the ill preparation of the observers that it seems to have taken them all by surprise, except the few who had observed the preceding transit. The cause of the appearance was first pointed out by Lalande, and is briefly this: when we look at a bright object on a dark ground, it looks a little larger than it really is, owing to the encroachment of the light upon the dark border. This encroachment, or irradiation, may arise from a number of causes—imperfections of the eye, imperfections of the lenses of the telescope when an instrument is used, and the softening effect of the atmosphere when we look at a celestial object near the horizon. To understand its effect, we have only to imagine a false edge painted in white around the borders of the bright object, the edge becoming narrower and darker where the bright object is reduced to a very narrow line. Thus, by painting around the borders of the light portions of Fig. 51, we have formed Fig. 53, and produced an appearance quite similar to that described by the observers of the transit. The better the telescope and the steadier the atmosphere, the narrower this border will be, and the more the planet will seem to preserve its true form, as in Fig. 52. In the observations of the recent transit of Venus with the improved instruments of the present time, very few of the more experienced observers noticed any distortion at all.

The results of the observations of 1769 were much more accordant than those of 1761, and seemed to indicate a parallax of about $8''.5$. Curious as it may seem, more than half a century elapsed after the transit before its results were completely worked up from all the observations in an entirely satisfactory manner. This was at length done by Encke, in

1824, for both transits, the result giving 8″.5776 for the solar parallax. Some suspicion, however, attached to some of the observations, which he was not at that time able to remove. In 1835, having examined the original records of the observations in question, he corrected his work, and found the following separate results from the two transits:

> Parallax from the observations of 1761............................8″.53
> Parallax from the observations of 1769............................8″.59
> Most probable result from both transits............................8″.571

The probable error of the result was estimated at 0″.037, which, though larger than was expected, was much less than the actual error has since proved to be. The corresponding distance of the sun is 95,370,000 miles, a classic number adopted by astronomers everywhere, and familiar to every one who has read any work on astronomy.

This result of Encke was received without question for more than thirty years. But in 1854 the celebrated Hansen, completing his investigations of the motions of the moon, found that her observed positions near her first and last quarters could not be accounted for except by supposing the parallax of the sun increased, and therefore his distance diminished, by about a thirtieth of its entire amount. The existence of this error has since been amply confirmed in several ways. The fact is, that although a century ago a transit of Venus afforded the most accurate way of obtaining the distance of the sun, yet the great advances made during the present generation in the art of observing, and the application of scientific methods, have led to other means of greater accuracy than these old observations. It is remarkable that while nearly every class of observations is now made with a precision which the astronomers of a century ago never thought possible, yet this particular observation of the interior contact of a planet with the limb of the sun has never been made with any thing like the accuracy which Halley himself thought he attained in his observation of the transit of Mercury two centuries ago.

The knowledge of this error in the fundamental astronomical unit gave increased interest to the transit of Venus which was to occur on December 8th, 1874. The rarity of the phenomenon was an advantage, in that it led to an amount of public interest being taken in it which could not have been excited by any other astronomical event, and thus secured from various governments the grants necessary to fit out the necessary parties of observation. Plans of observation began to be worked out very far in advance. In 1857, Professor Airy sketched a general plan of operations for the observation of the transits, and indicated the regions of the globe in which he considered the observations should be made. In 1870, before any steps whatever were taken in this country, he had advanced so far in his preparations as to have his observing huts all ready, and his instruments in process of construction. In 1869, the Prussian Government appointed a commission, consisting of six or eight of its most eminent astronomers, to devise a plan of operations, and report it to the Government with an estimate of the expenses. About the same time the Russian Government began making extensive preparations for observing the transit from a great number of stations in Siberia.

Active preparations for the observations in question were commenced by the United States Government in 1871. An account of the method of observation adopted by the Commission to whom the matter was intrusted may not be devoid of interest. The observations of the older transits having failed in giving results of the accuracy now required, it became necessary to improve upon the system then adopted. In this system, the parallax depended entirely on observations of contacts, the uncertainty of which we have already shown. Besides this uncertainty, Halley's method was open to the objection that, unless both contacts were observed at each station, the path of Venus could not be determined, and no result could be deduced. It was therefore proposed by De l'Isle early in the last century, that the observers should determine the longitudes of their stations, in order that, by means of it, they could find the actual intervals between the moments at which any given contact was seen at the different stations. This method was an improvement on Halley's, in that it diminished the chances of total failure. Still, it depended entirely upon making an accurate observation of the moment of contact, and was liable to fail from any accident which might interfere with such an observation—a passing cloud, or a disarrangement of some of the instruments of observation. Besides, it was not yet certain whether the observations could be made with the necessary accuracy. It was, therefore, desirable that, instead of depending on contacts alone, some method should be adopted of finding the position of Venus on the face of the sun as often as possible during the four hours which she should occupy in passing. The easiest and most effective way of doing this seemed to be to take photographs of the sun with Venus on his disk, which photographs could be brought home, compared, and measured at leisure.

This mode of astronomical measurement has been brought to great perfection in this country by Mr. L. M. Rutherfurd and others, and has been found to give results exceeding in accuracy any yet attained by ordinary eye observations. The advantages of the photographic method are so obvious that there could be no hesitation about employing it, and, so far as is known, it was applied by every European nation which sent out parties of observation. But there is a great and essential difference between the methods of photographing adopted by the Americans and by most of the Europeans. The latter seem to have devoted all their attention to the problem of securing a good sharp photograph, taking it for granted that when this photograph was measured there would be no further difficulty. But the measurement at home is necessarily made in inches and fractions, while the distance we must know is to be found in minutes and seconds of angular measure. If we have a map by measurements on which we desire to know the exact distance of two places, we must first know the exact scale on which the map is laid down, with a degree of accuracy corresponding to that of our measures. Just so with our photographs taken at various parts of the globe. We must know the scale on which the images are photographed before we can derive any conclusions from our measures. While the determination of this scale with sufficient precision for ordinary purposes is quite easy, this is by no means the case with a problem where so much accuracy was required, so that here lay the greatest difficulty which the photographic method offered.

In the mode of photographing adopted by the Americans this difficulty was met by using a telescope of great length —nearly forty feet. So long a telescope would be too unwieldy to point at the sun; it was therefore fixed in a horizontal position, the rays of the sun being thrown into it by a mirror. The scale of the picture was determined by actually measuring the distance between the object-glass and the photograph-plate. Each station was supplied with special apparatus by which this measurement could be made within the hundredth of an inch. Then, knowing the position of the optical centre of the glass, it is easy to calculate exactly how many inches any given angle will subtend on the photograph-plate. The following brief description of the apparatus will be readily understood by reference to the figures:

The object-glass and the support for the mirror are mounted on an iron pier extending four feet into the ground, and firmly embedded in concrete. The mirror is in a frame at the end of an inclined cast-iron axis, which is turned with a very slow motion by a simple and ingenious piece of clockwork. The inclination of the axis and the rate of motion are so adjusted that, notwithstanding the diurnal motion of the sun — or, to speak more accurately, of the earth — the sun's rays will always be reflected in the same direction. This result is not attained with entire exactness, but it is so near that it will only be necessary for an assistant to touch the screws of the mirror at intervals of fifteen or twenty minutes during the critical hours of the transit. The reflector is simply a piece of finely polished glass, without any silvering whatever.

It only reflects about a twentieth of the sun's light; but so intense are his rays that a photograph can be taken in less than the tenth of a second. The polishing of this mirror was the most delicate and difficult operation in the construction of the apparatus, as the slightest deviation from perfect flatness would be fatal. For instance, if a straight edge laid upon the glass should touch at the edges, but be the hundred-thousandth of an inch above it at the centre, the reflector would be useless. It might have seemed hopeless to seek for such a degree of accuracy, had it not been for the confidence of the Commission in the mechanical genius of Alvan Clark & Sons, to whom the manufacture of the apparatus was intrusted. The mirrors were tested by observing objects through a telescope, first directly, and then by reflection from the mirror. If they were seen with equally good definition in the two cases, it would show that there were no irregularities in the surface of the mirror; while if it were either concave or convex, the focus of the telescope would seem shortened or lengthened. The first test was sustained perfectly, while the

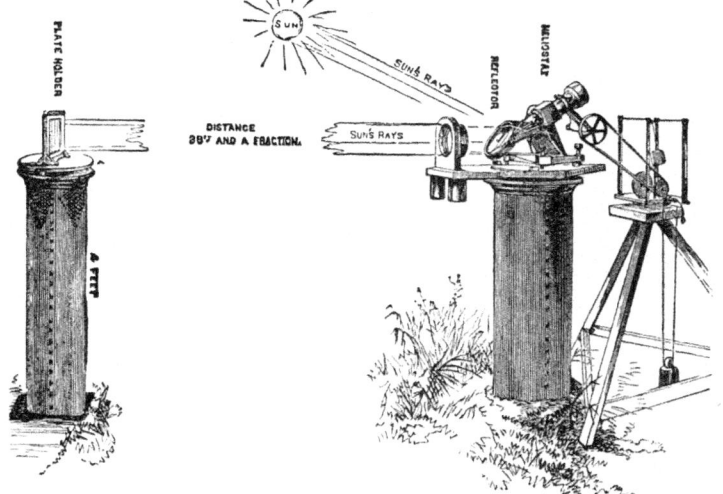

Fig. 54.—Method of photographing the transit of Venus used by the French and American observers, and by Lord Lindsay.

circles of convexity or concavity indicated by the changes of focus of the photographic telescope were many miles in diameter.

Immediately in front of the mirror is the object-glass. The curves of the lenses of which it is formed are so arranged that it is not perfectly achromatic for the visual rays, but gives the best photographic image. Thirty-eight feet and a fraction from the glass is the focus, where an image of the sun about four and a quarter inches in diameter is formed. Here another iron pier is firmly embedded in the ground for the support of the photographic plate-holder. This consists of a brass frame seven inches square on the inside, revolving on a vertical rod, which passes through the iron plate on top of the pier. Into this frame is cemented a square of plate-glass, just as a pane of glass is puttied in a window. The glass is divided into small squares by very fine lines about one-five-hundredth of an inch thick, which were etched by a process invented and perfected by Mr. W. A. Rogers, of the Cambridge Observatory. The sensitive plate goes into the other side of the frame, and when in position for taking the photograph, there is a space of about one-eighth of an inch between the ruled lines and the plate. The former are, therefore, photographed on every picture of the sun which is taken, and serve to detect any contraction of the collodion film on the glass plate.

The rod on which the plate-holder turns, and the frame itself, are perforated from top to bottom by a vertical opening one-sixth of an inch in diameter. Through the centre of this opening, passing between the ruled plate and the photograph plate, hangs a plumb-line of very fine silver wire. In every picture of the sun this plumb-line is also photographed, and this marks a truly vertical line on the plate very near the middle vertical etched line. A spirit-level is fixed to the top of the frame, and serves to detect any changes in the inclination of the ruled lines to the horizon.

One of the most essential features of the arrangement is that the photographic object-glass and plate-holder are on the same level, and in the meridian of the transit instrument with

188 PRACTICAL ASTRONOMY.

which the time is determined. The central ruled line on the plate-holder is thus used as a meridian mark for the transit. The great advantage of this arrangement is, that it permits the angle which the line joining the centres of the sun and Venus makes with the meridian to be determined with the greatest precision by means of the image of the plumb-line which is photographed across the picture of the sun.*

Although the contact observations were not wholly relied on, they were by no means neglected. On the contrary, the greatest pains were taken to avoid the sources of error which caused so much trouble in 1769. To learn what these errors probably were, and to practise the observers in making their observations so as to avoid them, an artificial planet was constructed to move over an artificial representation of a portion of the solar disk by clock-work. The apparatus was mounted on the top of a building about 3300 feet distant, in order to give the effect of atmospheric undulations and softening of the edges of the planet. The planet was represented by a black disk one foot in diameter, which made its apparent magnitude the same as that of Venus in transit. The sun was represented by a white screen behind the artificial Venus, the portions of the edge of the disk where Venus entered and left being formed by the sloping edges of a black triangle, as shown in the figure. There was no need of a representation of the entire sun. The motion was so regulated that the time occupied by the disk in passing from external to

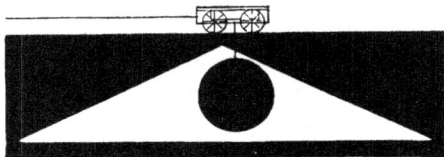

Fig. 55.—Artificial transit of Venus.

---

* The method of photographing the sun by a fixed horizontal telescope with a reflector in front of it is believed to have been first proposed in France by Captain Laussedat. It was independently invented by the late Professor Winlock, who put it into actual operation at the Harvard College Observatory in 1869, and, so far as the author is aware, was the first one to do so. It was employed not only by the American observers, but by the French, and by Lord Lindsay, M.P., of Scotland. The latter gentleman fitted out a finely equipped expedition at his own expense to observe the transit of Venus at the Mauritius.

internal contact, and the angle its motion made with the edges of the triangle, were the same as they would be in the actual transit as viewed from some point where it occurred near the zenith. The disk was put at such a height that it was only about three minutes from internal contact at ingress to internal contact at egress, instead of four hours.

The observations of this instrument have thrown much light on the question of the black drop, and the distortion of the planet seen in former transits of Venus, which have been already described. What is perhaps yet better, it has enabled us to account for a number of puzzling and discordant appearances described by the observers. Father Hell's black drop, seen before the limbs were in contact; the formation of internal contact by a fine line of light, though the cusps were blunt, as seen at Hudson Bay; Captain Cook's "atmosphere" around Venus, and his curious black piece cut out of the edge of the sun, may all be said to have been identified nearly enough to judge what the appearances really were which were so variously described. In looking at the artificial planet near the moment of internal contact, when the air is not still, the first thing which the observer sees is that there is really no constant shape to those parts of Venus and the sun which are approaching each other; but that, owing to the undulations of the air, they assume all sorts of shapes in rapid succession, so that different observers may give different descriptions of the appearances presented, though looking at the very same object. In the varied forms which may be seen, we recognize all the peculiar appearances described by the observers of the transit of 1769.

At each American station the scientific corps consisted of a chief of party, an assistant astronomer, and three photographers. The instruments at all the stations were precisely similar, and the operations and observations the same at all. This system was adopted to secure two great advantages: first, to run the least risk of entire failure from bad weather; and, second, to have all the observations strictly comparable. Much pains and trouble were devoted to these objects. To appreciate their importance, we must remember that, in order to deduce the parallax from the observations at any two stations, it is essential that the difference between observations should be due only to parallax, and that in every other respect they should be exactly the same; because, if there are other differences which we cannot certainly allow for, our calculation of the parallax will be wrong. It is also necessary that we compare the same kind of observations in order to get the parallax. To show how the chances of failure are lessened, suppose we have two stations in each hemisphere, in one of which eye observations are made, while in the other photographs are taken. Then, if the photographs in one hemisphere and the eye observations in the other are lost by clouds, or any other cause, everything will be lost, although one station in each hemisphere is successful, because the eye observations in the one hemisphere cannot be compared with the photographs in the other. It being decided, for these reasons, to have the same system of observations at all the stations, it became necessary to confine the choice of stations to points where the entire transit would be visible.

One of the most important features of the preparations, which distinguishes them from the preparations to observe the former transits, was the previous training of the observers. All the members of the observing parties assembled at Washington to practise together before leaving to make the observations. They took all their multitudinous instruments and apparatus out of their boxes, mounted them, and proceeded to practise with them in the same way they were to be used at the stations. Photographs of the sun were taken from day to day in the same way as on the 8th of December, and each chief of party was instructed in all the delicate operations necessary to secure the entire success of his operations.

To know where a party could be sent, it had first to be known when and where the transit would be visible. We give a small map of the world showing this at a glance. Could we have seen the planet Venus from the Eastern States on the afternoon of December 8th, 1874, we should have seen

SOLAR PARALLAX FROM TRANSITS OF VENUS. 191

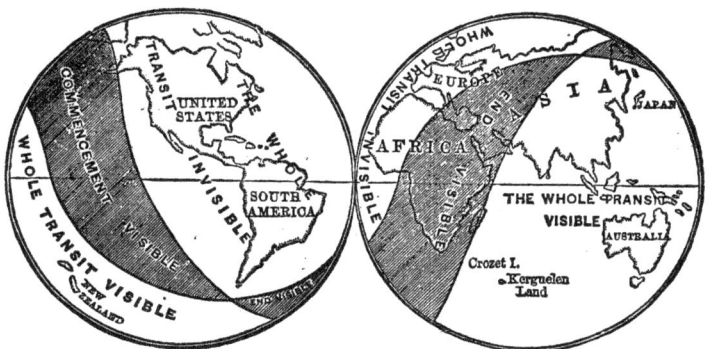

FIG. 56.—Map of the earth, showing the areas of visibility of the transit of 1874.

her approaching nearer and nearer the sun as the latter approached the horizon. In San Francisco, where sunset is three hours later than here, she would have been so near the sun as almost to seem to touch it. About an hour later she actually reached the solar disk. The sun was then shining on the whole Pacific Ocean, except that portion nearest the American coast, and on Eastern Asia, Australia, and the Indian and Antarctic oceans to the south pole. Venus was about four and a half hours passing over the face of the sun, and during this time the latter had set across the entire northern portion of the Pacific Ocean, and had risen as far west as Moscow and Vienna, from which cities the planet might have been seen to leave the disk just as the sun rose.

In the northern hemisphere suitable stations were easily found, as we have the whole of China, Japan, and Northern India. But in the southern hemisphere great difficulties were encountered, owing to the want of habitable stations in the regions which were astronomically the most favorable. Observations cannot be made from the deck of a ship; astronomers must have solid ground for their instruments. The south pole would have been the best station of all, if some antarctic Kane or Hall could take a party thither. The antarctic continent and the neighboring islands were not to be thought of, because a party could neither be landed nor subsisted there;

and if they could, the weather would probably have prevented any observations from being taken. The chance of having a clear sky on the eventful 8th of December was, indeed, one of the most important considerations on which the choice of a station had to depend. Information from every available source, official and private, respecting the meteorology of the various possible stations, was therefore sought. Where there was any American consul or consular agent, he was applied to through the State Department to have meteorological observations made during the months of November and December, 1872 and 1873. A sealing ship belonging to the firm of Williams, Haven, & Co., of New London, made observations at Heard's Island, in the Southern Indian Ocean. From all these reports, as well as from the printed reports issued by various authorities, it was found that the chances of good weather were much better in the northern than in the southern hemisphere. In consequence, instead of sending an equal number of parties north and south, it was determined to send three to the northern and five to the southern hemisphere.

The stations which the American parties finally occupied, with the names of the chiefs of party, are as follows:

NORTHERN HEMISPHERE.

Wladiwostok, Siberia............Professor ASAPH HALL, U. S. N.
Pekin, China......................Professor J. C. WATSON.
Nagasaki, Japan.................Professor GEORGE DAVIDSON, U. S. Coast Survey.

SOUTHERN STATIONS.

Kerguelen Island................Commander G. P. RYAN, U. S. N.
Hobart-town, Tasmania........Professor W. HARKNESS, U. S. N.
Campbelltown, Tasmania*.....Captain C. W. RAYMOND, Engineer Corps, U. S. A.
Queenstown, New Zealand....Professor C. H. F. PETERS.
Chatham Island..................EDWIN SMITH, Esq., U. S. Coast Survey.

The southern parties were all carried to their respective stations by the U. S. steamer *Swatara*, Captain Ralph Chandler, U. S. N., commanding.

---

* Captain Raymond's party was designed for the Crozet Islands, but the *Swatara* failed to effect a landing there.

The only thing which seriously interfered with the observations was the weather. Some photographs were obtained at every station, but the full number at none. Altogether, there were only about half the expected number obtained. No contacts at all were observed at Hobart-town or Chatham Island, but one or more were observed at each of the remaining six stations. Pekin was, however, the only one at which all four were observed. Among the parties sent out by other nations, the most fortunate, as regards weather, were the Germans, who were successful at all six of their stations. The English, French, and Russians were, on the average, about as successful as the Americans.

If the observations on the transit of 1874 had been made in the same way as those of the transit of 1769, they could be very speedily worked up, and we should soon expect to see the solar parallax deduced from the combination of them all. But the investigation and measurement of the photographs is so laborious an operation that the American results can hardly be published before 1878. The definitive value of the parallax must then be deduced, not from the observations of any one nation, but so far as possible from the combination of those of all nations. We must, therefore, wait for the final publication and discussion of all the observations before the definitive value of the parallax can be announced.

Under these circumstances, the question whether it is worth while to send out parties to observe the transit of 1882 will soon be a subject of discussion among astronomers, the answer to which will depend very largely on the success of the efforts made in 1874. On this success we cannot pronounce a final judgment until all the observations are worked up. The reason why doubt still remains on this point is that the sun is a very difficult object either to observe or to photograph with accuracy, owing to the action of his rays on the atmosphere. The air near the ground becomes heated, and thus causes the limb of the sun to undulate to a degree which sometimes renders its exact definition out of the question, while the outline of Venus undulates in the same way. Another difficulty is, that the irregularity in the transparency of the atmosphere, owing to clouds and vapors, renders the photographic representation of the limb of the sun quite uncertain, and thus requires all measures to be made from the sun's centre. Now, we cannot say how far these difficulties have been surmounted by the methods of observation adopted until we finally compare all the observations, and see how consistent they are with each other; and this cannot be done for several years.

The region of visibility of the transit of 1882 will be quite different from that of 1874, as it will include the whole American continent, except some portions in or near the arctic circle. The beginning will be visible over a large part of Africa, and the end over most of the Pacific Ocean. The most favorable northern stations for its observation are in the Eastern and Middle States.

§ 4. *Other Methods of determining the Sun's Distance, and their Results.*

The methods of determining the astronomical unit which we have described rest entirely upon measures of parallax, an angle which hardly ever exceeds 20″, and which it is therefore exceedingly difficult to measure with the necessary accuracy. If there were no other way than this of determining the sun's distance, we might despair of being sure of it within 200,000 miles. But the refined investigations of modern science have brought to light other methods, by at least two of which we may hope, ultimately, to attain a greater degree of accuracy than we can by measuring parallaxes. Of these two, one depends on the gravitating force of the sun upon the moon, and the other upon the velocity of light.

Parallactic Equation of the Moon.—The motion of the moon around the earth is largely affected by the gravitating force of the sun, or, to speak more exactly, by the difference of the gravitating force of the sun upon the moon and upon the earth. A part of this difference depends upon the proportion between the respective distances of the moon and the sun, so that when this force is known, the proportion can be deter-

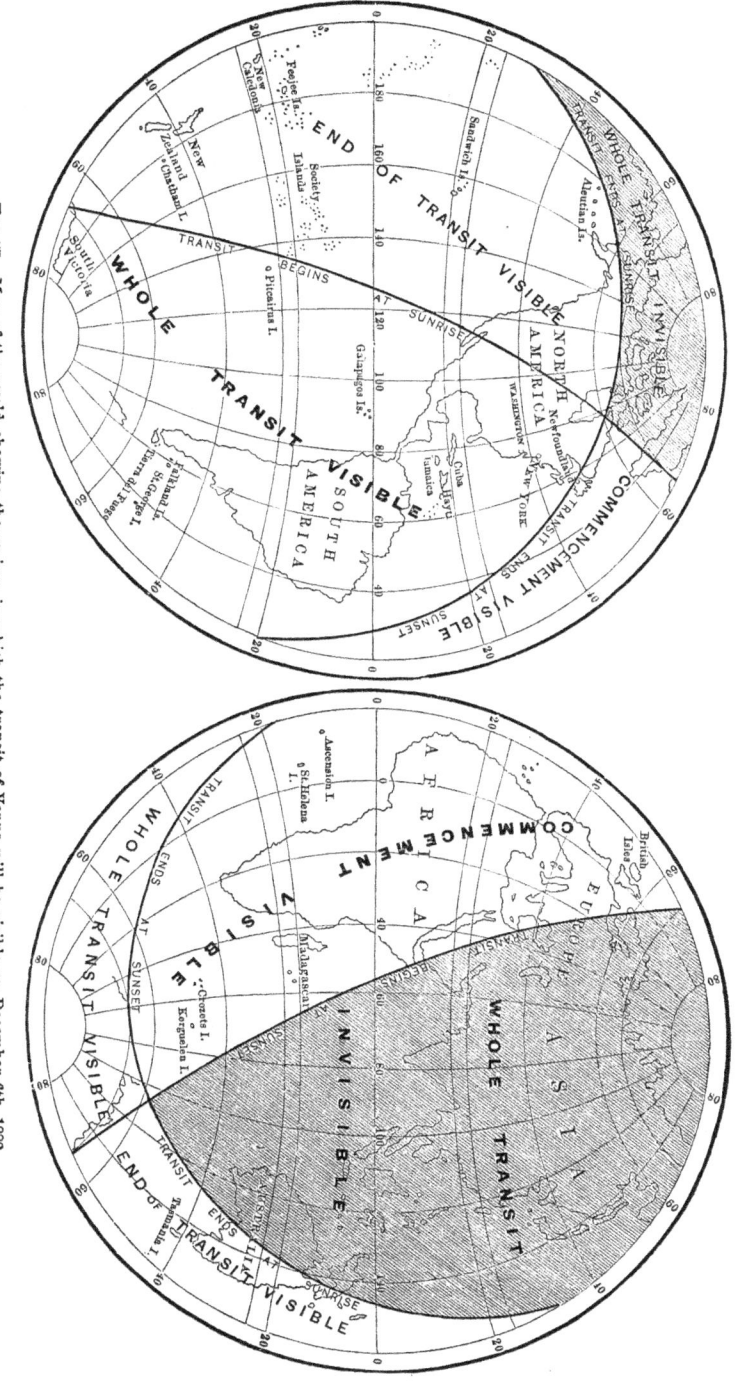

Fig. 57.—Map of the world, showing the regions in which the transit of Venus will be visible on December 6th, 1882.

METHODS OF DETERMINING THE SUN'S DISTANCE. 197

mined. The distance of the moon being known with all necessary precision, we have only to multiply it by the proportion thus obtained to get the distance of the sun. The force in question shows itself by producing a certain inequality in the moon's motion, by which she falls two minutes behind her mean place near the first quarter, and is two minutes ahead near her last quarter. In determining this inequality, we have to measure an angle about six times as great as the average of the planetary parallaxes on which the sun's distance depends; so that, if we could measure both angles with the same precision, the error, by using the moon, would be only onesixth as great as in direct measures of parallax. But it seems as if nature had determined to allow mankind no royal road to a knowledge of the sun's distance. It is the position of the moon's centre which we require for the purpose in question, and this can never be directly fixed. We have to make our observations on the limb or edge of the moon, as illuminated by the sun, and must reduce our observations to the moon's centre, before we can use them. The worst of the matter is, that one limb is observed at the first quarter, and another at the third quarter, so that we cannot tell with absolute certainty how much of the observed inequality is real, and how much is due to the change from one limb to the other. So great is the uncertainty here that, previous to 1854, it was supposed that the inequality in question was about 122″, agreeing with the theoretical inequality from Encke's erroneous value of the solar parallax. Hansen then found that it was really about 4″ greater, and thus was led to the conclusion that the parallax of the sun must be increased, and his distance diminished, by one-thirtieth of the whole amount.

It is quite likely that by adopting improved modes of observation, it will be found that the sun's distance can be more accurately measured in this way than through the parallaxes of the planets. Some pains have already been taken to determine the exact amount of the inequality from observations, the result being 125″.5. The entire seconds may here be relied on, but the decimal is quite uncertain. We can only say that we are pretty surely within three or four tenths of a second of the truth. From this value the parallax of the sun is found to be 8″.83, with an uncertainty of two or three hundredths of a second.

Sun's Distance from the Velocity of Light.—There is an extraordinary beauty in this method of measuring the sun's distance, arising from the contrast between the simplicity of the principle and the profoundness of the methods by which alone the principle can be applied. Suppose we had a messenger whom we could send to and fro between the sun and the earth, and who could tell, on his return, exactly how long it took him to perform his journey; suppose, also, we knew the exact rate of speed at which he travelled. Then, if we multiply his speed by the time it took him to go to the sun, we shall at once have the sun's distance, just as we could determine the distance of two cities when we knew that a train running thirty miles an hour required seven hours to pass between them. Such a messenger is light. It has been found practicable to determine, experimentally, about how fast light travels, and to find from astronomical phenomena how long it takes to come from the sun to the earth. How these determinations are made will be shown in the next chapter; here we shall stop only to give results. It is found by Foucault's experiment that light travels about 185,200 miles per second; and it is known from a study of several astronomical phenomena that it passes from the sun to the earth in 498 seconds. The product of these numbers gives a distance of 92,230,000 miles, a result, however, which is uncertain by $\frac{1}{200}$ of its entire amount, or nearly half a million of miles, owing to the uncertainty in each of the factors. This result was reached in 1862, and was one of the first confirmations of the increased value of the solar parallax found by Hansen. But since that time a redetermination of the velocity of light has been made by Cornu, of Paris, by a method soon to be described, with a different result. He finds a velocity of 300,400 kilometres or 186,670 miles per second, making the distance of the sun 92,960,000 miles, and its parallax 8″.794. This discrepancy

METHODS OF DETERMINING THE SUN'S DISTANCE.

is not yet explained, and the truth can be reached only by a repetition of one or both of the experiments.

These two methods of determining the distance of the sun may fairly be regarded as equal in accuracy to that by transits of Venus when they are employed in the best manner. There are also two or three minor methods which, though less accurate, are worthy of mention. One of the most ingenious of these was first applied by Leverrier. It is known from the theory of gravitation that the earth, in consequence of the attraction of the moon, describes a small monthly orbit around the common centre of gravity of these two bodies, corresponding to the monthly revolution of the moon around the earth, or, to speak with more precision, around the same common centre of gravity. If we know the mass (or weight) of the moon relatively to that of the earth, and her distance, we can thus calculate the radius of the little orbit referred to. In round numbers, it is 3000 miles. This monthly oscillation of the earth will cause a corresponding oscillation in the longitude of the sun, and by measuring its apparent amount we can tell how far the sun must be placed to make this amount correspond to, say 3000 miles. Leverrier found the oscillations in arc to be $6''.50$. From this he concluded the solar parallax to be $8''.95$. But Mr. Stone,* of Greenwich, found two errors in Leverrier's computation,† and, when these are corrected, the result is reduced to $8''.85$.

Another recondite method has been employed by Leverrier. It is founded on the principle that when the relative masses of the sun and earth are known, their distance can be found by comparing the distance which a heavy body will fall in one second at the surface of the earth with the fall of the latter towards the sun in the same time. The mass of the earth was found by its disturbing action on the planets Venus and Mars, as explained in the chapter on Gravitation. Leverrier

---

* Mr. E. J. Stone was then first assistant at the Royal Observatory, Greenwich, but has been Astronomer Royal at the Cape of Good Hope since 1870.
† "Monthly Notices of the Royal Astronomical Society," vol. xxvii., p. 241, and vol. xxviii., pp. 22, 23.

concluded that this method gave the value of the solar parallax as 8″.86. But one of his numbers requires a small correction, which reduces it to 8″.83. Another determination of the mass of the earth relative to that of the sun has recently been made by Von Asten, of Pulkowa, from the action of the earth upon Encke's comet. The solar parallax thence resulting is 9″.009, the largest recent value; but the anomalies in the apparent motions of this comet are such that very little reliance can be placed upon this result.

Yet another method of determining the solar parallax has been proposed and partially carried out by Dr. Galle.* It consists in measuring the parallax of some of the small planets between Mars and Jupiter at the times of their nearest approach to the earth, by observations in the northern and southern hemispheres. The least distance of the nearest of these bodies from us is little less than that of the sun, so that in this respect they are far less favorable than Venus and Mars. But they have the great advantage of being seen in the telescope only as points of light, like stars, and, in consequence, of having their position relative to the surrounding stars determined with greater precision than can be obtained in the case of disks like those of Venus and Mars. Observations of Flora were made in this way at a number of observatories in both hemispheres during the opposition of 1874, from which Dr. Galle has deduced 8″.875 as the value of the solar parallax.

Most Probable Value of the Sun's Parallax.—From the general accordance of the various methods we have described, it would appear that the solar parallax must lie between pretty narrow limits, probably between 8″.82 and 8″.86, and that the distance of the sun in miles probably lies between the limits 92,200,000 and 92,700,000. Of the distance of the sun, we may say with a reasonable approach to certainty that it is 92,000,000 and some fraction of another million; and

---

* Dr. J. G. Galle, now director of the observatory at Breslau, Eastern Prussia. He was formerly assistant at the Observatory of Berlin, where he became celebrated as the optical discoverer of the planet Neptune.

if we should guess that fraction to be 400,000, we should probably be within 200,000 miles of the truth. This is all we can say of the sun's distance until the results of the transits of Venus are obtained, when we may hope to find the uncertainty brought between yet narrower limits.

In many recent works the distance in question will be found stated at 91,000,000 and some fraction. This arises from the circumstance that into several of the first determinations by the new methods small errors and imperfections crept, which, by a singular coincidence, all tended to make the parallax too great, and therefore the distance too small. For instance, Hansen's original computations from the motion of the moon led him to a parallax of $8''.96$. Revising his calculations, he reduced it to $8''.917$. When his lunar tables, published in 1857, came to be compared with observations, it was found that his parallactic inequality was undoubtedly too great by one second or more. When this is corrected, the parallax is reduced about a tenth of a second more.

The observations of Mars, in 1862, as reduced by Winnecke and Stone, first led to a parallax of $8''.92$ to $8''.94$. But in these investigations only a small portion of the observations was used. When the great mass remaining was joined with them, the result was $8''.85$.

The early determinations of the time required for light to come from the sun were founded on the extremely uncertain observations of eclipses of Jupiter's satellites, and were five to six seconds too small. The time, 493 seconds, being used in some computations instead of 498 seconds, the distance of the sun from the velocity of light was made too small.

In both of Leverrier's methods some small errors of computation have been found, the effect of all of which is to make his parallax too great. Correcting these, and making no change in any of his data, the results are respectively $8''.85$ and $8''.83$.

§ 5. *Stellar Parallax.*

It is probable that no one thing tended more strongly to impress the minds of thoughtful men in former times with the belief that the earth was immovable than did the absence of stellar parallax. We may call to mind that the annual parallax of the fixed stars arises from the change in their direction produced by the motion of the earth from one side of its orbit to the other. One of the earliest forms in which we may suppose this parallax to have been looked for is shown in Fig. 58. Suppose AB to be the earth's orbit with the sun,

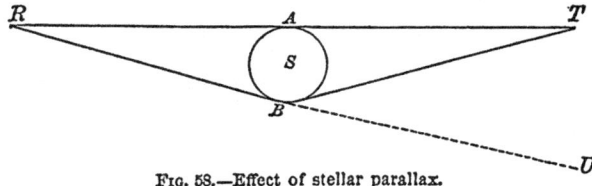

Fig. 58.—Effect of stellar parallax.

S, near its centre, and RT two stars so situated as to be directly opposite each other when the earth is at A; that is, when the direction of each star is 90° distant from that of the sun. Then it is clear that, after six months, when the earth is at B, the stars will no longer be opposite each other, the point U, which is opposite R, making the angle TBU, with the direction of T. The stars will all be displaced in the same direction that the sun is in from the earth. When it was found that the most careful observations showed no such displacement, the conclusion that the earth did not move seemed inevitable. We have seen how Tycho was led in this way to reject the doctrine of the earth's motion, and favor a system in which the sun moved around it. In this Tycho was followed by the ecclesiastical astronomers who lived during the seventeenth century, and who, finding no parallax whatever to any of the stars, were led to reject the Copernican system.

The telescope furnishing so powerful an auxiliary in measuring small angles, it was natural that the defenders of the Copernican system should be anxious to employ it in detecting the annual parallax of the stars. But the earlier observers had very imperfect notions of the mechanical appliances necessary to do this with success, and, in consequence, the invention of the telescope did not result in any immediate improvement in the methods of celestial measurement. A step was taken in 1669 by Hooke, of England, who was among the first to see how the telescope was to be applied in the measurement of the apparent distances of the stars from the zenith. He fixed a telescope thirty-six feet long in his house, in a vertical position, the object-glass being in an opening in the roof, while the eye-piece was in one of the lower rooms. A fine plumb-line hung down from the object-glass to a point below the eye-piece, which gave a truly vertical line from which to measure. The star selected for observation was γ Draconis, because it was comparatively bright, and passed over the zenith of London. His mode of observation was to measure the distance of the image of the star from the plumb-line from day to day at the moment of its passing the meridian. He had made but four observations when his object-glass was accidentally broken, and the attempt ended without leading to any result whatever.

Between 1701 and 1704, Roemer, then of Copenhagen, attempted to determine the sum of the double parallaxes of Sirius and α Lyræ by the principle shown in Fig. 58. These stars lie somewhere near the opposite quarters of the celestial sphere, and the angle between them will vary from spring to autumn by nearly double the sum of their parallaxes. The angle was measured by the transit instrument and the astronomical clock, by noting the time which elapsed between the transit of Sirius over the meridian, and that of α Lyræ. This time was found to be, on the average,

	Hrs.	Min.	Sec.
In February, March, and April....................	11	54	59.7
In September and October.........................	11	54	55.4
Difference..			4.3

Here was a difference of four seconds of time, or a minute of angle, which was then very naturally attributed to the motion of the earth, and which was afterwards printed in a dissertation entitled "Copernicus Triumphans." It is now known that there is no such parallax as this to either of these stars, and

Peters* has shown that the difference which was attributed to parallax by the enthusiastic Danish astronomers really arose, in great part, from the diurnal irregularity in the rate of their clock, caused by the action of the diurnal change of temperature upon the uncompensated pendulums. In the spring the interval of time measured elapsed during the night, Sirius passing the meridian in the evening, and α Lyræ in the morning. The cold of night made the clocks go too fast, and so the measured interval came out too great. In the autumn Sirius passed in the morning, and α Lyræ in the evening; the clock was going too slow on account of the heat of the day, and the interval came out too small.

Among the numerous other vain efforts made by the astronomers of the last century to detect the stellar parallax, that of Bradley is worthy of note, owing to the remarkable discovery of the aberration of light to which it led. The principle of his instrument was the same as that of Hooke, the zenith distance of the star γ Draconis at the moment of its passing the meridian being determined by the inclination of a telescope to a fine plumb-line. The instrument thus used, which has become so celebrated in the history of astronomy, has since been known as Bradley's zenith sector. In accuracy it was a long step in advance of any which preceded it, so that by its means Bradley was able to announce with certainty that the star in question had no parallax approaching a single second. But he found another annual oscillation of a very remarkable character, arising from the progressive motion of light, which will be described in the next chapter. It has frequently happened in the history of science that an investigation of some cause has led to discoveries in a different direction of an entirely unexpected character.

It would be tedious to describe in detail all the efforts made by astronomers, during the last century and the early part of the present one, to detect the stellar parallax. It will

---

* C. A. F. Peters, then of the Pulkowa Observatory, and now editor of the *Astronomische Nachrichten*.

be sufficient to say, in a general way, that they depended on absolute measures; that is, the astronomer endeavored, generally by a divided circle, to determine from day to day the zenith distance at which the star passed the meridian. The position of the zenith was determined in various ways—sometimes by a fine plumb-line, sometimes by the level of quicksilver. What is required is the angle between the plumb-line and the line of sight from the observer to the star. The same result can be obtained by observing the angle between a ray coming directly from a star and the ray which, coming from the star, strikes the surface of a basin of quicksilver, and is reflected upwards. Whatever method is used, a large angle has to be measured, an operation which is always affected by uncertainty, owing to the influences of varying temperatures and many other causes upon the instrument. The general result of all the efforts made in this way was that while several of the brighter stars seemed to some astronomers to have parallaxes, sometimes amounting to two or three seconds, though generally not much exceeding a second, yet there was no such agreement between the various results as was necessary to inspire confidence. As a matter of fact, we now know that these results were entirely illusory, being due, not to parallax, but to the unavoidable errors of the instruments used.

Struve was the first one to prove conclusively that the parallaxes even of the brighter stars were so small as to absolutely elude every mode of measurement before adopted. In principle his method was that employed by Roemer, the sum of the parallaxes of stars twelve hours distant in right ascension being determined by the annual change in the intervals between their times of transit over the meridian. But he made the great improvement of selecting stars which could be observed as they passed the meridian below the pole, as well as above it, so that a short time before or after observing the transit of a star he could turn his transit instrument below the pole, and observe the transit of the opposite star from west to east. Thus he was not under the necessity of depending on the rate of his clock for more than an hour or two, while Roemer had to depend on it for twelve hours. The result of Struve was that the average parallax of the twenty-five brightest stars within 45° of the pole could not much, if at all, exceed a single tenth of a second.

Such was the general state of things up to the year 1835. It was then decided by Struve and Bessel, in lieu of attempting to determine zenith distances, to adopt the method of relative parallaxes. The idea of this method really dates almost from the invention of the telescope. It was considered by Galileo and Huyghens that where a bright and a faint star were seen side by side in the field of view of a telescope, the latter was probably vastly more distant than the former, and that consequently they would change their relative position as the earth moved from one side of the sun to the other. If, for instance, one star was three times the distance of the other, its apparent motion produced by parallax would be only a third that of the other, and there would remain a relative parallax equal to two-thirds that of the brighter star, which could be detected by measuring the angular distance of the two stars as seen in the telescope from day to day throughout the year. The drawback to which this method is subject is the impossibility of determining how many times farther the one star is than the other; in fact, it may be that the smaller star is really no farther than the large one. No doubt it was this consideration which deterred the astronomers of the last century from trying this very simple method.

The astronomers of the last generation found cases in which there could be little doubt that a star was much nearer to us than the small stars which surrounded it in the field of the telescope. For instance, the star 61 Cygni, or rather the pair of stars thus designated, are found not to occupy a fixed position in the celestial sphere, like the surrounding small stars, but to be moving forward in a straight line at the rate of six seconds per year. This amount of proper motion was so unusual as to make it probable that the star must be one of the nearest to us, although it was only of the sixth magnitude. It was therefore selected by Bessel for the investigation of its parallax relative to two other stars in its neighborhood. The instrument used was the heliometer, an instrument which, as now made, admits of great precision, but which was then liable to small uncertainties from various causes. His early attempts to detect a parallax failed as completely as had those of former observers. He recommenced them in August, 1837, his first series of measures being continued until October, 1838. The result of this series was the detection of a parallax of about three-tenths of a second ($0''.3136$). He then took down his instrument, made some improvements in it, and commenced a second series, which he continued until July, 1839; and his assistant, Schlüter, until March, 1840. The final value of the parallax deduced by Bessel from all these observations was $0''.35$. The reality of this parallax has been well established by subsequent investigators, only it has been found to be a little larger. From a combination of all the results, Auwers, of Berlin, finds the most probable parallax to be $0''.51$.

The star selected by Struve for the measure of relative parallax was the bright one α Lyræ. This has not only a sensible proper motion, but is of the first magnitude; so that there is every reason to believe it to be among those which are nearest to us. The comparison was made with a single very small star in the neighborhood, the instrument used being the nine-inch telescope of the Dorpat Observatory. The observations extended from November, 1835, to August, 1838. The result was a relative parallax of a quarter of a second. Subsequent investigations have reduced this parallax to two-tenths of a second, so that although α Lyræ is nearly a hundred times as bright as either of the pair of stars 61 Cygni, it is more than twice as far from us.

So far as is known, and, beyond all reasonable doubt, in reality, the nearest fixed star is α Centauri, in the southern hemisphere. This fact was discovered by Henderson, the English Astronomer Royal at the Cape of Good Hope, about the same time that Struve and Bessel were making their first measures of parallaxes. The observations on which it was founded were made with the mural circle of the Cape Observatory, and were therefore absolute measures of zenith distance, instead of comparisons with surrounding stars, like the measures of Struve and Bessel. From a discussion of his own observations, and a very careful series by his successor, Henderson found the parallax of the pair of stars which compose α Centauri to be 0″.91.* This parallax corresponds to the distance of 226,000 astronomical units,† or more than twenty millions of millions of miles. Yet it is not only the nearest star, but so far the nearest that no other is known to be within nearly double the distance.

The most elaborate measures of stellar parallax made in recent times are those by Dr. Brünnow, formerly director of the observatory at Ann Arbor, Michigan. On his appointment to the post of Astronomer Royal for Ireland, Dr. Brünnow employed the equatorial telescope of the Dunsink Observatory in such determinations with great success. The results of his measures, with those of other astronomers, are given in the Appendix to the present work.

The recent researches of various observers have resulted in showing that there are about a dozen stars visible in our latitudes of which the parallax ranges from a tenth to half a second. Part of these are small stars, supposed to be near us from their large proper motion, while others are stars of the far brighter classes. It is, however, remarkable that among the thirteen stars of the first magnitude visible in our latitudes, less than half have been found to have any measurable parallax, even when the greatest refinements have been applied in the observations. For the most part, the stars with a decided parallax are not of a conspicuous magnitude. The two stars next in distance to α Centauri are 61 Cygni, of the fifth magnitude, and one in Ursa Major without a name, and too small

---

* The mean of all the measures of the parallax of this pair of stars hitherto made, gives 0″.93 as their most probable parallax, corresponding to a distance of 221,000 astronomical units.

† The astronomical unit is the distance of the earth from the sun, about 92½ millions of miles.

to be seen without a telescope. The parallax of the latter has been found by Professor Winnecke* to be 0″.501, which is nearly the same as that of 61 Cygni. The question of the average distance of the stars of the first magnitude must therefore be regarded as still unsolved. We can only say that the parallax of at least half of them is probably less than the tenth of a second, and, therefore, the distance greater than two million radii of the earth's orbit.†

In these measurements of the annual parallax of the fixed stars, it sometimes happens that the astronomer finds his observations to give a *negative* parallax. To understand what this means, we remark that a determination of the distance of a star is made by determining its directions, as seen from opposite points of the earth's orbit. If we draw a line from each of these points, in the observed direction of the star, the point in which the lines meet marks the position of the star. A negative parallax shows that the two lines, instead of converging to a point, actually diverge, so that there is no possible position of the star to correspond to the observations. Such a paradoxical result can arise only from errors of observation.

---

* Dr. A. Winnecke, formerly assistant at the Pulkowa Observatory, and now director of the observatory at Strasburg.

† A list of the stars of which the parallaxes have been determined will be found in the Appendix.

CHAPTER IV.

THE MOTION OF LIGHT.

INTIMATELY connected with celestial measurements are the curious phenomena growing out of the progressive movement of light. It is now known that when we look at a star we do not see the star that now is, but the star that was several years ago. Though the star should suddenly be blotted out of existence, we should still see it shining for a number of years before it would vanish from our sight. We should see an event that was long past, perhaps one that was past before we were born. This non-coincidence of the time of perception with that of occurrence is owing to the fact that light requires time to travel. We can see an object only by light which emanates from it and reaches our eye, and thus our sight is behind time by the interval required for the light to travel over the space which separates us from the object.

It was by observations of the satellites of Jupiter that it was first found that celestial phenomena were thus seen behind time. These bodies revolve round Jupiter much more rapidly than our moon does around the earth, the inner satellite making a complete revolution in eighteen hours. Owing to the great magnitude of Jupiter and his shadow, this satellite, as also the two next outside of it, are eclipsed at every revolution. The accuracy with which the times of disappearance in the shadow could be observed, and the consequent value of such observations for the determination of longitudes, led the astronomers of the seventeenth century to make tables of the times of occurrence of these eclipses. In attempting to improve the tables of his predecessors, it was found by Roemer (then of Paris, though a Dane by birth) that the times of the

THE MOTION OF LIGHT. 211

eclipses could not be represented by an equable motion of the satellites. He could easily represent the times of the eclipses when Jupiter was in opposition to the sun, and therefore the earth nearest to Jupiter. But then, as the earth receded from Jupiter in its annual course round the sun, the eclipses were constantly seen later, until, when it was at its greatest distance from Jupiter, the times appeared to be 22 minutes late. Such an inequality, Roemer concluded, could not be real; he therefore attributed it to the fact that it must take time for light to come from Jupiter to the earth, and that this time is greater the more distant the earth is from the planet. He therefore concluded that it took light 22 minutes to cross the orbit of the earth, and, consequently, 11 minutes to come from the sun to the earth.

The next great step in the theory of the progressive motion of light was made by the celebrated Bradley, afterwards Astronomer Royal of England, to whose observations at Kew on the star γ Draconis with his zenith sector, in order to determine the parallax of the star, allusion has already been made. The effect of parallax would have been to make the declination greatest in June and least in December; while in March and September the star would occupy an intermediate or mean position. But the actual result of the measures was entirely different, and exhibited phenomena which Bradley could not at first account for. The declinations of June and December were the same, showing no effect of parallax. But, instead of remaining the same the rest of the year, the declination was some forty seconds greater in September than in March, when the effect of parallax should be the same. Thus, the star had a regular annual oscillation; but instead of its apparent motion in this little orbit being opposite to that of the earth in its annual orbit, as required by the laws of relative motion, it was constantly at right angles to it.

After long consideration, Bradley saw the cause of the phenomenon in the progressive motion of light combined with the motion of the earth in its orbit. In Fig. 59 let S be a star, and OT a telescope pointed at it. Then, if the telescope is not in motion, the ray *SOT* emanating from the star, and entering the centre of the object-glass, will pass down near the right-hand edge of the eye-piece, and the star will appear in the right of the field of view. But, instead of being at rest, all our telescopes are carried along with the earth in its orbit round the sun at the rate of nearly nineteen miles a second. Suppose this motion to be in the direction of the arrow; then, while the ray is passing down the telescope, the latter moves a short distance, so that the ray no longer strikes the right-hand edge of the eye-piece, but some point farther to the left, as if the star were in the direction *S'*, and the ray followed the course of the dotted line. In order to see the star centrally, the eye end of the telescope must be dropped a little behind, so that, instead of pointing in the direction *S*, it will really be pointing in the direction *S'*, shown by the dotted ray. This will then represent the apparent direction of the star, which will seem displaced in the direction in which the earth is moving.

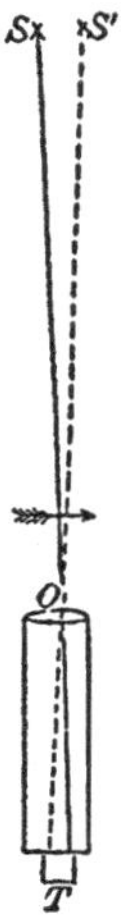

FIG. 59. — Aberration of light.

The phenomenon is quite similar to that presented by the apparent direction of the wind on board a steamship in motion. If the wind is really at right angles to the course of the ship, it will appear more nearly ahead to those on board; and if two ships are passing each other, they will appear to have the wind in different directions. Indeed, it is said to have been through noticing this very result of motion on board a boat on the Thames, that the cause of the phenomenon he had observed was suggested to Bradley.

The displacement of the stars which we have explained is called the *Aberration of Light*. Its amount depends on the ratio of the velocity of the earth in its orbit to the velocity of light. It can be determined by observing the declination of a star at the proper seasons during a number of years, by which the annual displacement will be shown. The value now most generally received is that determined by Struve at

THE MOTION OF LIGHT. 213

the Pulkowa Observatory, and is 20″.445. Though this is the most reliable value yet found, the two last figures are both uncertain. We can say little more than that the constant probably lies between 20″.43 and 20″.48, and that, if outside these limits at all, it is certainly very little outside.

This amount of aberration of each star shows that light travels 10,089 times as fast as the earth in its orbit. From this we can determine the time light takes to travel from the sun to the earth entirely independent of the satellites of Jupiter. The earth makes the circuit of its orbit in $365\frac{1}{4}$ days. Then light would make this same circuit in $\frac{365.25}{10089}$ of a day, which we find to be 52 minutes $8\frac{1}{2}$ seconds. The diameter of the earth's orbit is found by dividing its circumference by 3.1416, and the mean distance of the sun is half this diameter. We thus find from the above amount of aberration that light passes from the sun to the earth in 8 minutes 18 seconds.

The question now arises, Does the same result follow from the observations of the satellites of Jupiter? If it does, we have a striking confirmation of the astronomical theory of the propagation of light. If it does not, we have a discrepancy, the cause of which must be investigated. We have said that the first investigator of the subject found the time required to be 11 minutes. This determination was, however, uncertain by several minutes, owing to the very imperfect character of the early observations on which Roemer had to depend. Early in the present century, Delambre made a complete investigation from all the eclipses of the satellites which had been observed between 1662 and 1802, more than a thousand in number. His result was 8 minutes 13.2 seconds.

There is a discrepancy of five seconds between this result of Delambre, obtained some seventy years ago, and the modern determinations of the aberrations of the fixed stars made by Struve and others. What is its cause? Probably only the errors of the observations used by Delambre. In this case, there would be no real difference. But some physicists and astronomers have endeavored to show that there is a real cause for such a difference, which they hold to indicate an error in the value of the aberration derived from observation arising in this way. It is known from experiment that light passes through glass or any other refracting medium more slowly than through a void. In observations with a telescope the light has to pass through the objective, and the time lost in doing so will make the aberration appear larger than it really is, and the velocity of light will appear too small. But the commonly received theory (that of Fresnel) is that this loss of time is compensated by the objective partially drawing the ray with it. Desirous of setting the question at rest, Professor Airy, a few years ago, constructed a telescope, which he filled with water, with which he observed the constant of aberration. The aberration was found to be the same as with ordinary telescopes, thus proving the theory of Fresnel to be correct, because on the other theory the aberration ought to have been much increased by the water.

Hence this explanation of the difference of the two results fails, and renders it more probable that there is some error in Delambre's result. A reinvestigation of all the observations of Jupiter's satellites is very desirable; but so vast is the labor that no one since Delambre has undertaken it. Mr. Glasenapp, a young Russian astronomer, has, however, recently investigated all the observations of Jupiter's first satellite made during the years 1848–1873, and found from these that the time required for light to pass from the sun to the earth is 8 minutes 20 seconds. Instead of being smaller than Struve's result, this is two seconds larger, and seven seconds larger than that of Delambre. It is therefore concluded that the difference between the results of the two methods arises entirely from the errors of the observations used by Delambre, and that Struve's time (498 seconds) is not a second in error.

Each of the two methods we have described gives us the time required for light to pass from the sun to the earth; but neither of them gives us any direct information respecting the velocity of light. Before we can determine the latter from the former, we must know what the distance of the sun is. Dividing this distance in miles by 498, we shall have the distance which light travels in a second. Conversely, if we can find experimentally how far light travels in a second, then by multiplying this distance by 498 we shall have the distance of the sun. But we need only reflect that the velocity of light is about 180,000 miles per second to see that the problem of determining it experimentally is a most difficult one. It is seldom that objects on the surface of the earth are distinctly seen at a greater distance than forty or fifty miles, and over such a distance light travels in the forty-thousandth part of a second. As might be expected, the earlier attempts to fix the time occupied by light in passing over distances so short as those on the surface of the earth were entire failures. The first of these is due to Galileo; and his method is worth mentioning, to show the principle on which such a determination can be made. He stationed two observers a mile or two apart by night, each having a lantern which he could cover in a moment. The one observer, A, was to cover his lantern, and the distant one, B, as soon as he saw the light disappear, covered his also. In order that A might see the disappearance of B's lantern, it was necessary that the light should travel from A to B, and back again. For instance, if it took one second to travel between the two stations, B would continue to see A's light an entire second after it was really extinguished; and if he then covered his lantern instantly, A would still see it during another second, making two seconds in all after he had extinguished his own, besides the time B might have required to completely perform the movement of covering his.

Of course, by this rough method Galileo found no interval whatever. An occurrence which only required the hundredth part of the thousandth of a second was necessarily instantaneous. But we can readily elaborate his idea into the more refined methods used in recent times. Its essential feature is that which must always be employed in making the determination; that is, it is necessary that the light shall be sent from one station to another, and then returned to the first one, where the double interval is timed. There is no possibility of comparing the times at two distant stations with the necessary precision. The first improvement we should make on Galileo's method would be to set up a mirror at the distant station, and dispense with the second lantern, the observer A seeing his own lantern by reflection in the mirror. Then, if he screened his lantern, he would continue to see it by reflection in the mirror during the time the light required to go and come. But this also would be a total failure, because the reflection would seem to vanish instantly. Our next effort would be to try if we could not send out a flash of light from our lantern, and screen it off before it got back again. An attempt to screen off a single flash would also be a failure. We should then try sending a rapid succession of flashes through openings in a moving screen, and see whether they could be cut off by the sides of the openings before their return. This would be effected by the contrivance shown in Fig. 60. We have here a wheel with spokes extending from its circumference, the distance between them being equal to their breadth. This wheel is placed in front of the lantern, L, so that the light from the latter has to pass between the spokes of the wheel in order to reach the distant mirror.

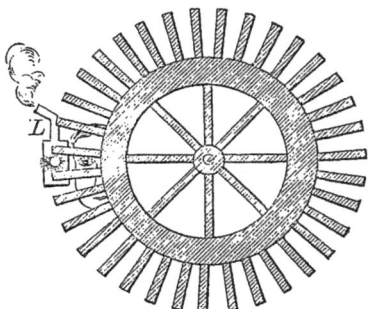

FIG. 60.—Revolving wheel, for measuring the velocity of light.

In the figure the reader is supposed to be between the wheel and the reflecting mirror, facing the former, so that he sees the light of the lantern, and also the eye of the observer, between the spokes. The latter, looking between the spokes, will see the light of the lantern reflected from the mirror. Now, suppose he turns the wheel, still keeping his eye at the same point. Then, each spoke cutting off the light of the lantern as it passes, there will be a succession of flashes of light which will pass through between the spokes, travel to the mirror, and thence be reflected back again to the wheel. Will they reach the eye of the observer behind the wheel? Evidently they will, if they return so quickly that a tooth has not had time to intervene. But suppose the wheel to turn so rapidly that a tooth just intervenes as the flash gets back to it. Then the observer will see no light in the mirror, because each successive flash is caught by the following tooth just before it reaches the observer's eye. Suppose, next, that he doubles the speed of his wheel. Then, while the flash is travelling to the mirror and back, the tooth will have passed clear across and out of the way of the flash, so that the latter will now reach the observer's eye through the opening next following that which it passed through to leave the lantern. Thus, the observer will see a succession of flashes so rapid that they will seem entirely continuous to the eye. If the speed of the wheel be again increased, the return flash will be caught on the second tooth, and the observer will see no light, while a still further increase of velocity will enable him to see the flashes as they return through the second interval between the spokes, and so on.

In principle, this is Fizeau's method of measuring the velocity of light. In place of spokes, he has exceedingly fine teeth in a large wheel. He does not look between the teeth with the naked eye, but employs a telescope so arranged that the teeth pass exactly through its focus. An arrangement is made by which the light passes through the same focus without reaching the observer's eye except by reflection from the distant mirror. The latter is placed in the focus of a second telescope, so that it can be easily adjusted to send the rays back in the exact direction from which they come. To find the time it takes the light to travel, it is necessary to know the exact velocity of the wheel which will cut off the return light entirely, and thence the number of teeth which pass in a second. Suppose, for instance, that the wheel had a thousand teeth, and the reflector was nine miles away, so that the light had to travel eighteen miles to get back to the focus of the telescope. Then it would be found that with a velocity of about five turns of the wheel per second, the light would be first cut off. Increasing the velocity, it would reappear, and would grow brighter until the velocity reached ten turns per second. It would then begin to fade away, and at fifteen turns per second would be again occulted, and so on. With the latter velocity, fifteen thousand teeth and fifteen thousand intervals would pass in a second, while two teeth and one interval passed during the time the light was performing its journey. The latter would, therefore, be performed in the ten-thousandth part of a second, showing the actual velocity to be 180,000 miles per second. The most recent determination made in this way is by M. Cornu, of Paris, who has made some improvements in the mode of applying it. His results will be described presently.

Ingenious and beautiful as this method is, I do not think it can be so accurate as another employed by Foucault, in which it is not a toothed wheel which revolves, but a Wheatstone mirror. To explain the details of the apparatus actually used would be tedious, but the principle on which the method rests can be seen quite readily. Suppose AB, Fig. 61, to represent a flat mirror, seen edgewise, revolving round an axis at X, and C a fixed concave mirror, so placed that the centre of its concavity shall fall on X. Let O be a luminous point, from which emanates a single ray of light, OX. This ray, meeting the mirror at X, is reflected to the concave mirror, C, which it meets at a right angle, and is therefore reflected directly back on the line from which it came, first to X, and then through the point O, from which it emanated, so that an eye stationed at E will see it returning exactly through the point O. No matter how the observer may turn the mir-

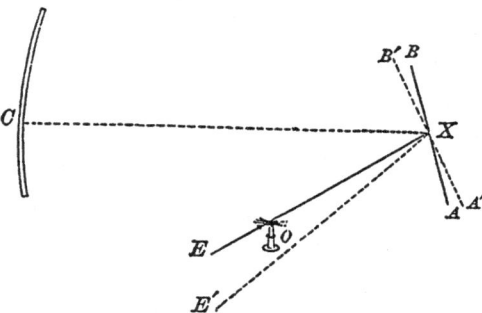

FIG. 61.—Illustrating Foucault's method of measuring the velocity of light.

ror AB, he cannot make the reflected ray deviate from this line: he can only make it strike a different point of the mirror C. If he turns AB so that after the ray is reflected from it, it does not strike C at all, then he will see no return ray. If the ray is reflected back at all, it will pass through O. This result is founded on the supposition that the mirror AB remains in the same position during the time the ray occupies in passing from X to C and back. But suppose the mirror AB to be revolving so rapidly that when the ray gets back to X, the mirror has moved to the position of the dotted line $A'B'$. Then it will no longer be reflected back through O, but will be sent in the direction E', the angle EXE' being double that through which the mirror has moved during the time the ray was on its passage. Knowing the velocity of the mirror, and the angle EXE', this time is easily found.

Evidently the observer cannot see a continuous light at E', because a reflection can be sent back only when the revolving mirror is in such a position as to send the ray to some point of the concave mirror, C. What will really be seen, therefore, is a succession of flashes, each flash appearing as the revolving mirror is passing through the position AB. But when the mirror revolves rapidly, these flashes will seem to the eye to form a continuous light, which, however, will be fainter than if the mirror were at rest, in the proportion which the arc of the concave mirror, C, bears to an entire circle. Beyond the enfeeblement of the light, this want of continuity is not productive of any inconvenience. It was thus found by Foucault that the velocity of light was 185,000 miles per second, a result which is probably within a thousand miles of the truth.

The preceding explanation shows the principle of the method, but not the details necessary in applying it. It is not practicable to isolate a single ray of light in the manner supposed in the figure, and therefore, without other apparatus, the light from O would be spread all over the space around E and E'. The desired result is obtained by placing a lens between the luminous point O and the revolving mirror in such a position that all the light falling from O upon the lens shall, after reflection, be brought to a focus upon the surface of the concave mirror, C. Then when the mirror AB is made to revolve rapidly, the return rays passing back through the lens on their return journey are brought to a focus at a point along-side O, and distant from it by an amount which is proportional to the time the light has required to pass from X to C and back again.

So delicate is this method, that the millionth of a second of time can be measured by it as accurately as a carpenter can measure the breadth of a board with his rule. Its perfection is the result of the combined genius of several men. The first idea of employing a revolving mirror in the measurement of a very minute interval of time is due to the late Sir Charles Wheatstone, who thus measured the duration of the electric spark. Then Arago showed that it could be applied to determine whether the velocity of light was greater in water or in air. Fizeau and Foucault improved on Arago's ideas by the introduction of the concave mirror, having its centre of curvature in the revolving mirror, and then this wonderful piece of apparatus was substantially complete. The last determination of the velocity of light with it was made by Foucault, and communicated to the French Academy of Sciences in 1862, with the statement that the velocity resulting from all his experiments was 298,000 kilometres (185,200 miles) per second.

The problem in question was next taken up by Cornu, of Paris, whose result has already been alluded to. Notwithstanding the supposed advantages of the Foucault-Wheatstone method, M. Cornu preferred that of Fizeau. His first results, reached in 1872, accorded quite well with those of Foucault just cited, indicating a small but somewhat uncertain increase. His experiments were repeated in 1874, and their results were communicated to the French Academy of Sciences in December of that year. In this last series of measurements his station was the observatory, and the distant mirror was placed on the tower of Montlhéry, at a distance of about fourteen English miles. The telescope through which

THE MOTION OF LIGHT. 221

the flashes of light were sent and received was twenty-nine feet long and of fourteen inches aperture. The velocity of the toothed wheel could be made to exceed 1600 turns a second, and by the electro-chronograph, on which the revolutions were recorded, the time could be determined within the thousandth of a second. At Montlhéry, the telescope, in the focus of which the reflecting mirror was placed, was six inches in aperture, and was held by a large cast-iron tube set in the masonry of the tower. At this distance M. Cornu was able, with the highest velocity of his revolving wheel, to make twenty of its teeth pass before the flashes of light got back, and to catch them, on their return, on the twenty-first tooth.

All the determinations, however, were not made with the wheel going at this rate, but with such different velocities that the rays were caught sometimes on one tooth and sometimes on another, from the fourth to the twenty-first. The following table shows the velocity of light in kilometres per second when the ray was caught on the fourth tooth, on the fifth, and so on to the twenty-first:

Tooth 4	300,130	Tooth 13	300,340
" 5	300,530	" 14	300,350
" 6	300,750	" 15	300,290
" 7	300,820	" 16	300,620
" 8	299,940	" 17	300,000
" 9	300,550	" 18	300,150
" 10	300,640	" 19	299,550
" 11	300,350	" 20	
" 12	300,500	" 21	300,060

M. Cornu hence concludes that the velocity of light in air is 300,330, and in a vacuum 300,400 kilometres per second. But Helmert, of Aix, has noticed a tendency in M. Cornu's numbers, as given above, to diminish as the velocity of the wheel is increased, and concludes that the true velocity to be derived from the measures is 299,990 kilometres. This result, though less than that derived by Cornu himself, is still nearly 2000 kilometres greater than that of Foucault.

CHAPTER V.

THE SPECTROSCOPE.

In one of Dr. Lardner's popular lectures on astronomy, delivered some thirty years ago, he introduced the subject of weighing the planets as one in which he could with difficulty expect his statements to be received with credulity. That men should measure the distances of the planets was a statement he expected his hearers to receive with surprise; but the step from measuring to weighing was so long a one, that it seemed to the ordinary mind to extend beyond all the bounds of possibility.

Had a hearer told the lecturer that men would also be able to determine the chemical constituents of the sun and stars, and to tell whether any of them did or did not contain iron, hydrogen, and other chemical elements, the lecturer would probably have replied that that statement quite exceeded the limits of his own credulity; that, while he himself saw clearly how the planets were measured and weighed, he looked upon the idea of determining their chemical constitution as a mere piece of pleasantry, or the play of an exuberant fancy. And yet, this very thing has, to a certain extent, been done by the aid of the spectroscope. The chemical constitution of matter in the state of gas or vapor can be detected almost as readily at the distance of the stars as if we had it in our laboratories. The difficulties which stand in the way do not arise from the distance, but from the fact that matter in the heavenly bodies seems to exist in some state which we have not succeeded in exactly reproducing in our laboratories. Like many other wonders, spectrum analysis, as it is called, is not at all extraordinary after we see how it is done. Indeed, the only wonder

THE SPECTROSCOPE. 223

now is how the first half of this century could have passed without physicists discovering it. The essential features of the method are so simple that only a knowledge of the elements of natural philosophy is necessary to enable them to be understood. We shall, therefore, briefly explain them.

It is familiarly known that if we pass the rays of the sun which enter a room by a small opening through a prism, the light is separated into a number of bright colors, which are spread out on a certain scale, the one end being red and the other violet, while a long range of intermediate colors is found between them. This shows that common white light is really a compound of every color of the spectrum. This compound is not like chemical compounds, made up of two or three or some limited number of simples, but is composed of an infinity of different kinds of light, all running into each other by insensible degrees; the difference, however, being only in color, or in the capacity of being refracted by the prism through which it passes. This arrangement of colors, spread out to our sight according to the refrangibility of the light which forms them, is called *the spectrum*. By the spectrum of any object is meant the combination of colors found in the light which emanates from that object. For instance, if we pass the light from a candle through a prism, so as to separate it into its component colors, and make the light thus separated fall on a screen, the arrangement of colors on the screen would be called the spectrum of the candle. If we look at a bright star through a prism, the combination of colors which we see is called the spectrum of the star, and so with any other object we may choose to examine.

As the experiment of forming a spectrum is commonly made, there is a slight mixing-up of light of the different colors, because light of the same degree of refrangibility, will fall on different parts of the screen according to the part of the prism it passes through. When the separation of the light is thus incomplete, the spectrum is said to be impure. In order to make any successful examination of the light which emanates from an object, our spectrum must be pure; that is,

224   PRACTICAL ASTRONOMY.

each point of the spectrum must be formed by light of one degree of refrangibility. To effect this in the most perfect way, the spectrum is not formed on a screen, but on the retina of the observer's eye. An instrument by which this is done is called a spectroscope.

The most essential parts of a spectroscope consist of a small telescope with a prism in front of the object-glass. The observer must adjust his telescope so that, removing the prism, and looking directly at the object, he shall obtain distinct vision of it. Then, putting the prism in its place, and turning the telescope to such an angle that the light which comes from the object shall, after being refracted by the prism, pass directly into the telescope, he looks into the latter. When the proper adjustments are made, he will see a pure spectrum of the object. In order that this experiment may succeed, it is essential that the object, when viewed directly, shall present the appearance of a point, like a star or planet. If it is an object which has a measurable surface, like the sun or moon, he will see either no spectrum at all or only a very impure one.

For this reason, a spectroscope which consists of nothing but a telescope and prism is not fitted for any purpose but that of trial and illustration. To fit it for general use, another object-glass, with a slit in its focus, is added. Fig. 62 shows the

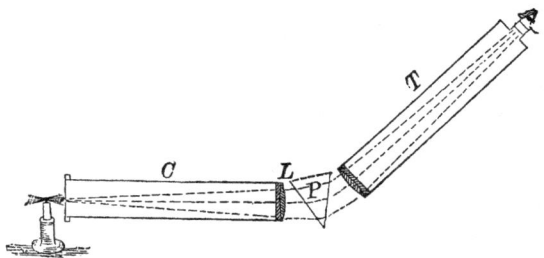

FIG. 62.—Course of rays through a spectroscope.

essential parts of a modern spectroscope. At the farther end of the second telescope, where the light enters, is a narrow slit, which can be opened or closed by means of a screw, and through which the light from the object is admitted. The rays of light following the dotted lines are made parallel by passing through the lens, L. They then fall on the prism, P, by which they are refracted, and from which they emerge parallel, except that the direction of the rays of different colors is different, owing to the greater or less degree of refraction produced by the prism. They then pass through the object-glass of the telescope, T, by which the rays of each color are brought to a focus at a particular point in the field of view, the red rays all coming together at the lower point, the violet ones at the upper point, and those of each intermediate color at their proper place along the line. The observer, looking into the telescope, sees the spectrum of whatever object is throwing its light through the slit.

If the object of which the observer wishes to see the spectrum is a flame, he places it immediately in front of the slit; and if it is an object of sensible surface, like the sun or moon, he points the collimator, C, directly at it, so that the light which enters the slit shall fall on the lens, L. But if it is a star, he cannot get light enough in this way to see it, and he must either remove his collimator entirely, or fasten his spectroscope to the end of a telescope, so that the slit shall be exactly in the focus. The latter is the method universally adopted in examining the spectrum of a star.

If, with this instrument, we examine the light which comes from a candle, from the fire, or from a piece of white-hot iron, we shall find it to be continuous; that is, there is no gap in the series of colors from one end to the other. But if we take the light from the sun, or from the moon, a planet, or any object illuminated by the sun, we shall find the spectrum to be crossed by a great number of fine dark lines, showing that certain kinds of light are wanting. It is now known that the particular kinds of light which originally belonged in these dark lines have been *culled out* by the gases surrounding the sun through which the light has passed. This culling-out is called *Selective Absorption*. It is found by experiment that each kind of gas has its own liking for light of peculiar degrees of refrangibility, and absorbs the light which belongs in the corresponding parts of the spectrum, letting all the other light pass.

Perhaps we may illustrate this process by a similar one which we might imagine mankind to perform. Suppose Nature should loan us an immense collection of many millions of gold pieces, out of which we were to select those which would serve us for money, and return her the remainder. The English rummage through the pile, and pick out all the pieces which are of the proper weight for sovereigns and half-sovereigns; the French pick out those which will make five, ten, twenty, or fifty franc pieces; the Americans the one, five, ten, and twenty dollar pieces, and so on. After all the suitable pieces are thus selected, let the remaining mass be spread out on the ground according to the respective weights of the pieces, the smallest pieces being placed in a row, the next in weight in an adjoining row, and so on. We shall then find a number of rows missing: one which the French have taken out for five-franc pieces; close to it another which the Americans have taken for dollars; afterwards a row which have gone for half-sovereigns, and so on. By thus arranging the pieces, one would be able to tell what nations had culled over the pile, if he only knew of what weight each one made its coins. The gaps in the places where the sovereigns and half-sovereigns belonged would indicate the English, that in the dollars and eagles the Americans, and so on. If, now, we reflect how utterly hopeless it would appear, from the mere examination of the miscellaneous pile of pieces which had been left, to ascertain what people had been selecting coins from it, and how easy the problem would appear when once some genius should make the proposed arrangement of the pieces in rows, we shall see in what the fundamental idea of spectrum analysis consists. The formation of the spectrum is the separation and arrangement of the light which comes from an object on the same system by which we have supposed the gold pieces to be arranged. The gaps we see in the spectrum tell the tale of the atmosphere through which the light has passed, as in the case of the coins they would tell what nations had sorted over the pile.

That the dark lines in the solar spectrum are picked out by the gases of the sun's atmosphere has long been surmised; indeed, Sir John Herschel seems to have had a clear idea of the possibility of spectrum analysis half a century ago. The difficulty was to find what particular lines any particular substance selects; since, to exert any selective action, a vastly greater thickness of gas is generally required than it is practicable to obtain experimentally. This difficulty was surmounted by the capital discovery of Kirchhoff and Bunsen, *that a glowing gas gives out rays of the same degree of refrangibility which it absorbs when light passes through it.* For example, if we put some salt into the flame of a spirit-lamp, and examine the spectrum of the light, we shall find a pair of bright-yellow lines, which correspond most accurately to a pair of black lines in the solar spectrum. These lines are known to be due to sodium, a component of common salt, and their existence in the solar spectrum shows that there is sodium in the sun's atmosphere. They are therefore called the sodium lines. By vaporizing various substances in sufficiently hot flames, the spectra of a great number of metals and gases have been found. Sometimes there are only one or two bright lines, while with iron the number is counted by hundreds. The quantity of a substance necessary to form these bright lines is so minute that the presence of some metals in a compound have been detected with the spectroscope when it was impossible to find a trace of them in any other way. Indeed, two or three new metals, the existence of which was before entirely unknown, first told their story through the spectroscope.

The general relations of the spectrum to the state of the substance from which the light emanated may be condensed into three rules, or laws, as follows:

1. The light from a glowing solid or liquid forms a continuous spectrum, in which neither bright nor dark lines are found. The spectrum is of the same nature, no matter how finely the substance may be divided.

2. If the light from the glowing solid passes through a gaseous atmosphere, the spectrum will be crossed by dark lines occupying those parts of the spectrum where the light culled out by the atmosphere belongs.

3. A glowing gas sends out light of the same degrees of refrangibility as belong to that which it absorbs, so that its spectrum consists of a system of bright lines occupying the same position as the dark lines it would produce by absorption.

If, then, on examining the spectrum of a star or other heavenly body, we find only bright lines with dark spaces between them, we may conclude that the body consists of a glowing gas, and we judge what the gas is by comparing the spectrum with those of various substances on the earth. If, on the other hand, the spectrum is a continuous one, except where crossed by fine dark lines, we conclude that it emanates from a glowing body surrounded by an atmosphere which culls out some of the rays of light.

It will be seen that the spectroscope gives us no definite information respecting the nature or composition of bodies in the solid state. If we heat any sort of metal white-hot, supposing only that it will stand this heat without being vaporized, we shall have a spectrum continuous from end to end, in which there will be neither bright nor dark lines to give any indications respecting the substance. In order, therefore, to detect the presence of any chemical element with this instrument, that element must be in the form of gas or vapor. Here we have one limitation to the application of the spectroscope to the celestial bodies. The tendency of bodies in space is to cool off, and when they have once become so cool as to solidify, the instrument in question can give us no further definite information respecting their constitution.

Even if the body be in the gaseous state, we cannot always rely on the spectroscope informing us with certainty of the nature of the gas. The light we analyze must either be emitted by the gas, the latter being so hot as to shine by its own light, or it must be transmitted through it. Thus, the application of spectrum analysis is confined to glowing gases and the atmospheres of the stars and planets, the application to the latter depending on the fact that the sunlight reflected from the surface of the planet passes twice through its atmosphere. Even in these cases the interpretation of its results is sometimes rendered difficult in consequence of the varied spectrum of the same gas at different temperatures and under different degrees of pressure. Under some conditions so many new lines are introduced into the spectrum of hydrogen that it can hardly be recognized. As a general rule, the greater the pressure, the greater the number of lines which appear; indeed, it has been found by Lockyer and Frankland that as the pressure and density of a gas are increased, its spectrum tends to become continuous. We must therefore regard the third of the above rules respecting spectrum analysis, or, rather, the general rule that a glowing gas gives a spectrum of bright lines, as not universally true. If we could, by artificially varying the temperature, pressure, and composition of gases, accurately reproduce the spectrum of a celestial body, the changes of the spectrum which we have mentioned would be a positive advantage; since they would enable us to determine, not merely the composition of a gaseous body, but its temperature and pressure. This is, however, a field in which success has not yet been reached.

The reader now understands that when the light from a celestial object is analyzed by the prism, and the component colors are spread out singly as on a sheet, the dark and bright lines which we see are the letters of the open book which we are to interpret so as to learn what they tell us of the body from which the light came, or the vapors through which it passed. When we see a line or a set of lines which we recognize as produced by a known substance, we infer the presence of that substance. The question may now be asked, How do we know but that the lines we observe may be produced by other substances besides those which we find to produce them in our laboratories? May not the same lines be produced by different substances? This question can be answered only by an appeal to probabilities. The evidence in the case is much the same as that by which, recognizing the picture of a friend, we conclude that it is not the picture of any one else. For anything we can prove to the contrary, another person might have exactly the same features, and might, therefore, make the very same picture. But, as a matter of fact, we know that practically no two men whom we have ever seen do look exactly alike, and it is extremely improbable that they ever would look so. The case is the same in spectrum analysis. Among the great number of substances which have been examined with the spectroscope, no two give the same lines. It is therefore extremely improbable that a given system of bright lines could be produced by more than one substance. At the same time, the evidence of the spectroscope is not necessarily conclusive in all cases. Should only a single line of a substance be found in the spectrum of a star or nebula, it would hardly be safe to conclude, from that alone, that the line was really produced by the known substance. Collateral evidence might, however, come in. If the same line were found both in the sunlight, and in that of a great number of stars, we should be justified in concluding that the lines were all produced by the same substance. All we can say in doubtful cases is, that our conclusions must be drawn with care and discrimination, and must accord with the probabilities of each special case.

PART III.—THE SOLAR SYSTEM.

CHAPTER I.

GENERAL STRUCTURE OF THE SOLAR SYSTEM.

HAVING, in the preceding parts, described the general structure of the universe, and the methods used by astronomers in measuring the heavens and investigating the celestial motions, we have next to consider in detail the separate bodies which compose the universe, and to trace the conclusions respecting the general order of creation to which this examination may lead us. Our natural course will be to begin with a general description of the solar system to which our earth belongs, considering, first, the great central body of that system, then the planets in their order, and, lastly, such irregular bodies as comets and meteors.

We have shown in the first part that the solar system was found by Copernicus, Kepler, and Newton to consist of the sun, as the great central body, with a number of planets revolving around it in ellipses, having the sun in one of their foci; the whole being bound together by the law of universal gravitation. Modern science has added a great number of bodies, and shown the system to be a much more complex one than Newton supposed. As we now know them, the bodies of the system may be classified as follows:

1. The sun, the great central body;
2. A group of four inner planets—Mercury, Venus, the Earth, and Mars;
3. A swarm of small planets or asteroids revolving outside the orbit of Mars (about 175 of them are now known);

4. A group of four outer planets—Jupiter, Saturn, Uranus, and Neptune;

5. A number of satellites of the planets, 18 being now known, of which all but one belong to the group of outer planets;

Fig. 63.—Relative size of sun and planets.

6. An unknown number of comets and meteors, revolving in very eccentric orbits.

The eight planets of groups 2 and 4 are called the major planets, to distinguish them from all others, which are smaller or less important.

GENERAL STRUCTURE OF THE SOLAR SYSTEM. 233

The range of size, distance, and mass among the bodies of the system is enormous. Neptune is eighty times as far from the sun as Mercury, and Jupiter several thousand times as heavy. It is, therefore, difficult to lay down a map of the whole system on the same scale. If the orbit of Mercury were represented with a diameter of one-fourth of an inch, that of Neptune would have a diameter of 20 inches.

With the exception of Neptune, the distances of the eight major planets proceed in a tolerably regular progression, the group of small planets taking the place of a single planet in the series. The progression is known as the law of Titius, from its first proposer, and is as follows: Take the series of numbers 0, 3, 6, 12, 24, 48, each one after the second being formed by doubling the one which precedes it. Add 4 to each of these numbers, and we shall have a series of numbers giving very nearly the relative distances of the planets from the sun. The following table shows the series of numbers thus formed, together with the actual distances of the planets expressed on the same scale, the distance of the earth being called 10:

Planet.	Numbers of Titius.	Actual Distance.	Error.
Mercury	0 + 4 = 4	3.9	0.1
Venus	3 + 4 = 7	7.2	0.2
Earth	6 + 4 = 10	10.0	0.0
Mars	12 + 4 = 16	15.2	0.8
Minor planets	24 + 4 = 28	20 to 35	
Jupiter	48 + 4 = 52	52.0	0.0
Saturn	96 + 4 = 100	95.4	4.6
Uranus	192 + 4 = 196	191.9	4.1
Neptune	384 + 4 = 388	300.6	87.4

It will be seen that before the discovery of Neptune the agreement was so close as to suggest the existence of an actual law of the distances. But the discovery of this planet in 1846 completely disproved the supposed law; and there is now no reason to believe that the proportions of the solar system are the result of any exact and simple law whatever. It is true that many ingenious people employ themselves from time to time in working out numerical relations between the distances of the planets, their masses, their times of rotation, and so on, and will probably continue to do so; because the number of such relations which can be made to come somewhere near to exact numbers is very great. This, however, does not indicate any law of nature. If we take forty or fifty numbers of any kind—say the years in which a few persons were born; their ages in years, months, and days at some particular event in their lives; the numbers of the houses in which they live; and so on—we should find as many curious relations among the numbers as have ever been found among those of the planetary system. Indeed, such relations among the years of the lives of great actors in the world's history will be remembered by many readers as occurring now and then in the public journals.

Range of Planetary Masses.—The great diversity of the size and mass of the planets is shown by the curious fact, that, considering the sun and the eight planets, the mass of each of the nine bodies exceeds the combined mass of all those which are smaller than itself. This is shown in the following simple calculation. Suppose the sun to be divided into a thousand millions of equal parts, one of which parts we take as the unit of weight: then, according to the best determinations yet made, the mass of each planet will be that used in the following calculation, in which each mass is added to the masses of all the planets which are smaller than itself, the planets being taken in the order of their masses, beginning with the smallest:

Mass of Mercury	200
Mass of Mars	339
Combined mass of Mercury and Mars	539
Mass of Venus	2,353
Combined mass of Mercury, Venus, and Mars	2,892
Mass of the Earth	3,060
Combined mass of the four inner planets	5,952
Mass of Uranus	44,250
Combined mass of five planets	50,202
Mass of Neptune	51,600
Combined mass of six planets	101,802
Mass of Saturn	285,580
Combined mass of seven planets	387,382
Mass of Jupiter	954,305
Combined mass of all the planets	1,341,687
Mass of the sun	1,000,000,000

ASPECTS OF THE PLANETS.

It will be seen that the combined mass of all the planets is less than $\frac{1}{700}$ that of the sun; that Jupiter is between two and three times as heavy as the other seven planets together; Saturn more than twice as heavy as the other six; and so on.

Aspects of the Planets.—The apparent motions of the planets are described in the first chapter of this work; and in the second chapter it is shown how these apparent motions result from the real motions as laid down by Copernicus. The best time to see one of the outer planets is when in opposition to the sun. It then rises at sunset, and passes the meridian at midnight. Between sunset and midnight it will be seen somewhere between east and south. During the three months following the day of opposition, the planet will rise from three to six minutes earlier every day. A month after opposition, it will be two to three hours high soon after sunset, and will pass the meridian between nine and ten o'clock at night; while three months after opposition, it will be on the meridian about six in the evening. Hence, knowing when a planet is in opposition, a spectator will know pretty nearly where to look for it. His search will be facilitated by the use of a star map showing the position of the ecliptic among the stars, because the planets are always very near the ecliptic. Indeed, if any bright star is not down on the map, he may feel sure that it is a planet.

In describing the individual planets, we give the times when they are in opposition, so that the reader may always be able to recognize them at favorable seasons, if he wishes to do so.

The arrangement of the planets, with their satellites, is as follows:

INNER GROUP....
- Mercury.
- Venus.
- Earth, with its moon.
- Mars.

The minor planets, or asteroids.

OUTER GROUP OF GREAT PLANETS.
- Jupiter, with 4 moons.
- Saturn, with rings and 8 moons.
- Uranus, with 4 moons.
- Neptune, with 1 moon.

236   THE SOLAR SYSTEM.

This arrangement is partly exhibited in the following plan of the solar system, showing the relations of the planetary orbits from the earth outward. The scale is too small to show the orbits of Mercury and Venus.

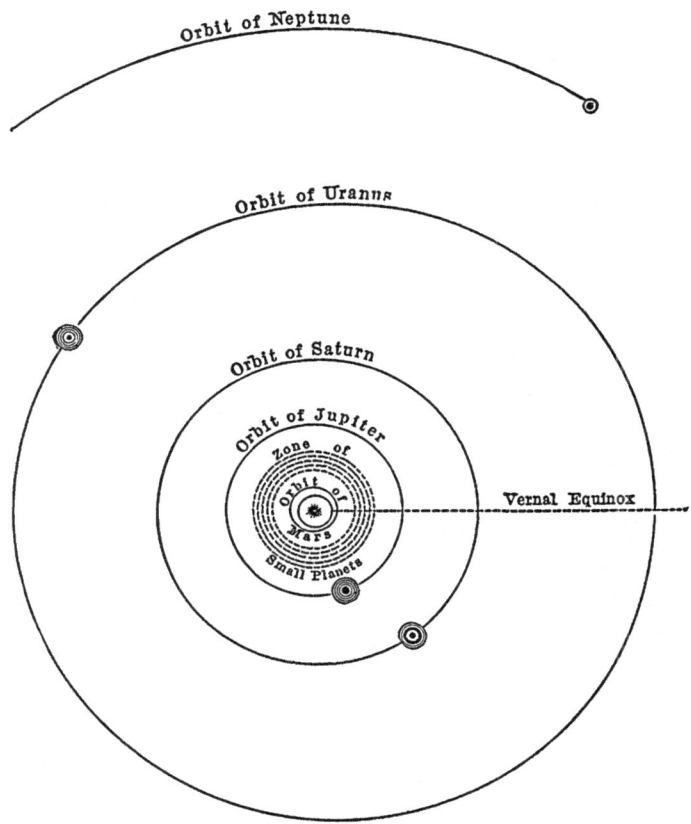

Fig. 64.—Orbits of the planets from the earth outward, showing their relative distances from the sun in the centre. The positions of the planets are near those which they occupy in 1877.

CHAPTER II.

THE SUN.

The sun presents to our view the aspect of a brilliant globe 32′, or a little more than half a degree, in diameter. To give precision to our language, the shining surface of this globe, which we see with the eye or with the telescope, and which forms the visible sun, is called the *photosphere*. Its light exceeds in intensity any that can be produced by artificial means, the electric light between charcoal points being the only one which does not look absolutely black against the unclouded sun. Our knowledge of the nature of this luminary commences with the invention of the telescope, since without this instrument it was impossible to form any conception of its constitution. The ancients had a vague idea that it was a globe of fire, and in this they were more nearly right than some of the moderns; but there was so entire an absence of all real foundation for their opinions that the latter are of little interest to any one but the historian of philosophy. We shall, therefore, commence our description of the sun with a consideration of the telescopic researches of recent times.

§ 1. *The Photosphere.*

To the naked eye the photosphere, or shining surface of the sun, presents an aspect of such entire uniformity that any attempt to gain an insight into its structure seems hopeless. But when we apply a telescope, we generally find it diversified with one or more groups of dark-looking spots; and if the vision is good, and we look carefully, we shall soon see that the whole bright surface presents a mottled appearance, looking like a fluid in which ill-defined rice-grains are suspended. Perhaps the most familiar idea of this appearance will be presented by saying that the sun looks like a plate of rice soup, the grains of rice, however, being really hundreds of miles in length. Some years ago Mr. Nasmyth, of England, examining the sun with high telescopic powers, announced that this mottled appearance seemed to him to be produced by the interlacing of long, narrow objects shaped like willow leaves, which, running and crossing in all directions, form a net-work, covering the entire photosphere. This view, though it has become celebrated through the very great care which Mr. Nasmyth devoted to his observations, has not been confirmed by subsequent observers.

Among the most careful and laborious telescopic studies of the sun recently made are those of Professor Langley.* He has a fine telescope at his command, in a situation where the air seems to be less disturbed by the sun's rays than is usual in other localities. According to his observations, when the sun is carefully examined, the mottling which we have described is seen to be caused by an appearance like fleecy clouds whose outlines are nearly indistinguishable. We may also discern numerous faint dots on the white background. Under high powers, used in favorable moments, the surface of any one of the fleecy patches is resolved into a congeries of small, intensely bright bodies, irregularly distributed, which seem to be suspended in a comparatively dark medium, and whose definiteness of size and outline, though not absolute, is yet striking, by contrast with the vagueness of the cloud-like forms seen before, and which we now perceive to be due to their aggregation. The "dots" seen before are considerable openings, caused by the absence of the white nodules at certain points, and the consequent exposure of the gray medium which forms the general background. These openings have been called pores. Their variety of size makes any measurements nearly valueless, though we may estimate in a very rough way the diameter of the more conspicuous at from $2''$ to $4''$.

---

* Professor S. P. Langley, Director of the Observatory at Allegheny, Pennsylvania.

In moments when the definition is very fine, the bright nodules or rice-grains are found to be made up of clusters of minute points of light or "granules," about one-third of a second in diameter. These have also been seen around the edges of the pores by Secchi, who estimated their magnitude as even less than that assigned by Langley. The fact that these points are aggregated into little clusters, which ordinarily present the appearance of rice-grains, gives the latter a certain irregularity of outline which has been remarked by Mr. Huggins. Thus, there appear to be three orders of aggregation in the brighter regions of the photosphere: cloud-like forms which can be easily seen at any time; rice-grains or nodules, into which these forms are resolved, and which can always be seen with a fair telescope under good definition; and granules which make up the rice-grains. This structure of the rice-grains has been seen only by Professor Langley.

If we carefully examine the sun with a very dark smoked glass, we shall find that the disk is brightest at the centre, shading off on all sides towards the limb. Careful comparisons of the intensity of radiation of different parts of the disk show that this diminution near the limb is common to all the rays, whether those of heat, of light, or of chemical action. The most recent measures of the heat rays were made by Langley by means of a thermo-electric pile, those of the light rays by Pickering,* and those of the chemical rays by Vogel.† The intensities of these several radiations at different distances from the centre of the disk as thus determined are shown in the table on the following page. The intensity at the centre is always supposed 100. The first column gives the distance from the centre in fractions of the sun's radius, which is supposed unity. Thus, the first line of the table corresponds to the centre; the last to the edge. Professor Langley's measures do not, however, extend to the extreme edge.

---

* Professor E. C. Pickering, director of the Harvard Observatory, Cambridge, Massachusetts.

† Dr. Hermann C. Vogel, formerly astronomer at Bothkamp, now of the Solar Observatory in Potsdam, Prussia.

240   THE SOLAR SYSTEM.

Distance from Centre of the Sun.	Heat Rays (Langley).	Light (Pickering).	Chemical Rays (Vogel).
.00	100	100	100
.125	....	99	100
.25	99	97	98
.375	....	94	95
.50	95	91	90
.625	....	86	81
.75	86	79	66
.85	....	69	48
.95	....	55	25
.96	62	....	23
.98	50	....	18
1.00	....	37	13

It will be seen that near the edge of the disk the chemical rays fall off most rapidly, the light rays next, and the heat rays least of all. Roughly speaking, each square minute near the limb of the sun gives about half as much heat as at the centre, about one-third as much light, and less than one-seventh as many photographic rays. Of the cause of this degradation of light and heat towards the limb of the sun no doubt has been entertained since it was first investigated. It is found in the absorption of the rays by a solar atmosphere. The sun being a globe surrounded by an atmosphere, the rays which emanate from the photosphere in a horizontal direction have a greater thickness of atmosphere to pass through than those which strike out vertically; while the former are those we see near the edge of the disk, and the latter near the centre. The different absorptions of different classes of rays correspond exactly to this supposition, it being known that the more refrangible or chemical rays are most absorbed by vapors, and the heat rays the least.

From this it follows that we get but a fraction—perhaps a small fraction—of the light and heat actually emitted by the sun; and that if the latter had no atmosphere, it would be much hotter, much brighter, and bluer in color, than it actually is. The total amount of absorption has been very differently estimated by different authorities, Laplace supposing it might be as much as eleven-twelfths of the whole amount. The smaller estimates are, however, more likely to be near the

THE PHOTOSPHERE. 241

truth, there being no good reason for holding that more than half the rays are absorbed. That is, if the sun had no atmosphere, it might be twice as bright and as hot as it actually is, but would not be likely to be three or four times so. Professor Langley suggests that the glacial epoch may have been due to a greater absorption of the sun's heat by its atmosphere in some past geological age.

A very important physical and astronomical problem is that of measuring the total amount of heat radiated by the sun to the earth during any period of time — say a day or a year. The question admits of a perfectly definite answer, but there are two difficulties in the way of obtaining it; one, to distinguish between the heat coming from the sun itself, and that coming from the atmosphere and surrounding objects; the other, to allow for the absorption of the solar heat by our atmosphere, which must be done in order to determine the total quantity emanating from the sun. The most successful experiments for this purpose are those of Pouillet and of Sir John Herschel. The results obtained by the former may be expressed thus: if the air were out of the way, and a sheet of ice were so held that the sun's rays should fall upon it perpendicularly, and be all absorbed, the ice would melt away at the rate of $14\frac{1}{2}$ inches in 24 hours. Since the sun is part of the time below the horizon, and is not perpendicular to more than a single point of the earth's surface when above it, the average amount of ice which would be melted over the whole earth is only a fraction of this, namely, 3.62 inches per day, or something more than 100 feet per year.

Attempts have been made to determine the temperature of the sun from the amount of heat which it radiates, but the estimates have varied very widely, owing to the uncertainty respecting the law of radiation at high temperatures. By supposing the radiation proportional to the temperature, Secchi[*] finds the latter to be several million degrees, while, by taking another law indicated by the experiments of Dulong and

---

[*] Father Angelo Secchi, Director of the Observatory at Rome.

Petit, others find a temperature not many times exceeding that of a reverberatory furnace. For the temperature of the photosphere, it seems likely that the lower estimates are more nearly right, being founded on an experimental law; but the temperature of the interior must be immensely higher.

§ 2. *The Solar Spots and Rotation.*

Even the poor telescopes made by the contemporaries of Galileo could hardly be directed to the sun many times without one or more spots being seen on his surface. Whatever credit may be due for a discovery which required neither industry nor skill should, by the rule of modern science already referred to, be awarded to Fabritius for the discovery of the solar spots. This observer, otherwise unknown in astronomy, made known the existence of the solar spots early in 1611— a year after Galileo began to scan the heavens with his telescope. His discovery was followed up by Galileo and Scheiner, by whom the first knowledge of the nature of the spots was acquired.

The first idea of Scheiner was that the spots were small planets in the neighborhood of the sun; but this was speedily disproved by Galileo, who showed that they must be on the surface of the sun itself. The idea of the sun being affected with any imperfection so gross as a dark spot was repugnant to the ecclesiastical philosophy of the times, and it is not unlikely that Scheiner's explanation was suggested by the desire to save the perfection of our central luminary.

A very little observation showed that the spots had a regular motion across the disk of the sun from east to west, occupying about 12 days in the transit. A spot generally appeared first on or near the east limb, and, after 12 or 14 days, disappeared at the west limb. At the end of another 14 days or more it reappeared at the east limb, unless in the mean time it had vanished from sight entirely. The spots were found not to be permanent objects, but to come into existence from time to time, and, after lasting a few days, weeks, or months, to disappear. But so long as they lasted, they always exhibited the motion just described, and it was thence inferred that the sun rotated on his axis in about 25 days.

The astronomers of the seventeenth and eighteenth centuries used a method of observing the sun which will often be found convenient for seeing the spots when one has not a telescope supplied with dark glasses at his disposal. Take an ordinary good spy-glass, or, indeed, a telescope of any size, and point

FIG. 65.—Man holding telescope, to show sun on screen.

it at the sun. To save the eyes, the right direction may be found by holding a piece of paper closely in front of the eyepiece: when the sun shines through the telescope on this paper, the pointing is nearly right. The telescope should be attached to some movable support, so that its pointing can be changed to the different directions of the sun, and should pass through a perforation in some sort of a screen, so that the sun cannot shine in front of the telescope except by passing through it. An opening in a window-shutter will answer a good purpose, only the rays must not have to pass through the glass of the window in order to reach the telescope. Draw out the eye-piece of the instrument about the eighth of an inch beyond the proper point for seeing a distant object. Then, holding a piece of white paper before the eye-piece at a distance of from 6 to 12 inches, an image of the sun will be thrown upon it. The distance of the paper must be adjusted to the distance the eye-piece is drawn out. The farther we draw out the eye-piece, the nearer the best image will be formed. Having adjusted everything so that the edge of the sun's image shall be sharply defined, one or more spots can generally be seen. This method, or something similar to it, is often used in observing eclipses and transits of Mercury, and is very convenient when it is desired to show an enlarged image of the sun to a number of spectators.

When powerful telescopes were applied to the sun, it was found that the spots were not merely the dark patches which they first appeared to be, but that they comprised two well-

Fig. 66.—Solar spot, after Secchi.

marked portions. The central part, called the *umbra* or *nucleus*, is the darkest, and is surrounded by a border, intermediate in tint between the darkness of the spot and the brill-

THE SOLAR SPOTS AND ROTATION. 245

iancy of the solar surface. This border is termed the *penumbra*. Ordinarily it appears of a uniform gray tint. But when carefully examined with a good telescope in a very steady atmosphere, it is found to be striated, looking, in fact, much like the bottom of a thatched roof, the separate straws being directed towards the interior of the spot. This appearance is shown in the figure.

The spots are extremely irregular in form and unequal in size. They are very generally seen in groups — sometimes two or more combined into a single one; and it frequently happens that a large one breaks up into several smaller ones. Their duration is also extremely variable, ranging from a few days to periods of several months.

Until about a century ago, it was a question whether the spots were not dark patches, like scoria, floating on the molten surface of the photosphere. Wilson, a Scotch observer, however, found that they appeared like cavities in the photosphere, the dark part being really lower than the bright surface around it. As a spot approached the edge of the disk, he found that the penumbra grew disproportionately narrow on the side nearest to the sun's centre, showing that this side of it was seen at a smaller angle than the other. This effect of perspective is shown in Fig. 67, where, near the sun's limb, the side of the penumbra nearest us is hidden by the photosphere. That the spots are cavities is also shown by the fact that when a large spot is exactly on the edge of the disk a notch is sometimes seen there. The shaded penumbra seems to form the sides of the cavity, while the umbra is the invisible bottom.

These observations gave rise to the celebrated theory of Wilson, which is generally connected with the name of Herschel, who developed it more fully. The interior of the sun is, by this theory, a cool, dark body, surrounded by two layers of clouds. The outer layer is intensely brilliant, and forms the visible photosphere, while the inner layer is darker, and forms the umbra around the spots. The latter are simply openings through these clouds, which form from time to

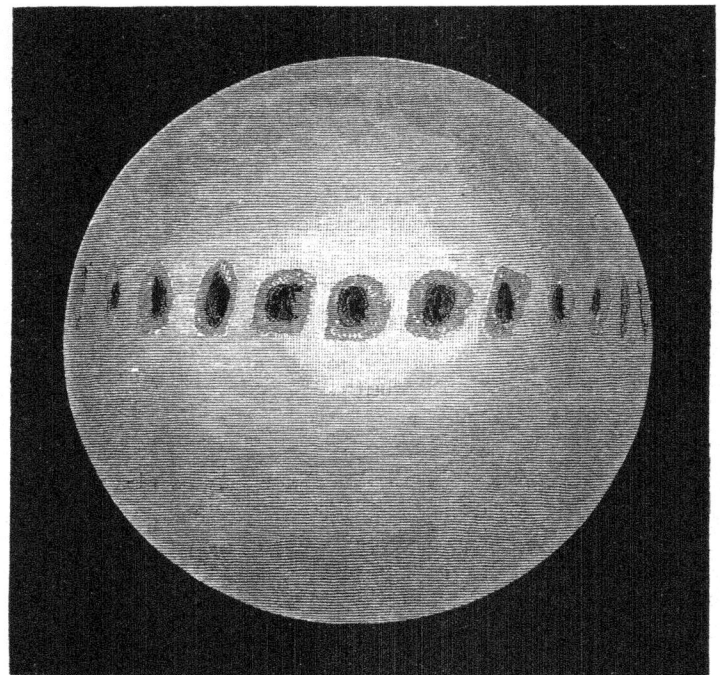

Fig. 67.—Changes in the aspect of a solar spot as it crosses the sun's disk, showing it to be a cavity in the photosphere.

time, and through which we see the dark body in the interior. Anxious that this body should serve some especial purpose in the economy of creation, they peopled it with intelligent beings, who were protected from the fierce radiation of the photosphere by the layer of cool clouds, but were denied every view of the universe without, except such glimpses as they might obtain through the occasional openings in the photosphere, which we see as spots.

Leaving out the fancy of living beings, this theory accounted very well for appearances. That the photosphere could not be absolutely and wholly solid, liquid, or gaseous seemed evident from the nature of the spots. If it were solid, the latter could not be in such a constant state of change as we see them; while if it were liquid or gaseous, these cavities could not continue for months, as they were sometimes seen to, because the liquid or gaseous matter would rush in from all sides, and fill them up. The only hypothesis that seemed left open to Herschel was that the photosphere consisted of clouds floating in an atmosphere. As the sides of the cavities looked comparatively dark, the conclusion seemed inevitable that the brilliancy of the photosphere was only on and near the surface; and as the bottom of the cavity looked entirely dark, the conclusion that the sun had a dark interior seemed unavoidable.

The discovery of the conservation of force, and of the mutual convertibility of heat and force, was fatal to this theory. Such a sun as that of Herschel would have cooled off entirely in a few days, and then we should receive neither light nor heat from it. A continuous flood of heat such as the sun has been radiating for thousands of years can be kept up only by a constant expenditure of force in some of its forms; but, on Herschel's theory, the supply necessary to meet this expenditure was impossible. Even if the heat of the photosphere could be kept up by any agency, it would be constantly conveyed to the interior by conduction and radiation; so that in time the whole sun would become as hot as the photosphere, and its inhabitants would be destroyed. In the time of Herschel it was not deemed necessary that the sun should be a very hot body, the heat received from his rays being supposed by many to be generated by their passage through our atmosphere. The photosphere was, therefore, supposed to be simply phosphorescent, not hot. This idea is still entertained by many educated men who have not made themselves acquainted with the laws of heat discovered during the present century. We may, therefore, remark that it is completely untenable. One of the best established results of these laws is that the surface of the sun is intensely hot, probably much hotter than any reverberatory furnace. The great question in the present state of science is, how the supply of heat is maintained against such immense loss by radiation.

§ 3. *Periodicity of the Spots.*

The careful observations of the solar spots which have been made during the last century seem to indicate a period of about eleven years in the spot-producing activity of the sun. During two or three years the spots are larger and more numerous than on the average; they then begin to diminish, and reach a minimum five or six years after the maximum. Another six years brings the return of the maximum. The intervals are, however, somewhat irregular, and further observations are required before the law of this period can be fixed with certainty. An idea of the evidence in favor of the period may be formed from some results of the observations of Schwabe, a German astronomer, who systematically observed the sun during a large part of a long life. One of his measures of the spot-producing power was the number of days on which he saw the sun without spots in the course of each year. The following are some of his results:

From 1828 to 1831, sun without spots on only 1 day.
In 1833, " " " 139 days.
From 1836 to 1840, " " " 3 days.
In 1843, " " " 147 days.
From 1847 to 1851, " " " 2 days.
In 1856, " " " 193 days.
From 1858 to 1861, " " " no day.
In 1867, " " " 195 days.

We see that the sun was remarkably free from spots in the years 1833, 1843, 1856, and 1867, about half the time no considerable spot being visible. This recurrence of the period has been traced back by Dr. Wolf, of Zurich, to the time of Galileo, and its average length is about 11 years 1 month. The years of fewest sun-spots during the present century were 1810, 1823, 1833, 1844, 1856, and 1867. Continuing the series, we may expect very few spots in 1878, 1889, etc. The years of greatest production of spots were 1804, 1816, 1829, 1837, 1848, 1860, and 1870, from which we may conclude that 1882, 1893, etc., will be years of numerous sun-spots.

The observations of Schwabe and the researches of Wolf seem to have placed the existence of this period beyond a doubt; but no satisfactory explanation of its cause has yet been given. When first noticed, its near approach to the period of revolution of Jupiter naturally led to the belief that there was a connection between the two, and that the attraction of the largest planet of the system produced some disturbance in the sun, which was greater in perihelion than in aphelion. But this connection seems to be disproved by the fact that the sun-spot period is at least six months, and perhaps a year, shorter than the revolution of Jupiter. It is therefore probable that the periodicity in question is not due to any action outside the sun, but is a result of some law of solar action of which we are as yet ignorant.

There are certain supposed connections of the sun-spot period with terrestrial phenomena which are of interest. Sir William Herschel collected quite a mass of statistics tending to show that there was an intimate connection between the number of sun-spots and the price of corn, the latter being low when there were few spots, and high when they were more numerous. His conclusion was that the fewer the spots, the more favorable the solar rays to the growth of the crops. This theory has not been confirmed by subsequent observation. There is, however, some reason to believe, from the researches of Professors Lovering and Loomis, that the frequency of auroras and of magnetic disturbances is subject to a period corresponding to that of sun-spots, these occurrences being most frequent when the spots are most numerous. Professor Loomis considers the coincidence to be pretty well proved, while Professor Lovering is more cautious, and waits for further research before coming to a positive conclusion. The occurrence of great auroras in 1859 and 1870–'71 was strikingly accordant with the theory.

§ 4. *Law of Rotation of the Sun.*

Between the years 1843 and 1861, a very careful series of observations of the positions and motions of the solar spots was made by Mr. Carrington, of England, with a view of deducing the exact time in which the sun rotates on his axis. These observations led to the remarkable result that the time of rotation shown by the spots was not the same on all parts of the sun, but that the equatorial regions seemed to perform a revolution in less time than those nearer the poles. Near the equator the period was about 25.3 days, while it was a day longer in 30° latitude. Moreover, the period of rotation seems to be different at different times, and to vary with the frequency of the spots. But the laws of these variations are not yet established. In consequence of their existence, we cannot fix any definite time of rotation for the sun, as we can for the earth and for some of the planets. It varies at different times, and under different circumstances, from 25 to 26½ days.

The cause of these variations is a subject on which there is yet no general agreement among those who have most carefully investigated the subject. Zöllner* and Wolf see in the general motions of the spots traces of currents moving from both poles of the sun towards the equator. The latter considers that the eleven-year spot-period is associated with a flood of liquid or gaseous matter thrown up at the poles of the sun about once in eleven years, and gradually finding its way to the equator. Zöllner adopts the same theory, and has submitted it to a mathematical analysis, the basis of which is that the sun has a solid crust, over which runs the fluid in which the spots are formed. The current springs up near the poles, and, starting towards the equator without any rotation, is acted on by the friction of the revolving crust. By this friction the crust continually tends to carry the fluid with it. The nearer the current approaches the equator, the more rapid the rotation of the crust, owing to its greater distance from the axis. The friction acts so slowly that the current reaches the equator before it takes up the motion of the crust. On this hypothesis, the crust of the sun really revolves in

---

* Dr. J. C. F. Zöllner, Professor in the University of Leipsic.

about 25 days; and the reason that the fluid which covers it revolves more slowly at a distance from the sun's equator is that it has not yet taken up this normal velocity of rotation.

This explanation of the seeming paradox that the equatorial regions of the sun perform their revolution in a shorter time than those parts nearer the poles, cannot be regarded as an established scientific theory. It is mentioned as being, so far as the writer is aware, the most completely elaborated explanation yet offered. It is possible that the spots have a proper motion of their own on the solar surface, and that this is the reason of the apparent difference in the time of rotation in different latitudes. Yet another theory of the subject is that of Faye,* who maintains that these differences in the rates of rotation are due to ascending and descending currents, as will be more fully explained in presenting his views. But we here touch upon questions which science is as yet far from being in a condition to answer.

§ 5. *The Sun's Surroundings.*

If the sun had never been examined with any other instrument than the telescope, nor been totally eclipsed by the intervention of the moon, we should not have formed any idea of the nature of the operations going on at his surface; but we might have been better satisfied that we had a complete knowledge of his constitution. Indeed, it is remarkable that modern science has shown us more mysteries in the sun than it has explained; so that we find ourselves farther than before from a satisfactory explanation of solar phenomena. When the ancients supposed the sun to be a globe of molten iron, they had an explanation which quite satisfied the requirements of the science of their times. The spots were no mystery to Galileo and Scheiner, being simply dark places in the photosphere. Herschel's explanation of them was quite in accord with the science of his time, and he may be regarded as the latest man who has held a theory of the physical constitution of the sun

---

* Mr. H. E. Faye, member of the French Academy of Sciences.

which was really satisfactory at the time it was propounded. We have shown how his theory was refuted by the discovery of the conservation of force; we have now to see what perplexing phenomena have been revealed in recent times.

Phenomena during Total Eclipses. — If, during the progress of a total eclipse, the gradually diminishing crescent of the sun is watched, nothing remarkable is seen until very near the moment of its total disappearance. But, as the last ray of sunlight vanishes, a scene of unexampled beauty, grandeur, and impressiveness breaks upon the view. The globe of the moon, black as ink, is seen as if it were hanging in mid-air, surrounded by a crown of soft, silvery light, like that which the old painters used to depict around the heads of saints. Besides this "corona," tongues of rose-colored flame of the most fantastic forms shoot out from various points around the edge of the lunar disk. Of these two appearances, the corona was noticed at least as far back as the time of Kepler; indeed, it was not possible for a total eclipse to happen without the spectators seeing it. But it is only within a century that the attention of astronomers has been directed to the rose-colored flames, although an observation of them was recorded in the Philosophical Transactions nearly two centuries ago. They are known by the several names of "flames," "prominences," and "protuberances."

The descriptions which have been given of the corona, although differing in many details, have a general resemblance. Halley's description of it, as seen during the total eclipse of 1715, is as follows:

"A few seconds before the sun was all hid, there discovered itself round the moon a luminous ring about a digit, or perhaps a tenth part of the moon's diameter, in breadth. It was of a pale whiteness, or rather pearl-color, seeming to me a little tinged with the colors of the iris, and to be concentric with the moon."

The more careful and elaborate observations of recent times show that the corona has not the circular form which was formerly ascribed to it, but that it is quite irregular in its outline. Sometimes its form is more nearly square than round, the corners of the square being about 45° of solar latitude, and the sides, therefore, corresponding to the poles and the equator of the sun. This square appearance does not, however, arise from any regularity of form, but from the fact that the corona seems brighter and higher half way between the poles and the equator of the sun than it does near those points.

Fig. 68.—Total eclipse of the sun as seen at Des Moines, Iowa, August 7th, 1869. Drawn by Professor J. R. Eastman. The letters, a, b, c, etc., mark the positions of the prominences.

These prominent portions sometimes seem like rays shooting out from the sun. The corona is always brightest at its base, gradually shading off toward the outer edge. It is impossible to say with certainty how far it extends, but there is no doubt that it has been seen as far as one semidiameter from the moon's limb.

The corona was formerly supposed to be an atmosphere either of the moon or of the sun. Thirty or forty years ago, the most plausible theory was that it was a solar atmosphere, and that the red protuberances were clouds floating in it. That the corona could be a lunar atmosphere was completely disproved by its irregular outline, for the atmosphere of a body like the moon would necessarily spread itself around in nearly uniform layers, and could not be piled up in some quarters, as the matter of the corona is seen to be. We shall soon see that there is no doubt about the corona being something surrounding the sun.

The question whether the red protuberances belong to the moon or the sun was settled during the total eclipse of 1860, which was observed in Spain. It was then proved by measures of their height above the limb of the moon that the latter did not carry them with her, but passed over them. This proved that they were fixed relatively to the sun.

At the time of this eclipse the spectroscope was in its infancy, and no one thought of applying it to the study of the corona and protuberances. The next considerable eclipse occurred eight years later, in July, 1868, and was visible in India and Siam. The spectroscope had, in the mean time, come into very general use, and expeditions were despatched from several European countries to India to make an examination of the spectra of the objects in question. The most successful observer was Janssen, of France, who took an elevated position in the interior, where the air was remarkably clear. When, on the eventful day, the last ray of sunlight was cut off by the advancing moon, an enormous protuberance showed itself, rising to a height of many thousand miles above the surface of the sun. The spectroscope was promptly turned upon it, and the practised eye of the observer saw in a moment that the spectrum consisted of the bright lines due to glowing hydrogen. The protuberance, therefore, did not consist of any substance shining merely by reflected sunlight, but of an immense mass of hydrogen gas, so hot as to shine by its own light. The theory of the cloud-like nature of the protuberances was overthrown in a moment.

This observation marks the commencement of a new era in solar physics, which, by a singular coincidence, was inaugurated independently by another observer. As Janssen looked at the lines which he was the first of men to see, it occurred to him that they were bright enough to be seen after the total phase of the eclipse had passed. He therefore determined to watch them, and find how long he could follow them. He kept sight of them, not only after the total phase had passed, but after the eclipse was entirely over. In fact, he found that with a sufficiently powerful spectroscope, he could see the spectral lines of the protuberances at any time when the air was perfectly clear, so that the varying forms of these remarkable objects which had hitherto been seen only during the rare moments of a total eclipse could be made a subject of regular observation.

But this great discovery was made in England, independently of the eclipse, by Mr. J. Norman Lockyer. This gentleman was an active student of the subject of spectroscopy; and it had occurred to him that the matter composing these protuberances, being so near the surface of the sun, must be hot enough, not only to shine by its own light, but to be quite vaporized, and, if so, its spectrum might be seen by means of the spectroscope. Finding that the instrument he possessed would show nothing, he ordered a more powerful one. But its construction was attended with so much delay that it was not ready till October, 1868. On the 20th of that month, he pointed it upon the margin of the sun, and found three bright lines in the spectrum, two of which belonged to hydrogen. Thus was realized an idea which he had formed two years before, but which he was prevented from carrying out by the want of a suitable instrument. His success was immediately communicated to the French Academy of Sciences, the news reaching that body on the very day that word was received from Janssen, in India, that he had also solved the same problem.

Following up his researches, Mr. Lockyer found that the protuberances arose from a narrow envelope surrounding the

Fig. 69.—Specimens of solar protuberances, as drawn by Secchi. The bright base in each figure represents the chromosphere from which the red flames rise.

whole surface of the sun, being, in fact, merely elevated portions of this envelope: that is to say, the sun is surrounded by an atmosphere composed principally of hydrogen gas, portions of which are here and there thrown up in the form of enormous tongues of flame, which, however, can never be seen except with the spectroscope, or during total eclipses. To this atmosphere Mr. Lockyer gave the name of the *chromosphere*. It had previously been seen and recognized by several observers during total eclipses, but nothing had been known respecting its nature.

The researches which we have described threw no light on the question of the corona, an object which seemed to have been almost lost sight of in the excitement caused by the discovery of the gaseous nature of the protuberances. Happily, only a year later, on August 7th, 1869, a total eclipse was visible in the United States. The shadow of the moon passed down the coast of Alaska, then entered into the interior, passing over the south-west portion of British America, entered the United States in the Territory of Nebraska, and passed over Iowa, Illinois, Kentucky, South-western Virginia, and North Carolina. This eclipse was observed very extensively by American astronomers, Professor Harkness, of the Naval Observatory, and Professor Young, of Dartmouth College, devoting especial attention to the spectroscopic observations. These observers found that the corona gave a very faint, continuous spectrum crossed by a single bright-green line, which was also seen in the spectrum of the protuberances. This solitary line was again seen during the eclipse of December 21st, 1870, in the Mediterranean; but it has not been certainly identified in the spectrum of any known terrestrial substance. There are several lines of iron in its neighborhood; but as this line stands alone, it does not seem likely that it can arise from the vapor of iron. All we can say is, that the substance which gives this line, and which seems to be the only gaseous element of the corona, is unknown, and may possibly be some gas much lighter than hydrogen which has not yet been discovered on the earth.

Continued observations of the spectra of the various gases surrounding the sun show a much greater number of lines than have ever been seen during total eclipses. Mr. Lockyer himself, by diligent observation extending over several years, found over a hundred. But the greatest advance in this respect was made by Professor C. A. Young. In 1871 an astronomical expedition was fitted out by the Coast Survey, for the purpose of learning by actual trial whether any great advantage would be gained by establishing an observatory on the most elevated point crossed by the Pacific Railway. This point was Sherman. The spectroscopic part of the expedition was intrusted to Professor Young. Although there was a great deal of cloudy weather, yet, when the air was clear, far less light was reflected from the sky surrounding the sun than at lower altitudes, which was a great advantage in the study of the sun's surroundings. Professor Young found no less than 273 bright lines which he was able to identify with certainty. The presence of many known substances, especially iron, magnesium, and titanium, is indicated by these lines; but there are also many lines which are not known to pertain to any terrestrial substance.

§ 6. *Physical Constitution of the Sun.*

Respecting the physical constitution of the sun, there are some points which may be established with more or less certainty, but the subject is, for the most part, involved in doubt and obscurity. Since the properties of matter are the same everywhere, the problem of the physical constitution of the sun is solved only when we are able to explain all solar phenomena by laws of physics which we see in operation around us. The fact that the physical laws operative on the sun must be at least in agreement with those in operation here, is not always remembered by those who have speculated on the subject. In stating what is probable, and what is possible, in the causes of solar phenomena, we shall begin on the outside, and go inwards, because there is less doubt about the operations which go on outside the sun than about those on his surface or in the interior.

As we approach the sun, the first material substance we meet with is the corona, rising to heights of five or ten, perhaps even fifteen, minutes above his surface, that is, to a height of from one to three hundred thousand miles. Of this appendage we may say with entire confidence that it cannot be an atmosphere in the sense in which that word is commonly used, that is, a continuous mass of elastic gas held up by its own elasticity. Of the two reasons in favor of this denial, one seems to me almost conclusive, the other entirely so. They are as follows:

1. Gravitation on the sun is about 27 times as great as on the earth, and any gas is there 27 times as heavy as here. In an atmosphere each stratum is compressed by the weight of all the strata above it. The result is, that as we go down by successive equal steps, the density of the atmosphere increases in geometrical progression. An atmosphere of the lightest known gas—hydrogen—would double its density every five or ten miles, though heated to as high a temperature as is likely to exist at the height of a hundred thousand miles above the sun's surface. But there is no approximation to such a rapid increase in the density of the corona as we go downwards. If we suppose the corona to be such an atmosphere, we must suppose it to be hundreds of times lighter than hydrogen.

2. The great comet of 1843 passed within three or four minutes of the surface of the sun, and therefore directly through the midst of the corona. At the time of nearest approach its velocity was 350 miles per second, and it went with nearly this velocity through at least 300,000 miles of corona, coming out without having suffered any visible damage or retardation. To form an idea what would have become of it had it encountered the rarest conceivable atmosphere, we have only to reflect that shooting-stars are instantly and completely vaporized by the heat caused by their encounter with our atmosphere at heights of from 50 to 100 miles; that is, at a height where the atmosphere entirely ceases to reflect the light of the sun. The velocity of shooting-stars is from 20 to 40 miles per second. Remembering, now, that resistance and heat increase at least as the square of the velocity, what would be the fate of a body, or a collection of bodies like a comet, passing through several hundred thousand miles of the rarest

atmosphere at a rate of over 300 miles a second? And how rare must such an atmosphere be when the comet passes not only without destruction, but without losing any sensible velocity! Certainly so rare as to be entirely invisible, and incapable of producing any physical effect.

What, then, is the corona? Probably detached particles partially or wholly vaporized by the intense heat to which they are exposed. A mere dust-particle in a cubic mile of space would shine intensely when exposed to such a flood of light as the sun pours out on every body in his neighborhood. The difficult question which we meet is, How are these particles held up? To this question only conjectural replies can be given. That the particles are not permanently held in one position is shown by the fact that the form of the corona is subject to great variations. In the eclipse of 1869, Dr. Gould thought he detected variations during the three minutes the eclipse lasted. The three conjectures that have been formed on the subject are:

1. That the matter of the corona is in what we may call a state of projection, being constantly thrown up by the sun, while each particle thus projected falls down again according to the law of gravitation. The difficulty we encounter here is that we must suppose velocities of projection rising as high as 200 miles per second constantly maintained in every region of the solar globe.

2. That the particles thrown out by the sun are held up a greater or less time by electrical repulsion. We know that atmospheric electricity plays an active part in terrestrial meteorology; and if electric action at the surface of the sun is proportional to those physical and chemical actions which we find to give rise to electrical phenomena here on the earth, the development of electricity there must be on an enormous scale.

3. That the corona is due to clouds of minute meteors circulating around the sun in the immediate vicinity of that luminary.

As already intimated, none of these explanations is much

PHYSICAL CONSTITUTION OF THE SUN. 261

better than a conjecture, though it is quite probable that the facts of the case are divided somewhere among them.

Next inside the corona lies the chromosphere. Here we reach the true atmosphere of the sun, rising in general a few seconds above his surface, but now and then projected upwards in immense masses which we might call flame, if the word were not entirely inadequate to convey any conception

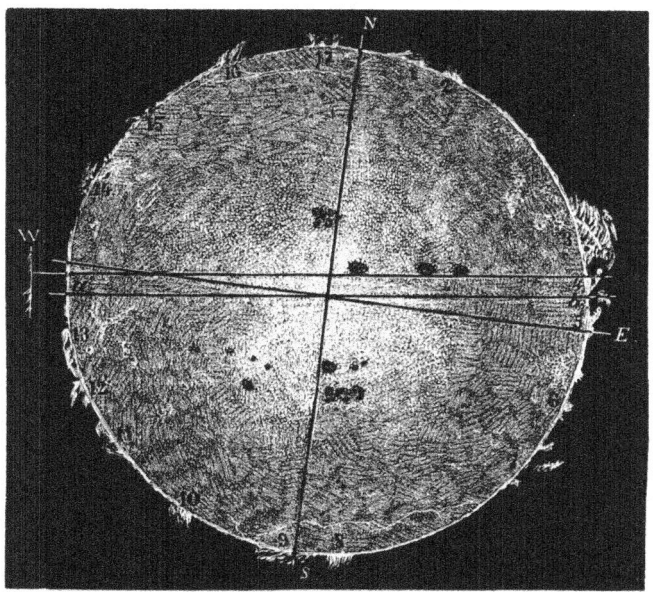

FIG. 70.—The sun, with its chromosphere and red flames, on July 23d, 1871, as drawn by Secchi. The figures mark the flames, 17 in number.

of the enormous scale on which thermal action is there carried on. What we call fire and flame are results of burning; but the gases at the surface of the sun are already so hot that burning is not possible. Hydrogen is the principal material of the upper part of the chromosphere; but, as we descend, we find the vapors of a great number of metals, including iron and magnesium. At the base, where the metals are most numerous, and the density the greatest, occurs the absorption of the solar rays which causes the dark lines in the spectrum already described (p. 225). This seems satisfactorily proved by an observation of Professor Young's during the eclipse of 1870, in Spain. At the moment of disappearance of the last rays of sunlight, when he had a glimpse of the base of the chromosphere, he saw all the spectral lines reversed; that is, they were bright lines on a dark ground. The vapors which absorb certain rays of the light which passes through them from the sun then emitted those same rays when the sunlight was cut off.

The most astonishing phenomena connected with the chromosphere are those outbursts of its matter which form the protuberances. The latter are of two classes—the cloud-like and the eruptive. The first class presents the appearance of clouds floating in an atmosphere; but as no atmosphere dense enough to sustain anything can possibly exist there, we find the same difficulty in accounting for them that we do in accounting for the suspension of the matter of the corona. In fact, of the three conjectural explanations of the corona, two are inadmissible if applied to the protuberances, since these cloud-like bodies sometimes remain at rest too long to be supposed moving under the influence of the sun's gravitation. This leaves the electrical explanation as the only adequate one yet brought forward. The eruptive protuberances seem to be due to the projection of hydrogen and magnesium vapor from the region of the chromosphere with velocities which sometimes rise to 150 miles a second. The eruption may continue for hours, or even days, the vapor spreading out into great masses thousands of miles in extent, and then falling back on the chromosphere.

Is it possible to present in language any adequate idea of the scale on which natural operations are here carried on? If we call the chromosphere an ocean of fire, we must remember that it is an ocean hotter than the fiercest furnace, and as deep as the Atlantic is broad. If we call its movements hurricanes, we must remember that our hurricanes blow only about a hundred miles an hour, while those of the chromosphere blow as far in a single second. They are such hurricanes as, "coming down upon us from the north, would, in thirty seconds after they had crossed the St. Lawrence, be in the Gulf of Mexico, carrying with them the whole surface of the continent in a mass, not simply of ruin, but of glowing vapor, in which the vapors arising from the dissolution of the materials composing the cities of Boston, New York, and Chicago would be mixed in a single indistinguishable cloud." When we speak of eruptions, we call to mind Vesuvius burying the surrounding cities in lava; but the solar eruptions, thrown fifty thousand miles high, would ingulf the whole earth, and dissolve every organized being on its surface in a moment. When the mediæval poets sung,

"Dies iræ, dies illa
Solvet sæclum in favilla,"

they gave rein to their wildest imagination, without reaching any conception of the magnitude or fierceness of the flames around the sun.

Of the corona and chromosphere the telescope ordinarily shows us nothing. They are visible only during total eclipses, or by the aid of the spectroscope. All we see with the eye or the telescope is the shining surface of the sun called the photosphere, on which the chromosphere rests. It is this which radiates both the light and the heat which reach us. The opinions of students respecting the constitution of the photosphere are so different that it is hardly possible to express any views that will not be challenged in some quarter. Although a contrary opinion is held by many, we may venture to say that the rays of light and heat seem to come, not from a gas, but from solid matter. This is indicated by the fact that their spectrum is continuous, and also by the intensity of the light, which far exceeds any that a gas has ever been made to give forth. It does not follow from this that the photosphere is a continuous solid or crust, since floating particles of solid matter will shine in the same way. The general opinion has been that the photosphere is of a cloud-like nature; that is, of minute particles floating in an atmosphere of heated gases. That it is not continuously solid like our earth seemed to be fully shown by the variations and motions of the spots, which have every appearance of going on in a fluid or gas. Indeed, of late, some of the most eminent physicists regard it as purely gaseous, the pressure making it shine like a solid.

But this theory is attended with a difficulty which has not been sufficiently considered. The photosphere is in striking contrast to the gaseous chromosphere, in being subject to no sensible changes of level. If it were gaseous, as supposed, the solid particles having no connection with each other, we should expect those violent eruptions which throw up the protuberances to carry up portions of it, so that it would now and then present an irregular and jagged outline, as the chromosphere does. But the most refined observations have never shown it to be subject to the slightest change of level, or deviation from perfect rotundity, except in the region of the spots, where its continuity seems to be broken by immense chasm-like openings.

The serene immobility of the photosphere, under such violent actions around it as we have described, lends some color to the supposition that it is a solid crust which forms around the glowing interior of the sun, or, at least, that it is composed of a comparatively dense fluid resting upon such a crust. The latter is the view of Zöllner, who considers some sort of an envelope between the exterior and the interior of the sun absolutely necessary to account for the eruptive protuberances. He places this solid envelope three or four thousand miles below the surface of the photosphere.

Inside the photosphere we have the enormous interior globe, 860,000 miles in diameter. The best-sustained theory of the interior is the startling one that it is neither solid nor liquid, but gaseous; so that our great luminary is nothing more than an immense bubble. The pressure upon the interior portions of this mass is such as to reduce it to nearly the density of a liquid; while the temperature is so high as to keep the substances in a state which is between the liquid and the gaseous, and in which no chemical action is possible. The strong point in support of this gaseous theory of the sun s interior is, that it is the only one which explains how the sun's light and heat are kept up. How it does this will be shown in treating of the laws which govern the secular changes of the universe at large.

§ 7. *Views of Distinguished Students of the Sun on the Subject of its Physical Constitution.*

The progress of our knowledge of the sun during the past ten years has been so rapid that only those can completely follow it who make it the principal business of their lives. For the same reason, the views respecting the sun entertained by those who are engaged in studying it must be modified and extended from time to time. The interest which necessarily attaches to the physical source of all life and motion on our globe renders the author desirous of presenting these views to his readers in their latest form; and, through the kindness of several of the most eminent investigators of solar physics now living, he is enabled to gratify that desire. The following statements are presented in the language of their respective authors, except that, in the case of Messrs. Secchi and Faye, they are translated from the French for the convenience of the English reader. It will be noticed that in some minor points they differ from each other, as well as from those which the author has expressed in the preceding section. Such differences are unavoidable in the investigation of so difficult a subject.

Views of the Rev. Father Secchi.—" For me, as for every one else, the sun is an incandescent body, raised to an enormous temperature, in which the substances known to our chemists and physicists, as well as several other substances still unknown, are in a state of vapor, heated to such a degree that its spectrum is continuous, either on account of the pressure to which the vapor is subjected, or of its high temperature. This incandescent mass is what constitutes the photosphere. Its limit is defined, as in the case of incandescent gases in general, by the temperature to which the exterior layer is reduced by its free radiation in space, together with the force of gravity exerted by the body. The photosphere presents itself as composed of small, brilliant granulations, separated by a dark net-work. These granulations are only the summits of the flames which constitute them, and which rise above the lower absorbing layer, which forms the net-work, as we shall soon more clearly see.

"Above the photospheric layer lies an atmosphere of a very complex nature. At its base are the heavy metallic vapors, at a temperature which, being less elevated, no longer permits the emission of light with a continuous spectrum, although it is sufficient to give direct spectra with brilliant lines, which may be observed, during total eclipses of the sun, at its limb. This layer is extremely thin, having a depth of only one or two seconds of arc. According to the law of absorption laid down by Kirchhoff, these vapors absorb the rays of the spectrum from the light of the photosphere which passes through them, thus giving rise to the breaks known as the Fraunhofer dark lines of the solar spectrum. These vapors are mixed with an enormous quantity of hydrogen. This gas is present in such a quantity that it rises considerably above the other layer, and forms an envelope rising to a height of from ten to sixteen seconds, or even more, which constitutes what we call the chromosphere. This hydrogen is always mixed with another substance, provisionally called *helium*, which forms the yellow line D_3 of the spectrum of the protuberances, and with another still rarer substance, which gives the green line 1474 K. This last substance rises to a much greater elevation than the hydrogen; but it is not so easily seen in the full sun as the latter. Probably there is some other substance not yet well determined. Thus, the substances which compose this solar envelope appear to be arranged in the order of their density; but still without any well-defined separation, the diffusion of the gases producing a constant mixture.

"This atmosphere becomes visible in total eclipses in the form of the corona. It is very difficult to fix its absolute height. The eclipses prove that it may reach to a height equal to the solar diameter in its highest portions.

"No doubt it extends yet farther, and it may well be connected with the zodiacal light. The visible layer of this atmosphere is not spherical; it is higher in middle latitudes, near forty-five degrees, than at the equator. It is still more depressed at the poles. At the base of the chromosphere, the hydrogen has the shape of small flames composed of very thin, close filaments which seem to correspond to the granulations of the photosphere. During periods of tranquillity the direction of these filaments is perpendicular to the solar surface; but during periods of agitation they are generally more or less inclined, and often directed systematically towards the poles.

"The body of the sun is never in a state of absolute repose. The various substances coming together in the interior of the body tend to combine, in consequence of their affinity, and necessarily produce agitations and interior movements of every kind and of great intensity. Hence the numerous crises which show themselves at the surface through the elevation of the lower strata of the atmosphere by eruptions, and often by actual explosions. Then the lower metallic vapors are projected to considerable heights, hydrogen especially, at an elevation visible in the spectroscope (in full sunlight) of one-fourth the solar diameter. These masses of hydrogen, leaving the photosphere at a temperature higher than that of the atmosphere, rise to the superior regions of the latter, remaining suspended, diffusing themselves at considerable elevations, and forming what are called the prominences or protuberances. The structure of the hydrogenous protuberances is entirely similar to that of fluid veins raising themselves from denser layers, and diffusing in the more rare ones: but their extreme variability, even at the base, and the rapid changes of the place of exit and diffusion, prove that they do not pass through any orifice in a solid resisting layer.

"These eruptions are often mixed with columns of metallic vapors of greater density, which do not attain the elevation of the hydrogen, and of which the nature can be recognized by the aid of the spectroscope: occasionally we see them falling back on the sun in the form of parabolic jets. The most common substances are sodium, magnesium, iron, calcium, etc.—indeed, the same substances which are seen to form the low, absorbing layer of the solar atmosphere, and which by their absorption produce the Fraunhofer lines. A rigorous and inevitable consequence of these conditions is the fact that when the mass thus elevated is carried by the rotation of the sun between the photosphere and the eye of the observer, the absorption becomes very sensible, and produces a dark spot on the photosphere itself. The metallic absorption lines are then really wider and more diffused in this region; and if the elevated mass is high and dense enough, we can even see the re-reversal of the lines already reversed; that is to say, we can see the bright lines of the substance itself on the background of the spot. This often happens for hydrogen, which rises to a great height, and also with sodium and magnesium, which metals have the rarest vapors. Here, then, we have the origin of the solar spots. They are formed by masses of absorbing vapors which, brought out from the interior of the sun, and interposed between the photosphere and the eye of the observer, prevent a large part of the light from reaching our eyes.

"But these vapors are heavier than the surrounding mass into which they have been thrown. They therefore fall by their own weight, and, tending to sink into the photosphere, produce in it a sort of cavity or basin filled with a darker and more absorbing mass. Hence the aspect of a cavity recognized in the spots. If the eruption is instantaneous, or of very short duration, this vaporous mass, fallen back on the photosphere, soon becomes incandescent, reheated, and dissolved, and the spot rapidly disappears; but the interior crises of the body of the sun may be continued a long time; and the eruption may maintain itself in the same place during two or more rotations of the sun. Hence the persistence of the spots; for the cloud can continue to form so long and so fast as the photosphere dissolves it, as happens with the jets of vapor from our volcanoes. The eruptions, when about to terminate, may be revived and reproduced several times near the same place, and give rise to spots very variable in form and position.

VIEWS ON THE PHYSICAL CONSTITUTION OF THE SUN. 269

"The spots are formed of a central region, called the nucleus, or umbra, and of a surrounding part less dark, called the penumbra. The latter is really formed of thin dark veils, and of filaments or currents of photospheric matter which tend to encroach upon the dark mass. These currents have the form of tongues, often composed of globular masses looking like strings of beads or willow leaves, and evidently are only the grains of the photosphere precipitating themselves towards the centre of the spot, and sometimes crossing it like a bridge.

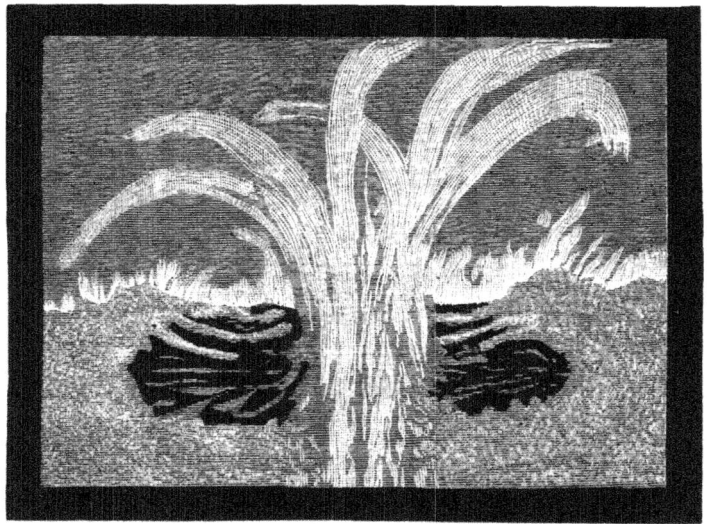

FIG. 71.—Illustrating Secchi's theory of solar spots.

"In each spot we must distinguish three periods of existence: the first, of formation; the second, of rest; the third, of extinction. In the first, the photospheric mass is raised and distorted by a great agitation, often in the nature of a vortex, which elevates it all around the flowing streams, and forms irregular elevations, either without penumbra or with a very irregular one. These irregular movements defy description: their velocities are enormous, and the agitated region extends itself over several square degrees; but this upturning soon comes to an end, and the agitation slowly subsides, and is succeeded by calm. In the second period, the agitated and elevated mass falls back again, and tends to combine in masses more or less circular, and to sink by its weight into the surface of the photosphere. Hence the depressed form of the photosphere, resembling a funnel, and the numerous currents which come from each point of the circumference to rush upon this obscure mass; but at the same time the contrast between it and the substance issuing still persists. The spot takes a nearly stable and circular form, a contrast which may last a long time—so long, in fact, as the interior actions of the solar globe furnish new materials. At length, the latter ceasing, the eruptive action languishes and is exhausted, and the absorbing mass invaded on all sides by the photosphere is dissolved and absorbed, and the spot disappears.

"The existence of these three phases is established by the comparative study of the spots and eruptions. When a spot is on the sun's border during its first period, although the dark region is invisible, its position is indicated by eruptions of metallic vapors, if the spot be considerable. On the darkest ones the vapors of sodium, iron, and magnesium are seen in the greatest quantity, and raised to great heights. A calm and circular spot is crowned by beautiful faculæ and jets of hydrogen and metallic vapors, very low, though quite brilliant. A spot which is on the point of closing up has no metallic jets, and at the utmost only a few small jets of hydrogen, and a more agitated and elevated chromosphere. Besides, observation teaches that the eruptions in general accompany the spots, and that they are deficient at times when the spots are wanting. Thus the solar activity is measured by the double activity of eruptions and spots which have a common source, and the spots are really only a secondary phenomenon, depending upon the eruptions and the more or less absorbing quality of the materials: if the erupted materials were not absorbent, we could see no spots at all.

"The eruptions composed simply of hydrogen do not produce spots; thus they are seen on all points of the disk, while the spots are limited to the tropical zones, where alone the metallic eruptions appear. The eruptions of simple hydrogen give rise to the faculæ. The greater brilliancy of the faculæ is due to two causes: the first is, the elevation of the photosphere above the absorbing stratum of vapor which is very thin (only one or two seconds of arc, as we have before said); this elevated region thus escapes the absorption of the lower stratum, and appears more brilliant. The other cause may be that the hydrogen, in coming out, displaces the absorbing stratum, and, taking the place of the metallic vapors, permits a better view of the light of the photosphere itself.

"Thus, in conclusion, the spots are a secondary phenomenon, but, nevertheless, inform us of the violent crises which prevail in the interior of the radiant globe. The frequency of the spots corresponding to the frequency of eruptions, the two phenomena, taken in connection, are the mark of solar activity. The spots occupy the zones on each side of the solar equator, and rarely pass beyond the parallel of thirty degrees. One or two seen at forty-five degrees are exceptions. That parallel is therefore the limit of greatest activity of the body. It is remarkable that the parallels of thirty degrees divide the hemispheres into two sectors of equal volume. Beyond these parallels we see faculæ, but not true spots—or, at most, only veiled spots indicative of a very feeble metallic eruption.

"Such a fluid mass, in which the parts are exposed to very different temperatures, could not subsist without an interior circulation. We do not yet know its laws; but the following facts are well enough established: the zones of spots are not fixed, but have a progressive motion from the equator towards the poles. The spots, arrived at a certain high latitude, cease to appear, but after some time reappear at lower latitudes, and afterwards go on anew. Between these phases of displacement there is commonly a minimum of spots. During periods of activity the protuberances have a dominant direction towards the pole, as also the flames of the chromosphere. This indicates a general movement of the photosphere from the equator to the poles. This movement is supported by the displacement of the zones of eruption and of the protuberances, which always seem to move towards the poles.

"Besides this movement in latitude, the photosphere has also a movement in longitude, which is greatest at the equator. Thus the time of rotation of the body is different upon different parallels, the minimum being at the equator. These phenomena lead to the conclusion that the entire mass is affected with a vortical motion which sets from the equator towards the poles, in a direction oblique to the meridians. The theory of these movements is still to be elaborated, and is, no doubt, connected with the primitive mode in which the sun was formed.

"The activity of the body is subject to considerable fluctuations: the best established period is one of eleven and one-third years, but the activity increases more rapidly than it diminishes—it increases about four years, and diminishes about seven. This activity is connected with the phenomena of terrestrial magnetism, but we cannot say in what way. We may suppose a direct electro-magnetic influence of the sun upon our globe, or an indirect influence due to the thermal action of the sun, which reacts upon its magnetism. It is, indeed, very natural to suppose that the ethereal mass which fills the spaces of our planetary system may be greatly altered and modified by the activity of the central body. But, whatever may be the cause of these changes of activity, we are completely ignorant of them. The action of the planets has been proposed as plausible, but it is far from being satisfactory. The true explanation is reserved for the science which shall reveal the nature of the connection which unites heat to electricity, to magnetism, and to the cause of gravity.

"Of the interior of the sun we have no certain information. The superficial temperature is so great, notwithstanding the continual loss of heat which it suffers, that we cannot suppose it less in the interior; and, consequently, no solid layer can exist there, except perhaps at depths where the pressure due to gravity equals or surpasses the molecular dilatation produced by temperature. However it may be, the layer accessible to the exploration of our instruments is, no doubt, fluid and gaseous, and we can thus explain the variations of the solar diameter established by certain astronomers. Notwithstanding these small fluctuations, the radiation of the body into its planetary system is nearly constant during widely separated periods, and especially is it so during the historic period. This constancy is due to several causes: first, to the enormous mass of the body, which can be cooled only very slowly, owing to its very high temperature; second, to the contraction of the mass, which accompanies the condensation consequent upon the loss of heat; third, to the emission of the heat of dissociation due to the production of chemical actions which may take place in the total mass.

"The origin of this heat is to be found in the force of gravity; for it is well proved that the solar mass, by contracting from the limits of the planetary system to its present volume, would produce, not only its actual temperature, but one several times greater. As to the absolute value of this temperature, we cannot fix it with certainty. Science not yet having determined the relation which exists between molecular living force (*vis viva*) and the intensity of radiation to a distance (which last is the only datum given by observation), we find ourselves in a state of painful uncertainty. Nevertheless, this temperature must be several million degrees of our thermometer, and capable of maintaining all known substances in a state of vapor.

"Rome, February 11th, 1877."

Views of M. Faye.—" In studying without any prepossession the movements of the spots, we find, with Mr. Carrington, that there exists a simple relation between their latitude and their angular velocity. Nevertheless, this law does not suffice to represent the observations with the exactitude which they admit of. It is still necessary to take account by calculation of a parallax of depth which I estimate at $\frac{1}{200}$ of the radius of the sun, and of certain oscillations of very small extent, and of long period, which the spots undergo perpendicular to their parallels. Then the observations are represented with great precision, from which I conclude that we have to deal with a quite simple mechanical phenomenon. The law in question can be expressed by the formula,

$$\omega = a - b \sin^2 \lambda;$$

ω being the angular velocity of a spot at the latitude λ, and a and b being constants, having the same value ($a=857'.6$ and $b=157'.3$) over the whole surface of the sun. These constants may vary slowly with the time, but I have not studied their variations.

"Admitting, as we shall see farther on, that the velocity of a spot is the same as the mean velocity of that zone of the photosphere in which it is formed, we see:

"1. That the contiguous strips of the photosphere are animated with a velocity of rotation nearly constant for each filament, at least during a period of several months or years, but varying with the latitude from one strip to another.

"2. That these strips move nearly parallel to the equator, and never give indications of currents constantly directed towards either pole, as in the upper regions of our atmosphere.

"3. That the spots are hollow, or at least that the black nucleus is perceptibly depressed in respect to the photosphere.

"The diminution in the rate of superficial rotation, more and more marked towards the poles, and the absence of all motion from the equator, can only proceed from the vertical ascent of materials rising incessantly from a great depth towards all points of the surface. It is sufficient that this depth goes on increasing from the equator towards the poles, following a law analogous to that of the rotation, in order that it may produce at the surface a retardation increasing with the latitude. This retardation is about two days in each rotation at forty-five degrees of latitude. The mass of the sun, being formed principally of metallic vapors condensable at a certain temperature, and that temperature being reached at a certain level in consequence of the exterior cooling, there ought to be established a double vertical movement of ascending vapors, which go to form a cloud of condensed matter susceptible of intense radiation, and of condensed products which fall back in the form of rain into the interior. The latter are stopped at the depth at which they meet a temperature high enough to vaporize them anew, and afterwards force them to reascend. As almost the entire mass of the sun partakes of this double movement, the heat radiated by the cloud will be borrowed from this mass, and not from a superficial layer, the temperature of which would rapidly fall, and which would soon condense into a complete crust. Hence the formation and support of the photosphere, and the constancy and long duration of its radiation, which is also partly fed by the slow contraction of the whole mass of the sun.

"The contiguous bands of the photosphere being animated with different velocities, there results a multitude of circular gyratory movements around a vertical axis extending to a great depth, as in our rivers and in the great upper currents of our atmosphere. These whirlpools, which tend to equalize the differences of velocity just spoken of, follow the currents of the photosphere in the same way that whirlpools, and the whirlwinds, tornadoes, and cyclones of our atmosphere follow the upper currents in which they originate. Like these, they are descending, as I have proved (against the meteorologists) by a special study of these terrestrial phenomena. They carry down into the depths of the solar mass the cooler materials of the upper layers, formed principally of hydrogen, and thus produce in their centre a decided extinction of light and heat as long as the gyratory movement continues. Finally, the hydrogen set free at the base of the whirlpool becomes reheated at this great depth, and rises up tumultuously around the whirlpool, forming irregular jets which appear above the chromosphere. These jets constitute the protuberances.

"The whirlpools of the sun, like those on the earth, are of all dimensions, from the scarcely visible pores to the enormous spots which we see from time to time. They have, like those of the earth, a marked tendency first to increase, and then to break up, and thus form a row of spots extending along the same parallel. The penumbra is due to a portion of the photosphere which forms around their conical surface at a lower level, on account of the lowering of the temperature produced by the whirlpool. Sometimes in this sort of luminous sheath we see traces of the whirling movement going on in the interior.

"It is more difficult to account for the periodicity of the spots. It seems to me that it must depend upon fluctuations in the form of the interior layer, to which the condensed matter of the photosphere falls in the form of rain. This flow of materials from above must alter, little by little, the velocity of rotation of this layer. If its compression is changed in the course of time, and if it becomes rounder, the variations in the superficial velocity of the photosphere, as well as the gyratory movements, will diminish in intensity and frequency.

"A time will at length arrive when the vertical movements which feed the photosphere will become more and more hindered. The cooling will then be purely superficial, and the surface of the sun will harden into a continuous crust.

"Paris, February, 1877."

Views of Professor Young.—" 1. It seems to me almost demonstrated, as a consequence of the low mean density of the sun and its great force of gravity, that the central portions of that body, and, in fact, all but a comparatively thin shell near the surface, must be in a gaseous condition, and the gases at so high a temperature as to remain for the most part dissociated from each other, and incapable of chemical interaction. Under the influence of the great pressure and high temperature, however, their density and viscosity are probably such as to render their mechanical behavior more like that of such substances as tar or honey than that of air, as we are familiar with it.

"2. The visible surface of the sun, the photosphere, is composed of clouds formed by the condensation and combination of such of the solar gases as are cooled sufficiently by their radiation into space. These clouds are suspended in the mass of uncondensed gases like the clouds in our own atmosphere, and probably have, for the most part, the form of approximately vertical columns, of irregular cross-section, and a length many times exceeding their diameter. The liquid and solid particles of which they are made up descend continually, their places being constantly supplied by fresh condensation from the ascending currents which rise between the cloud-columns. From the under-surface of the photosphere there must be an immense precipitation of what may be called solar 'rain and snow,' which descends into the gaseous core, and by the internal heat is re-evaporated, decomposed, and restored to its original gaseous condition; the heat lost by the surface radiation being replaced mainly by the mechanical work due to the gradual diminution of the sun's bulk, and the thickening of the photosphere. I do not know any means of determining the thickness of the photospheric shell, but, from the phenomena of the spots, judge that it can hardly be less than ten thousand miles, and that it may be much more.

"3. The weight of the cloud-shell, and the resistance offered to the descending products of condensation, act to produce on the enclosed gaseous core a constricting pressure, which forces the gases upwards through the intervals between the clouds with great velocity; so that jets or blasts of heated gas continually ascend all over the sun's surface, the same material subsequently redescending in the cloud-columns, partly condensed into solid or liquid particles, and partly uncondensed, but greatly cooled. It seems also not unlikely that in the upper part of the channels through which the ascending currents rush, there may often occur the mixture of different gases cooled by expansion to temperatures sufficiently below the dissociation point to allow of their explosive combination.

"4. The 'chromosphere' is simply the layer of uncondensed gases which overlies the photosphere, though separated from it by no definite surface. The lower portion of the chromosphere is rich in all the vapors and gases which enter into the sun's composition; but at a comparatively small height the denser and less permanent gases disappear, leaving in the upper regions only hydrogen and some other substances not as yet identified. The dark lines of the solar spectrum originate mainly in the absorption produced by the denser gases which bathe the photospheric clouds, and these metallic vapors are only occasionally carried into the upper regions by ascending jets of unusual violence. When this occurs, it is almost invariably in connection with a solar spot. The prominences are merely heated masses of the hydrogen and other chromospheric gases, carried to a considerable height by the ascending currents, and apparently floating in the 'coronal atmosphere,' which interpenetrates and overtops the chromosphere.

"5. I do not know what to make of the corona. Its spectrum proves that a considerable portion of its light comes from some exceedingly rare form of gaseous matter, which cannot be identified with anything known to terrestrial chemistry; and this gas, whatever it may be, exists at a height of not less than a million of miles above the solar surface, constituting the 'coronal atmosphere.' Another portion of its light appears to be simply reflected sunshine. But by what forces the peculiar radiated structure of the corona is determined, I have no definite idea. The analogies of comets' tails and auroral streamers both appear suggestive; but, on the other hand, the spectra of the corona, the aurora borealis, the comets, and the nebulæ are all different—no two in the least alike.

"6. As to sun-spots, there can be no longer any doubt, I think, that they are cavities in the upper surface of the photosphere, and that their darkness is due simply to the absorbing action of the gases and vapors which fill them. It is also certain that very commonly, if not invariably, there is a violent uprush of hydrogen and metallic vapors all around the outer edge of the penumbra, and a considerable depression of the chromosphere over the centre of the spot; probably, also, there is a descending current through its centre. As to the cause of the spots, and the interpretation of their telescopic details, I am unsatisfied. The theory of Faye appears to me, on the whole, the most reasonable of all that have yet been proposed; but I cannot reconcile it with the want of systematic rotation in the spots, or their peculiar forms. Still, it undoubtedly has important elements of truth, and may perhaps be modified so as to meet these difficulties. As to the periodicity of the spots,

I am unable to think it due in any way to planetary action; at least, the evidence appears to me wholly insufficient as yet; but I have no hypothesis to offer. Nor have I any theory to propose to account for the certain connection between disturbances of the solar surface and of terrestrial magnetism.

"7. As to the temperature of the sun's surface, I have no settled opinion, except that I think it must be much higher than that of the carbon points in the electric light. The estimates of those who base their calculations on Newton's law of cooling, which is confessedly a mere approximation, seem to me manifestly wrong and exaggerated; on the other hand, the very low estimates of the French physicists, who base their calculations on the equation of Dulong and Petit, seem to me hardly more trustworthy, since their whole result depends upon the accuracy of a numerical exponent determined by experiment at low temperatures and under circumstances differing widely from those of the sun's surface. The process is an unsafe extrapolation. The sensible constancy of the solar radiation seems to be fairly accounted for on the hypothesis of slow contraction of the sun's diameter.

"8. I look upon the accelerated motion of the sun's equator as the most important of the unexplained facts in solar physics, and am persuaded that its satisfactory elucidation will carry with it the solution of most of the other problems still pending.

"Such, in brief, are my 'opinions;' but many of them I hold with little confidence and tenacity, and anxiously await more light, especially as regards the theory of the sun's rotation, the cause and constitution of the spots, and the nature of the corona. The only peculiarity in my views lies, I think, in the importance I assign to the effects of the descending products of condensation, which I conceive to form virtually a sort of constricting skin, producing pressure upon the gaseous mass beneath, something as the film of a bubble compresses the enclosed air. To the pressure thus produced I ascribe mainly the eruptive phenomena of the chromosphere and prominences.

"Dartmouth College, March, 1877."

Views of Professor Langley.—"It seems to me that we have now evidence on which to pass final adverse judgment on views which regard the photosphere as an incandescent liquid, or the spots as analogous either to scoriac matter, on the one hand, or to clouds above the luminous surface, on the other. According to direct telescopic evidence, the photosphere is purely vaporous, and I consider these upper vapors to be lighter than the thinnest cirri of our own sky. The observation of faculæ allies them and the whole 'granular' cloud structure of the surface most intimately with chromospheric forms, seen by the spectroscope, and associates both with the idea of an everywhere-acting system of currents which transmit the internal heat, generated by condensation, to the surface, and take back the cold, absorbent matter. This vertical circulation goes to a depth, I think, sensible even by comparison with the solar diameter. It coexists with approximately horizontal movements observed in what may be called the successive upper photospheric strata in the vicinity of spots. The spots give evidence of cyclonic action such as could only occur in a fluid. Their darkness is due to the presence, in unusual depth, of the same obscuring atmosphere which forms the gray medium in which the luminous photospheric forms seem suspended, and which we here look through, where it fills openings in the photospheric stratum, down to regions of the solar interior made visible by the dim light of clouds of luminous vapor, precipitated in lower strata where the dew-point has been altered by changed conditions of temperature and pressure. All observation and all legitimate inference go to show that the sun is gaseous throughout its mass, though by this it is not meant to deny the probable precipitation of cooling photospheric vapors in something analogous to rain; a condition perhaps necessary to the maintenance of the equilibrium of the interchange of cold and heated matter between exterior and interior; nor is it meant that the conditions of a perfect fluid are to be expected, where these are essentially modified (if by no other cause) by the viscosity due to extreme heat. The temperature of the sun is, in my view, necessarily

VIEWS ON THE PHYSICAL CONSTITUTION OF THE SUN. 281

much greater than that assigned by the numerous physicists, who maintain it to be comparable with that obtainable in the laboratory furnace; but we cannot confidently assign any upper limit to it until physics has advanced beyond its present merely empirical rules connecting emission and temperature; for this, and not the lack of accurate data from physical astronomy, is the source of nearly all the obscurity now at-

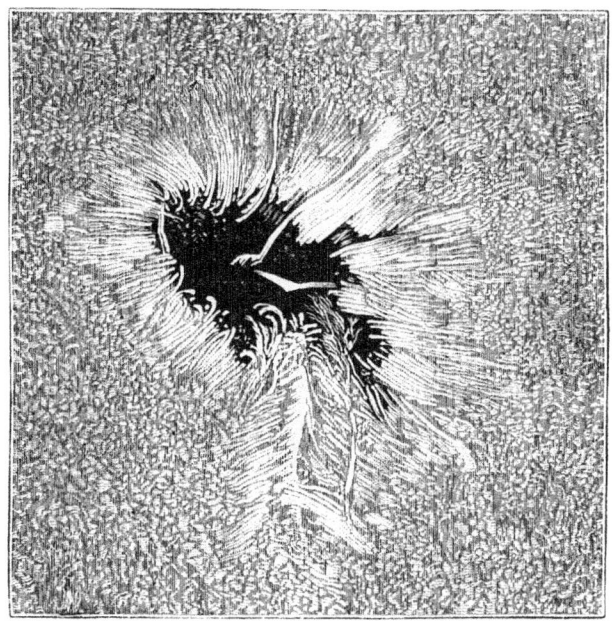

FIG. 72.—Solar spot, after Langley.

tending this important question. No theory of the solar constitution which is free from some objection has yet been proposed; but if the master-key to the diverse problems it presents has not been found, it is still true, I think, that the one which unlocks most is that of M. Faye.

"Of the potential energy of the sun, we may say that we believe it to be sufficient for a supply of the present heat during periods to be counted by millions of years. But what immediately concerns us is the constancy of the rate of conversion of this potential into actual radiant energy, as we receive it, for on this depends the uniformity of the conditions under which we exist. Now, this uniformity in turn depends on the equality of the above-mentioned interchanges between the solar surface and the interior, an equality of whose constancy we know nothing save by limited experience. The most important statement with reference to the sun, perhaps, which we can make with certainty is even a negative one. It is that we have no other than empirical grounds, in the present state of knowledge, for believing in the uniformity of the solar radiation in prehistoric periods and in the future.

"The above remarks, limited as they are, appear to me to cover nearly all the points as to the sun's physical constitution (outside of the positive testimony of the spectroscope) on which we are entitled to speak with confidence, even at the present time."

CHAPTER III.

THE INNER GROUP OF PLANETS.

§ 1. *The Planet Mercury.*

MERCURY is the nearest known planet to the sun, and the smallest of the eight large planets. Its mean distance from the sun is 40 millions of miles, and its diameter about one-third that of the earth. It was well known to the ancients, being visible to the naked eye at favorable times, if the observer is not in too high a latitude. The central and northern regions of Europe are so unfavorably situated for seeing it that it is said Copernicus died without ever having been able to obtain a view of it. The difficulty of seeing it arises from its proximity to the sun, as it seldom sets more than an hour and a half after the sun, or rises more than that length of time before it. Hence, when the evening is sufficiently advanced to allow it to be seen, it is commonly so near the horizon as to be lost in the vapors which are seen in that direction. Still, by watching for favorable moments, it can be seen several times in the course of the year in any part of the United States. The following are favorable times for seeing it after sunset:

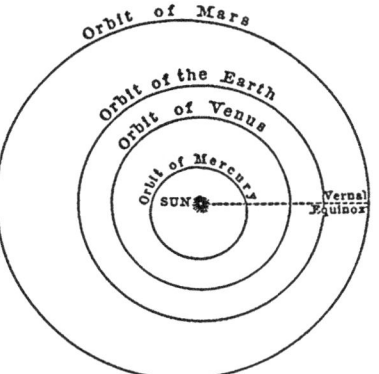

FIG. 73.—Orbits of the four inner planets, illustrating the eccentricity of those of Mercury and Mars.

1877..............................May 3d, August 26th, December 25th.
1878..............................April 14th, August 9th, December 9th.
1879..............................March 28th, July 23d, November 21st.

The corresponding times in subsequent years may be found by subtracting 18 days from the dates for each year; that is, they will occur 18 days earlier in 1879 than in 1878; 18 days earlier in 1880 than in 1879, and so on. It is not necessary to look on the exact days we have given, as the planet is generally visible for fifteen or twenty days at a time. Each date given is about the middle of the period of visibility, which extends a week or ten days on each side. The best time for looking is in the evening twilight, about three-quarters of an hour after sunset, the spring is in this respect much more favorable than autumn.

Aspect of Mercury.—Mercury shines with a brilliant white light, brighter than that of any fixed star, except, perhaps, Sirius. It does not seem so bright as Sirius, because it can never be seen at night except very near the horizon. Owing to the great eccentricity of its orbit and the great variations of its distance from the earth, its brilliancy varies considerably; but the favorable times we have indicated are near those of greatest brightness.

Viewed with a telescope under favorable conditions, Mercury is seen to have phases like the moon. When beyond the sun, it seems round and small, being only about 5″ in diameter. When seen to one side of the sun, near its greatest apparent angular distance, it appears like a half-moon. When nearly between the sun and earth, its diameter is between 10″ and 12″, but only a thin crescent is visible. The manner in which these various phases are connected with the position of the planet relative to the earth and sun is the same as in the case of Venus, and will be shown in the next section.

Rotation, Figure, Atmosphere, etc.—About the beginning of the present century Schröter, the celebrated astronomer of Lilienthal, who made the telescopic study of the planets a speciality, thought that at times, when Mercury presented the aspect of a crescent, the south horn of this crescent seemed blunted at certain intervals. He attributed this appearance to the shadow of a lofty mountain, and by observing the times of its return was led to the conclusion that the planet revolved on its axis in 24 hours 5 minutes. He also estimated the height of the mountain at twelve miles. But the more powerful instruments of modern times have not confirmed these conclusions, and they are now considered as quite doubtful, if not entirely void of foundation. That is, we must regard the time of rotation of Mercury on its axis, and, of course, the position of that axis, as not known with certainty, but as perhaps very nearly 24 hours.

The supposed atmosphere of Mercury, the deviation of its body from a spherical form, and many other phenomena which observers have described, must be received with the same scepticism. No deviation from a spherical form can be considered as proved, the discordance of the measures showing that the supposed deviations are really due to errors of observation. So, also, the appearances which many observers have attributed to an atmosphere are all to be regarded as optical illusions, or as due to the imperfections of the telescope made use of. From measures of its light at various phases Zöllner has been led to the conclusion that Mercury, like our moon, is devoid of any atmosphere sufficiently dense to reflect the light of the sun. If this doubt and uncertainty seems surprising, it must be remembered that the nearness of this planet to the sun renders it a very difficult object to observe with accuracy. We must look at it either in the daytime, when the air is disturbed by the sun's rays, or in the early evening, when the planet is very near the horizon, and therefore in an unfavorable situation.

Transits of Mercury.—Transits of this planet across the face of the sun are much more frequent than those of Venus, the average interval between successive transits being less than ten years, and the longest interval thirteen years. These transits are always looked upon with great interest by astronomers, on account of the questions to which they have given rise. From the earliest ages in which it was known that Mercury moved around the sun, it was evident that it must sometimes pass between the earth and the sun; but its diameter is too small to admit of its being seen in this position with the naked eye.

The first actual observation of Mercury projected on the face of the sun was made by Gassendi, on November 7th, 1631. His mode of observation was that already described for viewing the solar spots, the image of the sun being thrown on a screen by means of a small telescope. He came near missing his observation, owing to his having expected that the planet would look much larger than it did. The imperfect telescopes of that time surrounded every brilliant object with a band of diffused light which greatly increased its apparent magnitude, so that Gassendi had no idea how small the planet really was.

Gassendi's observation was hardly accurate enough to be of any scientific value at the present time. It was not till 1677 that a really good observation was made. Halley, of England, in that year was on the island of St. Helena, and, being provided with superior instruments, was fortunate enough to make a complete observation of a transit of Mercury over the sun which occurred on November 7th. We have already mentioned the great accuracy which he attributed to his observation, and the phenomenon of the black drop which he was the first to see.

The following are the dates at which it has been calculated that transits of Mercury will occur during the remainder of the present century. The first transit will be visible over the whole United States, and the second on the Pacific coast.

1878..................May 6th.	1894..................November 10th.
1881..................November 7th.	1901..................November 4th.
1891..................May 9th.	

§ 2. *The Supposed Intra-Mercurial Planets.*

At the present time the greatest interest which attaches to transits of Mercury arises from the conclusion which Leverrier has drawn from a profound comparison of transits observed before 1848 with the motion of Mercury as determined from the theory of gravitation. This comparison indicates, according to Leverrier, that the perihelion of Mercury moves more rapidly by 40″ a century than it ought to from the gravitation of all the known planets of the system. He accounted for this motion by supposing a group of small planets between Mercury and the sun, and the question whether such planets exist, therefore, becomes important.

Apparent support to Leverrier's theory is given by the fact that various observers have within the past century recorded the passage over the disk of the sun of dark bodies which had the appearance of planets, and which went over too rapidly or disappeared too suddenly to be spots. But when we examine these observations, we find that they are not entitled to the slightest confidence. There is a large class of recorded astronomical phenomena which are seen only by unskilful observers, with imperfect instruments, or under unfavorable circumstances. The fact that they are not seen by practised observers with good instruments is sufficient proof that there is something wrong about them. Now, the observations of intra-Mercurial planets belong to this class. Wolf has collected nineteen observations of unusual appearances on the sun, extending from 1761 to 1865, but, with two or three exceptions, the observers are almost unknown as astronomers. In at least one of these cases the observer did not profess to have seen anything like a planet, but only a cloud-like appearance. On the other hand, for fifty years past the sun has been constantly and assiduously observed by such men as Schwabe, Carrington, Secchi, and Spoerer, none of whom have ever recorded anything of the sort. That planets in such numbers should pass over the solar disk, and be seen by amateur observers, and yet escape all these skilled astronomers, is beyond all moral probability.

In estimating this probability we must remember that a real planet appearing on the sun would be far more likely to be recognized by a practised than by an unpractised observer, much as a new species of plant or animal is more likely to be recognized by a naturalist than by one who is not such. One not accustomed to the close study of the solar spots might have some difficulty in distinguishing an unusually round spot from a planet. He is also liable to be deceived in various ways.* For instance, the sun, by his apparent diurnal motion, presents different parts of the edge of his disk to the horizon in the course of a day; he seems, in fact, in the northern hemisphere to turn round in the same direction with the hands of a watch. Hence, if a spot is seen near the edge of his disk it will seem to be in motion, though really at rest. On the other hand, should an experienced observer see a planet projected on the sun's face, he could hardly fail to recognize it in a moment; and should any possible doubt exist, it would be removed by a very brief scrutiny.

The strongest argument against these appearances being planets is, that the transit of a planet in such a position could not be a rare phenomenon, but would necessarily repeat itself at certain intervals, depending on its distance from the sun and the inclination of its orbit. For instance, supposing an inclination of 10°, which is greater than that of any of the principal planets, and a distance from the sun one-half that of Mercury, the planet would pass over the face of the sun, on the average, about once a year, and its successive transits would occur either very near the same day of the year, or on a certain day of the opposite season. The supposed transits to which we have referred occur at all seasons, and if we suppose them real, we must suppose, as a logical consequence, that the transits of these several planets are repeated many times a year, and yet constantly elude the scrutiny of all good observers, though occasionally seen by unskilled ones. This is a sufficient *reductio ad absurdum* of the theory of their reality.

It is very certain, then, that if the motion of the perihelion of Mercury is due to a group of planets, they are each so small as to be invisible in transits across the sun. They must

---

* Some readers may recall Butler's sarcastic poem of the "Elephant in the Moon," as illustrative of the possibility of an observer being deceived by some peculiarity of his telescope. In one instance, about thirty years since, a telescopic observation of something which we now know must have been flights of distant birds over the disk of the sun was recorded, and published in one of the leading astronomical journals, as a wonderful transit of meteors. The publication was probably not seriously intended, the description being a close parallel to that of the satirical poet. See *Astronomische Nachrichten*, No. 549.

also be so small as to be invisible during total eclipses of the sun, because they have always failed to show themselves then. But to produce the observed effect on Mercury, their total mass must be three or four times that of Mercury. Being so small individually, and so large in the aggregate, their number must be counted by thousands; and if seen at all, they will be seen only as a cloud-like mass. Now, in the zodiacal light we have such a mass, and the question arises whether the matter which reflects this light can be that which affects the motions of Mercury. Although the affirmative of this question involves nothing intrinsically improbable, it cannot be accepted without further investigation. The delicate point involved is, that unless we suppose the hypothetical group of planetoids to move nearly in the plane of the orbit of Mercury, they must change the node of that planet as well as its perihelion. Now, the observations discussed by Leverrier do not show any motion of the node above that due to the action of the known planets. We thus reach the enforced conclusion that if the motion of the perihelion is due to the cause assigned by Leverrier, the planetoids which cause it must, in the mean, move in nearly the same plane with Mercury. But it has not yet been shown that the axis of the zodiacal light deviates from the ecliptic by so great an angle as the orbit of Mercury, namely 7°. A great deal of research—more, in fact, than is likely to be applied to the subject during the present generation—will be required before the question can be settled.

§ 3. *The Planet Venus.*

The planet Venus moves around the sun about half-way between the orbits of Mercury and the earth, its mean distance from the sun being 67 millions of miles. Its orbit is more nearly circular than that of any of the other principal planets. It is very nearly the size of the earth, its diameter being little, if any, more than four per cent. less than that of our globe. Next to the sun and moon, it is the most brilliant object in the heavens, sometimes casting a very distinct shadow. It never recedes more than about 45° from the sun, and is, therefore, seen by night only in the western sky in the evening, or the eastern sky in the morning, according as it is east or west of the sun. There is, therefore, seldom any difficulty in recognizing it. When at its greatest brilliancy, it can be clearly seen by the naked eye in the daytime, provided that one knows exactly where to look for it. It was known to the ancients by the names of *Hesperus* and *Phosphorus*, or the evening and the morning star, the former name being given when the planet, being east of the sun, was seen in the evening after sunset, and the latter when, being to the west of the sun, it was seen in the east before sunrise. It is said that before the birth of exact astronomy *Hesperus* and *Phosphorus* were supposed to be two different bodies, and that it was not until their motions were studied, and the one was seen to emerge from the sun's rays soon after the other was lost in them, that their identity was established.

Aspect of Venus.—To the unaided eye Venus presents the appearance of a mere star, distinguishable from other stars only by its intense brilliancy. But when Galileo examined this planet with his telescope, he found it to exhibit phases like those of the moon. Desiring to take time to assure himself of the reality of his discovery, without danger of losing his claim to priority through some one else in the mean time making it independently, he published the following anagram, in which it was concealed:

"Hæc immatura a me jam frustra leguntur o. y."
(These unripe things are now vainly gathered by me).

By transposing the letters of this sentence he afterwards showed that they could be made into the sentence,

"Cynthiæ figuras æmulatur mater amorum"
(The mother of the loves imitates the phases of Cynthia).

That the disk of Venus was not round was first noticed by Galileo in September, 1610. A computation of its position at that time shows that it must have been a little gibbous, more than half of its face being illuminated; but after a few months it changed into a crescent. Therefore Galileo could not have found it necessary to wait long before explaining his anagram.

The variations of the aspect and apparent magnitude of Venus are very great. When beyond the sun, it is at a distance of 160 millions of miles, and presents the appearance of a small round disk 10" in diameter. When nearest the earth, it is only 25 millions of miles distant; and if its whole face were visible, it would be more than 60" in diameter.

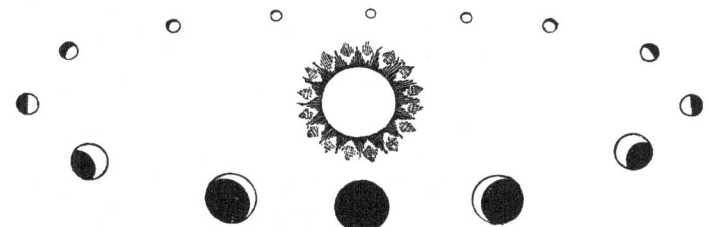

Fig. 74.—Phases of Venus, showing apparent figure and magnitude of the bright and dark portions of the planet in various points of its orbit.

But, being then on the same side of the sun with us, its dark hemisphere is turned towards us, except, perhaps, an extremely thin crescent of the illuminated hemisphere. Between these two positions it goes through all the intermediate phases, the universal rule of which is that the nearer it is to the earth, the smaller the proportion of its apparent disk which is illuminated; but the larger that disk would appear could the whole of it be seen. Its greatest brilliancy occurs between the time of its greatest elongation from the sun and its inferior conjunction.

Supposed Rotation of Venus.—The earlier telescopists naturally scrutinized the planets very carefully, with a view of finding whether there were any inequalities or markings on their surfaces from which the time of rotation on their axes could be determined. In April, 1667, Cassini saw, or thought he saw, a bright spot on Venus, by tracing which for several successive evenings he found that the planet revolved in between 23 and 24 hours. Sixty years later Blanchini, an Italian astronomer, whose telescope is shown on page 112, supposed that he found seven spots on the planet, which he considered to be seas. By watching them from night to night, he concluded that it required more than 24 days for Venus to revolve on its axis. This extraordinary result was criticised by the second Cassini, who showed that Blanchini, only seeing the planet a short time each evening, and finding the spots night after night in nearly the same position, concluded that it had moved very little from night to night; whereas, in fact, it had made a complete revolution, and a little more. At the end of 24 days it would be seen in its original position, but would have made 25 revolutions in the mean time, instead of one only, as Blanchini supposed. This would make the time of rotation 23 hours $2\frac{1}{2}$ minutes, while Cassini found 23 hours 15 minutes from his father's observations.

Between 1788 and 1793 Schröter applied to Venus a mode of observation similar to that he used to find the rotation of Mercury. Watching the sharp horns when the planet appeared as a crescent, he thought that one of them was blunted at certain intervals. Attributing this appearance to a high mountain, as in the case of Mercury, he found a time of rotation of 23 hours 21 minutes.

On the other hand, Herschel was never able to see any permanent markings on Venus. He thought he saw occasional spots, but they varied so much and disappeared so rapidly that he could not gather any evidence of the rotation of the planet. He therefore supposed that Venus was surrounded by an atmosphere, and that whatever markings might be occasionally seen were due to clouds or other varying atmospheric phenomena.

In 1842, De Vico, of Rome, came to the rescue of the older astronomers by publishing a series of observations tending to show that he had rediscovered the markings found by Blanchini more than a century before. He deduced for the time of rotation of the planet 23 hours 21 minutes 22 seconds.

The best-informed astronomers of the present day look with suspicion on nearly all these observations, being disposed to sustain the view of Herschel, though on grounds entirely different from those on which he founded it. It is certain that there are plenty of observers of the present day, with instruments much better than those of their predecessors, who have never been able to see any permanent spots. The close agreement between the times of rotation found by the older observers is indeed striking, and might seem to render it certain that they must have seen spots which lasted several days. It must also be admitted in favor of these observers that a fine steady atmosphere is as necessary for such observations as a fine telescope, and it is possible that in this respect the Italian astronomers may be better situated than those farther north. But the circumstance that the deduced times of rotation in the cases both of Mercury and Venus differ so little from that of the earth is somewhat suspicious, because if the appearance were due to any optical illusion, or imperfection of the telescope, it might repeat itself several days in succession, and thus give rise to the belief that the time of rotation was nearly one day. The case is one on which it is not at present possible to pronounce an authoritative decision; but the balance of probabilities is largely in favor of the view that the rotatation of Venus on its axis has never been seen or determined by any of the astronomers who have made this planet an object of study.*

Atmosphere of Venus.—The appearance of Venus when nearly between us and the sun affords very strong evidence of the existence of an atmosphere. The limb of the planet farthest from the sun is then seen to be illuminated, so that it appears as a complete circle of light. If only half the globe of the planet were illuminated by the sun, this appearance could never present itself, as it is impossible for an observer to see more than half of a large sphere at one view. There is no

---

* The latest physical observations on Venus with which I am acquainted are those of Dr. Vogel at Bothkamp, in Part II. of the "Bothkamp Observations" (Leipzig, Engelmann, 1873). The result to which these observations point is that the atmosphere of Venus is filled with clouds so dense that the solid body of the planet can not be seen, and no time of rotation can be determined.

known way in which the sun can illuminate so much more than the half of Venus as to permit a complete circle of light to be seen except by the refraction of an atmosphere.

The appearance to which we allude was first noticed by David Rittenhouse, of Philadelphia, while observing the transit of Venus on June 3d, 1769. When Venus had entered about half-way upon the sun's disk, so as to cut out a notch of the form of a half-circle, that part of the edge of the planet which was off the disk appeared illuminated so that the outline of the entire planet could be seen. As this appearance was not confirmed by other observers, it seems to have excited no attention. But it was found by Mädler in 1849 that when Venus was near inferior conjunction, the visible crescent extended through more than a half-circle. This showed that more than half the globe of Venus was illuminated by the sun, and Mädler, computing the refractive power of the atmosphere which would be necessary to produce this effect, found that it would exceed that of our own atmosphere; the horizontal refraction being 44', whereas on the earth it is only 34'. He therefore concluded that Venus was surrounded by an atmosphere a little more dense than that of the earth.

The next important observation of the kind was made by Professor C. S. Lyman, of Yale College. In December, 1866, Venus was very near her node at inferior conjunction, and passed unusually near the line drawn from the earth to the sun. Examining the minute crescent of the planet with a moderate-sized telescope, he found that he could see the entire circle of the planet's disk, an exceedingly thin thread of light being stretched round the side farthest from the sun. So far as known, this was the first time that the whole circle of Venus had been seen in this way since the time of Rittenhouse. It is remarkable that both observations should have been made by isolated observers in America.

Notwithstanding the concurrent testimony of Rittenhouse, Mädler, and Lyman, the bearing of their observations on what was to be expected during the transit of Venus in December, 1874, was entirely overlooked. Accordingly, many of the observers were quite taken by surprise to find that when Venus was partly on and partly off the sun, the outline of that part of her disk outside the sun could be distinguished by a delicate line of light extending around it. In some cases the time of internal contact at egress of the planet was missed, through the observer mistaking this line of light for the limb of the sun.

That no one but Rittenhouse saw this line of light during the transit of 1769 is to be attributed to the low altitude of the planet at most of the stations, and to the imperfect character of many of the instruments used. It is also to be remarked that the observers of that time had an erroneous notion of the appearance which would be presented by an atmosphere of Venus. It was supposed that the atmosphere would give the planet a nebulous border when on the sun, caused by the partial absorption of the light in passing through it. Captain Cook, at Otaheite, made separate observations of the contacts of the supposed atmosphere and of the planet with the limb of the sun. In fact, however, it would not be possible to see any indications of an atmosphere under such circumstances, for the reason that the light passing through its denser portions would be refracted entirely out of its course, so as not to reach an observer on the earth at all.

The spectroscope shows no indication that the atmosphere of Venus exerts any considerable selective absorption upon the light which passes through it. No new and well-marked spectral lines are found in the light reflected from the planet, nor has the spectrum been certainly found to differ from the regular solar spectrum, except, perhaps, that some of the lines are a little stronger. This would indicate that the atmosphere in question does not differ in any remarkable degree from our own, or, at least, does not contain gases which exert a powerful selective absorption on light.

Supposed Visibility of the Dark Hemisphere of Venus.—Many astronomers of high repute have seen the dark atmosphere of Venus slightly illuminated, the planet presenting the appearance known as "the old moon in the new moon's arms," which may be seen on any clear evening three or four days after the change of the moon. It is well known that in the case of the moon her dark hemisphere is thus rendered visible by the light reflected from the earth. But in the case of Venus, there is no earth or other body large enough to shed so much light on the dark hemisphere as to make it visible. There being no sufficient external source of light, it has been attributed to a phosphorescence of the surface of the planet. If the phosphorescence were always visible under favorable circumstances, there would be no serious difficulty in accepting this explanation. But, being only rarely seen, it is hard to conceive how any merely occasional cause could act all at once over the surface of a planet the size of our globe, so as to make it shine. Indeed, one circumstance makes it extremely difficult to avoid the conclusion that the whole appearance is due to some unexplained optical illusion. The appearance is nearly always seen in the daytime or during bright twilight—rarely or never after dark. But such an illumination would be far more easily seen by night than by day, because during the day an appearance easily seen at night might be effaced by the light of the sky. If, then, the phenomenon is real, why is it not seen when the circumstances are such that it should be most conspicuously visible? This is a question to which no satisfactory answer has been given, and until it is answered we are justified in considering the appearance to be purely optical.

Supposed Satellite of Venus.—No better illustration of the errors to which observations with imperfect instruments are liable can be given than the supposed observations of a satellite of Venus, made when the telescope was still in its infancy. In 1672, and again in 1686, Cassini saw a faint object near Venus which exhibited a phase similar to that of the planet. But he never saw it except on these two occasions. A similar object was reported by Short, of England, as seen by him on October 23d, 1740. The diameter of the object was a third of that of Venus, and it exhibited a similar phase. Several other observers saw the same thing between 1760 and 1764.

One astronomer went so far as to compute an orbit from all the observations; but it was an orbit in which no satellite of Venus could possibly revolve unless the mass of the planet were ten times as great as it really is. A century has now elapsed without the satellite having been seen, and the fact that during this century the planet has been scrutinized with better telescopes than any which were used in the observations referred to affords abundant proof that the object was entirely mythical.

How the observers who thought they saw the object could have been so deceived it is impossible, at this distance of time, to say with certainty. Had they been inexperienced, we could say with some confidence that they were misled by the false images produced to some extent in every telescope by the light reflected from the cornea of the eye against the nearest surface of the eye-piece, and thence back again into the eye. Similar images are sometimes produced by the reflection of light between the surfaces of the various lenses of the eye-piece. They are well known to astronomers under the name of "ghosts;" and one of the first things a young observer must learn is to distinguish them from real objects. They may also arise from a slight maladjustment of the lenses of the eye-piece, and if, proceeding from this cause, they are produced only when the actual object is in the centre of the field, they may, for the moment, deceive the most experienced observer.* If, in an ordinary achromatic telescope, in which the interior curvatures of the lenses are the same, the latter are not exactly at the same distance all the way round, a ghost will be seen along-side of every bright object in all positions. It is probable that all the observations alluded to were the results of some sort of derangements in the telescope, producing false images by reflection from the glasses.

---

* One of the eye-pieces of the great Washington telescope shows a beautiful little satellite along-side the planet Uranus or Neptune when the image of the planet is brought exactly in the centre of the field of view, but it disappears as soon as the telescope is moved. The writer was deceived by this appearance on two occasions while scrutinizing these planets for close satellites.

§ 4. *The Earth.*

Our earth is the third planet in the order of distance from the sun, and slightly the largest of the inner group of four. Its mean distance from the sun is about $92\frac{1}{2}$ millions of miles; but it is a million and a half less than this mean on January 1st of every year, and as much greater on July 1st. That is, its actual distance varies from 91 to 94 millions of miles. As already remarked, these numbers are uncertain by several hundred thousand miles.

Much of what we may call the astronomy of the earth—such as its figure and mass, the length of the year, the obliquity of the ecliptic, the causes of the changes in the seasons and in the length of the days—has already been treated in the chapter on gravitation, so that we have little of a purely astronomical character to add here. The features of its surface and the phenomena of its atmosphere belong rather to geography and meteorology than to astronomy. But its constitution gives rise to several questions in the treatment of which astronomical considerations come into play. Prominent among these is that of the state of the great interior mass of our globe, whether solid or liquid. It is well known that wherever we descend into the solid portions of the earth, we find a rise in temperature, going on uniformly with the depth, at a rate which nowhere differs greatly from 1° Fahrenheit in 50 feet. This rise of temperature has no connection with the sea-level, but is found at all points of the surface, no matter how elevated they may be. Wherever a difference of temperature like this exists, there is necessarily a constant transfer of heat from the warmer to the cooler strata by conduction. In this way, the inequality would soon disappear by the warmer strata cooling off, if there were not a constant supply of heat inside the earth. The rise of temperature, therefore, cannot be something merely superficial, but must continue to a great depth. If we trace to past times the conditions which must have existed in order that the increase might show itself at the present time, we shall find it almost certain that, a thousand years ago, the whole earth was red-hot at a distance of ten or fifteen miles below its surface; because otherwise its interior could not have furnished the supply of heat which now causes the observed increase. This being the case, it is probably red-hot still, since it would be absurd to expect a state of things like this to be merely temporary. In a word, we have every reason to believe that the increase of say 100° a mile continues many miles into the interior of the earth. Then we shall have a red heat at a distance of 12 miles, while, at the depth of 100 miles, the temperature will be so high as to melt most of the materials which form the solid crust of the globe.

We are thus led to the theory, very generally received by geologists, that the earth is really a sphere of molten matter surrounded by a comparatively thin solid crust, on which we live. This crust floats, as it were, on the molten interior. It must be confessed that geological facts are, on the whole, favorable to this view. Observations on the pendulum have been supposed to show that the specific gravity of the earth under the great mountain chains is generally less than in the adjoining plains, which is exactly the result that would flow from the theory. The heavier masses, pressing upon the interior fluid, would tend to elevate the surrounding lighter masses, and when the two were in equilibrium, the latter would be the higher, as a floating block of pine wood will rise higher out of the water than a block of oak. Boiling springs in many parts of the globe show that there are numerous hot regions in the earth's interior, and this heat cannot be merely local, because then it would soon be dissipated. But the geologist finds the strongest proof of the theory in volcanoes and earthquakes. The torrents of lava which have been thrown out of the former through thousands of years show that there are great volumes of molten matter in the earth's interior,

FIG. 75.—Showing thickness of the earth's crust according to the geological theory of a molten interior. The circle is thicker in proportion than the solid crust.

while the latter show this interior to be subject to violent changes which a solid could not exhibit.

But mathematicians have never been able entirely to reconcile the theory in question with the observed phenomena of precession, nutation, and tides. To all appearance, the earth resists the tide-producing action of the sun and moon exactly as if it were solid from centre to circumference. Sir William Thomson has shown that if the earth were less rigid than steel, it would yield so much to this action that the tides would be much smaller than on a perfectly rigid earth; that is, the attraction of the bodies in question would draw the earth itself out into an ellipsoidal form, instead of drawing merely the waters of the ocean. Earth and ocean moving together, we could see no tides at all. If the earth were only a thin shell floating on a liquid interior, the tides would be produced in the latter; the thin shell would bend in such a way that the tides in the ocean would be nearly neutralized. Again, the question has arisen whether the liquid interior would be affected by precession; whether, in fact, the crust would not slip over it, so that in time the liquid would rotate in one direction, and the crust in another. Altogether, the doctrine of the earth's fluidity is so fraught with difficulty that, notwithstanding the seeming strength of the evidence in its favor, it must be regarded as at least very doubtful. It may be added that no one denies that the interior of our planet is intensely hot— hot enough, in fact, to melt the rocks at its surface—but it is supposed that the enormous pressure of the outer portions tends to keep the inner part from melting. Nor is it questioned by Sir William Thomson that there are great volumes of melted matter in the earth's interior from which volcanoes are fed; but he maintains that, after all, these volumes are small compared with that of the whole earth.

Refraction of the Atmosphere.—If a ray of light pass through our atmosphere in any other than a vertical direction, it is constantly curved downwards by the refractive power of that medium. The more nearly horizontal the course of the ray, the greater the curvature. In consequence of this, all the heavenly bodies appear a little nearer the zenith, or a little higher above the horizon, than they actually are. The displacement is too small to be seen by the naked eye except quite near the horizon, where it increases rapidly, amounting to more than half a degree at the horizon itself. Consequently, at any point where we have a clear horizon, as on a prairie, or the sea-shore, the whole disk of the sun will be seen above the horizon when the true direction is below it. A slight increase is thus given to the length of the day. The sun in our latitudes always rises three or four minutes sooner, and sets three or four minutes later, than he would if there were no atmosphere. At the time of the equinoxes, if we suppose the day to begin and end when the centre of the sun is on the horizon, it is not of the same length with the night, but is six or eight minutes longer. If we suppose the day to begin with the rising of the sun's upper limb, and not to end till the same limb has set, then we must add some three minutes more to its length.

If, standing on a hill, we watch the sun rise or set over the ocean, one effect of refraction will be quite clearly visible. When his lower limb almost seems to touch the water, it will be seen that the form of his disk is no longer round, but elliptical, the horizontal diameter being greater than the vertical. The reason of this is that the lower limb is more elevated by refraction than the upper one, and thus the vertical diameter is diminished.

In practical astronomy, all observations of the altitude of the heavenly bodies above the horizon must be corrected for refraction, the true altitude being always less than that observed. Very near the zenith the refraction is about $1''$ for every degree, or $\frac{1}{3600}$ part the distance from the zenith. But it increases at first in the proportion of the tangent of the zenith distance, so that at $45°$, or half-way between the zenith and the horizon, it amounts to $60''$; at the horizon it is $34'$.

The Aurora Borealis.—This phenomenon, though so well known, is one of which great difficulty has been found in giving a satisfactory explanation. That it is in some way con-

302 THE SOLAR SYSTEM.

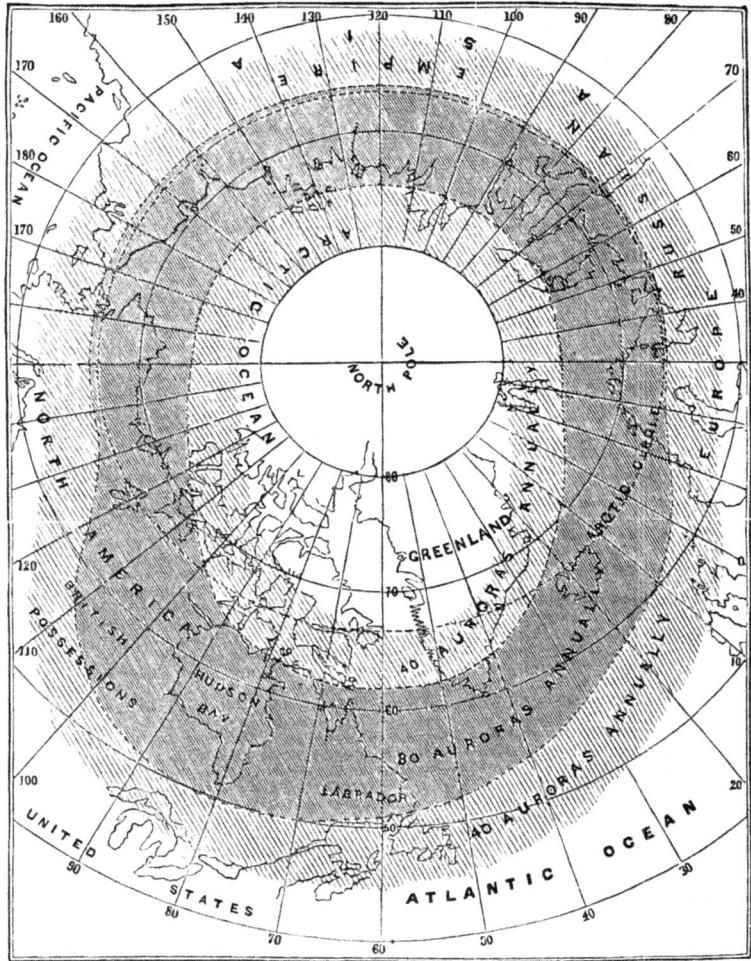

Fig. 76.—Distribution of auroras, after Loomis. The darker the color, the more frequently auroras are seen.

nected with the pole of the earth is shown by the fact that its frequency depends on the latitude. In the equatorial regions of our globe it is quite rare, and increases in frequency as we go north. But the region of greatest frequency seems

THE EARTH. 303

to be, not the poles, but the neighborhood of the Arctic Circle, from which it diminishes towards both the north and the south. This is shown more exactly in Professor Loomis's auroral map, of which we give a copy on the preceding page. A close study of the aurora indicates that its connection is not with the geographical, but with the magnetic pole. Two distinct kinds of light are seen in the aurora; or we might say that the light assumes two distinct forms, of which sometimes the one and sometimes the other preponderates. They are as follows:

1. The cloud-like form. This consists of a large irregular patch of light, frequently of a red or purple tinge. It is seen in every direction, but more frequently in or near the northern horizon, where it assumes the form of an arch or crown of light. The two ends of the arch rest on the horizon, one on each side of the north point. The middle of the arch rises a few degrees above the horizon.

Fig. 77.—View of aurora.

2. The streamer or pillar form. This form consists of long streamers or pillars, which extend in the direction of the dipping magnetic needle. They look curved or arched, like the celestial sphere on which they are projected, but they are really straight. They are in a state of constant motion. Some-

times they are spread out in the form of an immense flag with numerous folds, dancing, quivering, and undulating, as if moved by the wind.

Electric Nature of the Aurora.—There is abundant evidence that the aurora is intimately connected with the electricity and magnetism of the earth. During a brilliant aurora such strong and irregular currents of electricity pass through the telegraph wires that it is difficult to send a despatch. Sometimes the current runs with such force that a message may be sent without a battery. The magnetic needle is also in a state of great agitation. Before the spectroscope came into use, these electric phenomena gave rise to the opinion that the aurora was due entirely to currents of electricity passing through the upper regions of the atmosphere from one pole to the other. But recent researches seem to show that, though this view may be partly true, it is far from the whole truth, and does not afford a complete explanation. The great height of the aurora and the nature of its spectrum both militate against it.

Height of the Aurora.—Several attempts have been made in recent times to determine the height of the aurora above the surface of the earth, by simultaneous observations of some prominent streamer or patch of light from several far-distant stations. The general result is that it extends to the height of from 400 to 600 miles. But the evidence of shooting-stars and meteors seems to indicate that the limit of the atmosphere is between 100 and 110 miles in height. If it extends above this, it must be too rare to conduct electricity long before it reaches the greatest height of the aurora; indeed, it is doubtful whether it does not attain this rarity at a height of 40 or 50 miles. If, then, the aurora really extends to the great height we have mentioned, and still exists in a gaseous medium, it seems difficult to avoid the conclusion that this medium is something far more ethereal than the gases which form our atmosphere. It would, however, be unphilosophical to assume the existence of such a medium without some other evidence in its favor than that afforded by the aurora. We must include the aurora among those things in which modern observations have opened up more difficulties than modern theories have explained.

Spectrum of the Aurora.—The spectrum of the aurora is so far from uniform as to be quite puzzling. There is one characteristic bright line in the green part of the spectrum, known as Ångstrom's line, from its first discoverer. This was the only line Ångstrom could see: he therefore pronounced the light of the aurora to be entirely of one color. Subsequent observers, however, saw many additional lines, but they were different in different auroras. Among those who have made careful studies of the aurora with the spectroscope are the late Professor Winlock, of Harvard University; Professor Barker, of Philadelphia; and Dr. H. C. Vogel, formerly of Bothkamp.

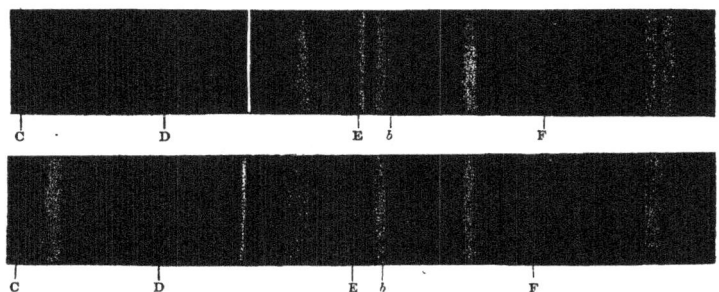

Fig. 78.—Spectrum of two of the great auroras of 1871, after Dr. H. C. Vogel.

Fig. 78 shows the spectra of two auroras, as drawn by Dr. Vogel. It will be seen that there is one fine bright line between D and E, which would fall in the yellowish-green part of the spectrum, while the others are all broad, ill-defined bands. Dr. Vogel notices a remarkable connection between these lines and several groups of lines produced by the vapor of iron, and inquires whether this vapor can possibly exist in the upper regions of our atmosphere. A more complete study of the spectra of vapors at different pressures and temperatures is necessary before we can form a decided opinion as to what the aurora really is.

306 THE SOLAR SYSTEM.

Of the supposed periodicity of the aurora, and its connection with sun-spots, we have already spoken. Granting the reality of this connection, we may expect that auroras will be very frequent between the years 1880 and 1884; and if this expectation is realized, little doubt of the connection will remain.

§ 5. *The Moon.*

The moon is much the nearest to us of all the heavenly bodies; no other, except possibly a comet, ever coming nearer than a hundred times her distance. Her mean distance is, in round numbers, 240,000 miles. Owing to the ellipticity of her orbit and the attractive force of the sun, it varies from ten to twenty thousand miles on each side of this mean in the course of each monthly revolution. The least possible distance is 221,000 miles; the greatest is 259,600 miles. It very rarely approaches either of these limits, the usual oscillation being about 13,000 miles on each side of the mean distance of 240,300. The diameter of the moon is 2160 miles, or somewhat less than two-sevenths that of the earth. Her volume is about one-fiftieth that of the earth, and if she were as dense as the latter, her mass would be in the same proportion.

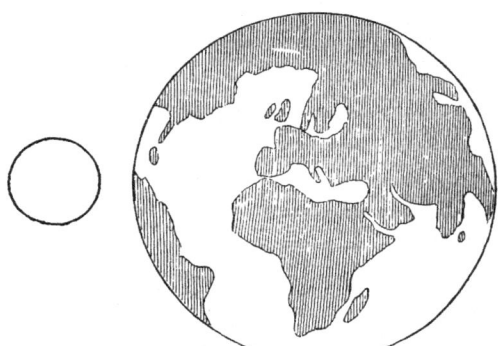

FIG. 79.—Relative size of earth and moon.

But her actual mass is only about one-eightieth that of the earth, showing that her density, or the specific gravity of the material of which she is composed, is little more than half that of our globe. Her weight is, in fact, about $3\frac{1}{2}$ times that of her bulk of water.

The most remarkable feature of the motion of the moon is, that she makes one revolution on her axis in the same time that she revolves around the earth, and so always presents the same face to us. In consequence, the other side of the moon must remain forever invisible to human eyes. The reason of this peculiarity is to be found in the ellipticity of her globe. That she should originally have been set in revolution on her axis with precisely the same velocity with which she revolved around the earth, so that not the slightest variation in the relation of the two motions should ever occur in the course of ages, is highly improbable. If such had been the state of things, the correspondence of the two motions could not have been kept up without her axial rotation varying; because, owing to the secular acceleration already described, the moon, in the course of ages, varies her time of revolution, and so the two motions would cease to correspond. But the effect of the attraction of the earth upon the slightly elongated lunar globe is such that if the two motions are, in the beginning, very near together, not only will the axial rotation accommodate itself to the orbital revolution around the earth, but as the latter varies, the former will vary with it, and thus the correspondence will be kept up.

Figure, Rotation, and Libration of the Moon.—Supposing the shape of the moon to be the same as if it were a fluid mass, or covered by an ocean, it will be an ellipsoid with three unequal axes. The shortest axis will be that around which it revolves, which is not very far from being perpendicular to the ecliptic. The next longest is that which lies in the direction in which the moon moves; while the longest of all is that which points towards the earth. The reason that the polar axis is the shortest is the same which makes the polar axis of the earth the shortest, that is, the centrifugal force generated by the revolution round that axis. If we considered only the action of this force, we should conclude that the moon, like the earth, was an oblate spheroid, the equator being a perfect circle. But the attraction of the earth upon the moon tends to elongate it in the direction of the line joining the two bodies, in the same way that the attraction of the moon upon the earth generates a tide-producing force which we have already explained. At the centre of the moon the attraction of the earth and the centrifugal force of the moon in its orbit exactly balance each other. But if we go to the farther side of the moon, the centrifugal force will be greater, owing to the larger orbit which that part of the moon has to describe, while the attraction of the earth will be less owing to the greater distance of the particles it attracts. Hence, that part of the moon tends to fly off from the centre and from the earth. On this side of the moon the case is reversed, the attractive force of the earth exceeding the centrifugal force of those parts of the moon, whence those parts are impelled by a force tending to draw them to the earth. The effect would be much the same as if a rope were fastened to this side of the moon, and constantly pulled towards the earth, while another were fastened to the opposite side, and as constantly pulled from the earth. Supposing the moon to be a liquid, so as to yield freely, it is clear that the effect of these forces would be to elongate her in the direction of the earth.

The deviations from a spherical form produced by these causes are very minute. Taking the results of Lagrange and Newton, the mean axis would be $46\frac{1}{2}$ feet longer than the shortest one, and the longest 186 feet longer than the mean one, or $232\frac{1}{2}$ feet longer than the shortest one.* These differences are so much smaller than the average height of the lunar mountains that the irregularities produced by the latter might entirely overpower them; but the correspondence between the motions of rotation and revolution of the moon shows that there must be, on the average, a real elongation in

---

* These numbers are, perhaps, not strictly correct. The extension of 186 feet was deduced by Newton from a comparison of the distorting powers of the centrifugal force of the earth with that of the force we have just described. He seems to have overlooked the fact that the small density of the moon will cause the elongation to be greater.

the direction of the earth. This correspondence is kept up by the slight additional attraction of the earth upon this extension of the moon towards the earth, combined with the additional centrifugal force of the extension on the other side. Although these forces are not by any means the same as the distorting forces already described, they may be represented in the same way by two ropes, one of which pulls the protuberance on this side towards the earth, while the other pulls the protuberance on the other side from it. If the two protuberances do not point exactly towards the earth, the effect of these two minute forces will be to draw them very slowly into line. Consequently, notwithstanding the slow variations to which the motion of the moon around the earth is subject in the course of ages, the attraction of the earth will always keep this protuberant face turned towards us. Human eyes will never behold the other side of the moon, unless some external force acts upon her so as to overcome the slight balancing force just described, and set her in more or less rapid motion on her axis. If it is disappointing to reflect that we are forever deprived of the view of the other side of our satellite, we may console ourselves with the reflection that there is not the slightest reason to believe that it differs in any respect from this side. The atmosphere with which it has been covered, and the inhabitants with which it has been peopled, are no better than the products of a poetic imagination.

The forces we have just described as tending to keep the same face of the moon pointed towards us would not produce this effect unless the adjustment of the two motions—that around the earth, and that on her axis—were almost perfect in the beginning. If her axial rotation were accelerated by so small an amount as one revolution in two or three years, there is every reason to believe that she would keep on revolving at the new rate, notwithstanding the force in question. The case is much like that of a very easy-turning fly-wheel, which is slightly weighted on one side. If we give the wheel a gentle motion in one direction or another, the weight will cause the wheel to turn till the heavy side is the lowest, and the wheel will then vibrate very slowly on one side and the other of this point. But if we give the wheel a motion rapid enough to carry its heavy side over the highest point, then the weight will accelerate the wheel while it is falling as much as it will retard it while rising; and if there were no friction, the wheel would keep on turning indefinitely. The question now arises, How does it happen that these two motions are so exactly adjusted to each other that not only is the longer axis of the moon pointed exactly towards the earth, but not the slightest swing on one side or the other can be detected? That this adjustment should be a mere matter of chance, without any physical cause to produce it, is almost infinitely improbable, while to suppose it to result from the mere arbitrary will of the Creator is contrary to all scientific philosophy. But if the moon were once in a partially fluid state, and rotated on her axis in a period different from her present one, then the enormous tides produced by the attraction of the earth, combined with the centrifugal force, would be accompanied by a friction which would gradually retard the rate of rotation, until it was reduced to the point of exact coincidence with the rate of revolution round the earth, as we now find it. We therefore see in the present state of things a certain amount of probable evidence that the moon was once in a state of partial fluidity.

The force we have just described as drawing the protuberant portion of the moon towards the earth is so excessively minute that it takes it a long time to produce any sensible effect; consequently, although the moon moves more rapidly in some points of her orbit than in others, the force in question produces no corresponding change in the moon's rotation. The protuberance does not, therefore, always point exactly at the earth, but sometimes a little one side, and sometimes a little the other, according as the moon is ahead of or behind her mean place in the orbit. The result is, that the face which the moon presents to us is not always exactly the same, there being a slight *apparent* (not real) oscillation, due to the real inequality in her orbital motion. This apparent swaying is called *libration*, and in consequence of it there is nearly six-tenths of the lunar surface which may, at one time or another, come into view from the earth.

The Lunar Day.—In consequence of the peculiarity in the moon's rotation which we have described, the lunar day is 29½ times as long as the terrestrial day. Near the moon's equator the sun shines without intermission nearly fifteen of our days, and is absent for the same length of time. In consequence, the vicissitudes of temperature to which the surface is exposed must be very great. During the long lunar night the temperature of a body on the moon's surface would probably fall below any degree of cold that we ever experience on the earth, while during the day it must become hotter than anywhere on our globe.

Astronomical phenomena, to an observer on the moon, would exhibit some peculiarities. The earth would be an immense moon, going through the same phases that the moon does to us; but instead of rising and setting, it would only oscillate back and forth through a few degrees. On the other side of the moon it would never be seen at all. The diurnal motion of the stars would take place in twenty-seven of our days, much as they do here every day, while, as we have said, the sun would rise and set in 29½ of our days.

Geography of the Moon.—With the naked eye it is quite readily seen that the brilliancy of the moon is far from uniform, her disk being variegated with irregular dark patches, which have been supposed to bear a rude resemblance to a human face. It is said to have been a fancy of some of the ancient philosophers that the light and dark portions were caused by the reflection of the seas and continents of the terrestrial globe, though it is hard to conceive of such an opinion being seriously entertained. The first rude idea of the real nature of the lunar surface was gained by Galileo with his telescope. He saw that the brighter portions of the disk were broken up with inequalities of the nature of mountains and craters, while the dark parts were, for the most part, smooth and uniform. Here he saw a striking resemblance to the geographical features of our globe, and is said to have suggested that the brighter and rougher portions might be continents, and the dark, smooth portions oceans. This view of the resemblance to terrestrial scenery is commemorated in Milton's description of Satan's shield:

> "Like the moon, whose orb
> Through optic glass the Tuscan artist views
> At evening, from the top of Fesolé,
> Or in Valdarno, to descry new lands,
> Rivers, or mountains in her spotty globe."

The opinion that the dark portions of the lunar disk were seas was shared by Kepler, Hevelius, and Ricciolus. The last two made maps of the moon in which they gave names to the supposed seas, which names the regions still bear, though they are strikingly fanciful. Among them are *Oceanus Procellarum* (the Ocean of Storms), *Mare Tranquillitatis* (Sea of Tranquillity), *Mare Imbrium* (Rainy Sea), etc. The names of great philosophers and astronomers were given to prominent features, craters, etc.

If this resemblance between the earth and moon had been established; if it had been found that our satellite really had seas and atmosphere, and was fitted for the support of organic life; still more, if any evidence of the existence of intelligent beings had been found, our interest in lunar geography would have been immensely heightened. But the more the telescope was improved, the more clearly it was seen that there was no similarity between lunar and terrestrial scenery. A very slight increase of telescopic power showed that there was no more real smoothness in the regions of the supposed seas than elsewhere. The inequalities were smaller and harder to see on account of the darkness of color; but that was all. The sun would have been brilliantly imaged back from the surfaces of the oceans in certain positions of the moon; but nothing of the kind was ever seen. The polariscope showed that the sun's rays did not pass through any liquid at the moon's surface. Positive evidence of an atmosphere was sought in vain. Supposed volcanoes were traced to bright spots, illuminated by light from the earth. Inequalities of surface there were; but in form they were wholly different from the mountains of the earth. So the beautiful fancies of

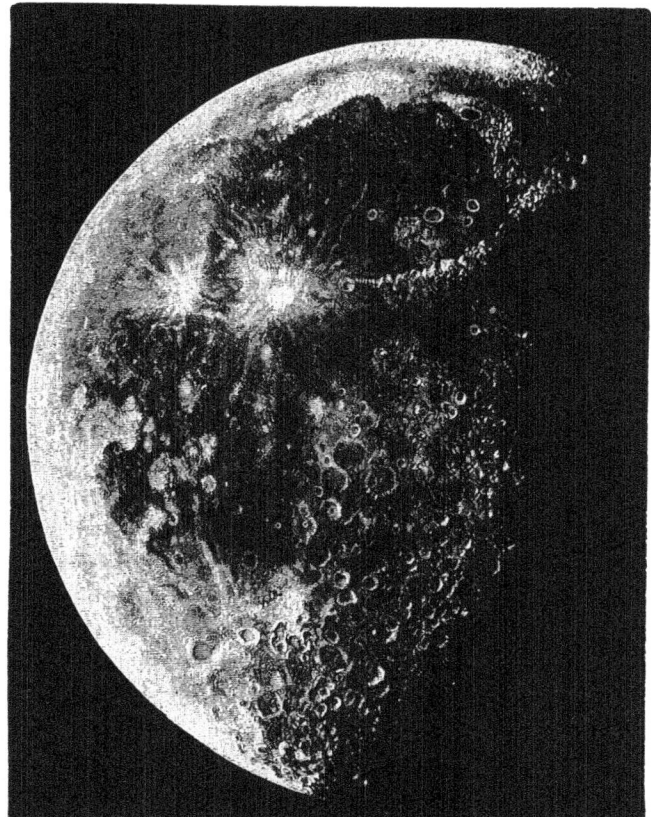

FIG. 80.—View of moon near the third quarter. From a photograph by Professor Henry Draper.

the earlier astronomers all faded away, leaving our satellite as lifeless as an arid rock.

As the moon is now seen and mapped, the difference between the light and dark portions is due merely to a difference in the color of the material, much of which seems to be darker than the average of terrestrial objects. The mountains consist, for the most part, of round saucer-shaped elevations, the interior being flat, with small conical mounds rising here and there. Sometimes there is a single mound in the centre. It is very curious that the figures of these inequalities in the lunar surface can be closely imitated by throwing pebbles upon the surface of some smooth plastic mass, as mud or mortar. They may be well seen during an eclipse of the sun, when the contrast between the smoothness of the sun's limb and the roughness of that of the moon cannot escape notice. Their appearance is most striking when the eclipse is annular or total. In the latter case, as the last streak of sunlight is disappearing, it is broken up into a number of points, which have been known as "Baily's beads," from the observer who first described them, and which are caused by the sun shining through the depressions between the lunar mountains.

To give the reader an idea what the formation of the lunar surface is, we present a view of the spot or crater "Copernicus," by Secchi, taken from the "Memoirs of the Royal Astronomical Society," vol. xxxii. The diameter of the central portion, so much like a fort, is about 45 or 50 miles.

Among the most curious and inexplicable features of the moon's surface are the long narrow streaks of white material which radiate from certain points, especially from the great crater Tycho. Some of these can be traced more than a thousand miles. The only way in which their formation has been accounted for is by supposing that in some former age immense fissures were formed in the lunar surface which were subsequently filled by an eruption of this white matter which forms the streaks.

Has the Moon an Atmosphere? — This question may be answered by saying that no evidence of a lunar atmosphere entitled to any weight has ever been gathered, and that if there is such an atmosphere, it is certainly not $\frac{1}{400}$ part the density of the earth's atmosphere. The most delicate known test of an atmosphere is afforded by the behavior of a star when in apparent contact with the limb of the moon. In this

THE MOON. 315

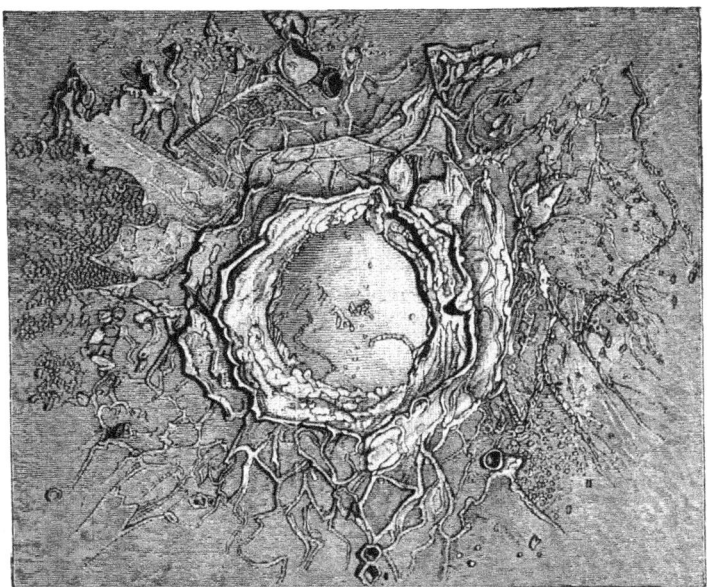

FIG. 81.—Lunar crater "Copernicus," after Secchi.

position the rays of light coming from the star would pass through the lunar atmosphere, and be refracted by twice the horizontal refraction of that atmosphere. The star would then be apparently thrown out of its true position in the direction from the moon's centre by the amount of this double refraction. But observations of stars in this position, at the moment when the limb of the moon passes over them, have never indicated the slightest displacement. It is certain that, had the displacement been decidedly in excess of half a second, it would have been detected; therefore, the double horizontal refraction of the lunar atmosphere, if any exist, must be as small as half a second.* The corresponding refraction of the earth's atmosphere is 4000 seconds. Therefore, the re-

---

* A similar test is afforded by the occultation of a planet, especially Saturn or Venus, the limb of which would be a little flattened as it touched the moon. The writer looked very carefully for this appearance during an unusually favorable occultation of Saturn which occurred on Aug. 6th, 1876, without seeing a trace of it.

fractive power of the lunar atmosphere cannot be much in excess of $\frac{1}{5000}$ that of the earth's, and certainly falls below $\frac{1}{2000}$.

Without an atmosphere no water or other volatile fluid can exist on the moon, because it would gradually evaporate and form an atmosphere of its own vapor. The evaporation would not cease till the pressure of the vapor became equal to its elastic force at the mean temperature of the moon. If this temperature were as low as the freezing-point, the pressure of an atmosphere of water vapor would be $\frac{1}{160}$ that of our atmosphere. So dense an envelope could not fail of detection with our present means of observation.

The question whether any change is taking place on the surface of the moon is one of interest. Hitherto, the preponderance of evidence has been against the idea of any change. It is true that a few years ago there was a great discussion in the astronomical world about a supposed change in the aspect of the spot Linnæus, which was found not to present the same appearance as on Beer and Mädler's map. But careful scrutiny showed that, owing to some peculiarity of its surface, this spot varied its aspect according to the manner in which it was illuminated by the sun, and these variations appear to be sufficient to account for the supposed change. To whatever geological convulsions the moon may have been subjected in ages past, it seems as if she had now reached a state in which no further change was to take place, unless by the action of some new cause. This will not seem surprising if we reflect what an important part the atmosphere plays in the changes which are going on on the surface of the earth. The growth of forests, the formation of deltas, the washing-away of mountains, the disintegration and blackening of rocks, and the decay of buildings, are all due to the action of air and water, the latter acting in the form of rain. Changes of temperature powerfully re-enforce the action of these causes, but are not of themselves sufficient to produce any effect. Now, on the moon, there being neither air, water, rain, frost, nor organic matter, the causes of disintegration and decay are all absent. A marble building erected upon the surface of the moon would remain century after century just as it was left. It is true that there might be bodies so friable that the expansions and contractions due to the great changes of temperature to which the surface of the moon is exposed would cause them to crumble. But whatever crumbling might thus be caused would soon be done with, and then no further change would occur.

Light and Heat of the Moon.—That the sun is many times brighter than the moon is evident to the eye; but no one judging by the unaided eye would suppose the disparity to be so great as it really is. It is found by actual trial that the light of the sun must be diminished several hundred thousand times before it becomes as faint as the full moon. The results of various experiments range between 300,000 and 800,000. Professor G. B. Bond, of Cambridge, found the ratio to be 470,000. The most careful determination yet made is by Zöllner, who finds the sun to give 619,000 times as much light as the full moon. This result is probably quite near the truth.

The moon does not shine by sunlight alone. Whenever the narrow crescent of the new moon is seen through a clear atmosphere, her whole surface may be plainly seen faintly illuminated. This appearance is known as " the old moon in the new moon's arms." The faint light thus shed upon the dark parts of the moon is reflected from the earth. An observer on the moon would see the earth in his sky as a large moon, much larger than the moon is seen by us. When it is new moon with us, it would be *full earth*, if we may be allowed the term, to an observer on this side of the moon. Hence, under those circumstances, most of the lunar hemisphere hidden by the sun is illuminated by earth-light, or by sunlight reflected by the earth, and is thus rendered visible. The case is the same as if an observer on the moon should see the dark hemisphere of the earth by the light of the full moon.

As the moon reflects the light of the sun, so also must she reflect his heat. Besides, she must radiate off whatever heat she absorbs from the sun. Hence, we must receive some heat from the moon, though calculation will show the quantity to be so small as to defy detection with the most delicate thermometer, the average quantity being only $\frac{1}{280000}$ part of that received from the sun. As the direct rays of the sun will not raise the black-bulb thermometer more than 50 or 60 degrees above the temperature of the air, those of the moon cannot raise it more than $\frac{1}{5000}$ of a degree. By concentrating the rays in the focus of a telescope of large aperture and comparatively short focal length, the temperature might be increased a hundred times or more; but even then we should only have an increase of $\frac{1}{50}$ of a degree. Even this increase might be unattainable, for the reason that the heat radiated by the moon would not pass through glass. It is, therefore, only since the discovery of thermo electricity and the invention of the thermo-electric pile that the detection of the heat from the moon has been possible. The detection is facilitated by using a reflecting telescope to concentrate the lunar rays, because the moon is not hot enough to radiate such heat as will penetrate glass. Lord Rosse and M. Marie-Davy, of Paris, have thus succeeded in measuring the heat emanating from the moon. The former sought not merely to determine the total amount of heat, but how much it varied from one phase of the moon to the other, and what portion of it was the reflected heat of the sun, and what portion was radiated by the moon herself, as if she were a hot body. He found that from new to full moon, and thence round to new moon again, the quantity of heat received varied in the same way with the quantity of light; that is, there was most at full-moon, and scarcely any when the moon was a thin crescent. That only a small proportion of the total heat emitted was the reflected heat of the sun, was shown by the fact that while 86 per cent. of solar heat passes through glass, only 12 per cent. of lunar heat does so. This absorption by glass is well known to be a property of the heat radiated by a body which is not itself at a high temperature. The same result was indicated in another way, namely, that while the sun is found by Zöllner to give 618,000 times as much light as the moon, it only gives 82,600 times as much heat. Thus both the ratio of solar to lunar heat, and the proportion of the latter which is absorbed by glass, agree in indicating that about six-sevenths of the heat received from the moon is radiated by the latter, owing to the temperature of her surface produced by the absorption of the sun's rays.

Lord Rosse was thus enabled to estimate the change of temperature of the moon's surface according as it was turned towards or from the sun, and found it to be more than 500° Fahrenheit. But there was no way of determining the temperatures themselves with exactness. Probably when the sun does not shine the temperature is two or three hundred degrees below zero, and therefore below any ever known on the earth; while under the vertical sun it is as much above zero, and therefore hotter than boiling water.

Effect of the Moon on the Earth.—We have already explained, in treating of gravitation, how the attraction of the moon causes tides in the ocean. This is one of the best-known effects of lunar attraction. It is known from theory that a similar tide is produced in the air, affecting the height of the barometer; but it is so minute as to be entirely masked by the changes constantly going on in the atmospheric pressure from other causes. There is also reason to believe that the occurrence of earthquakes may be affected by the attraction of the moon; but this is a subject which needs further investigation before we can pronounce with certainty on a law of connection.

Thus far there is no evidence that the moon directly affects the earth or its inhabitants in any other way than by her attraction, which is so minute as to be entirely insensible except in the ways we have described. A striking illustration of the fallibility of the human judgment when not disciplined by scientific training is afforded by the opinions which have at various times obtained currency respecting a supposed influence of the moon on the weather. Neither in the reason of the case nor in observations do we find any real support for such a theory. It must, however, be admitted that opinions of this character are not confined to the uneducated. In scientific literature several papers are found in which long series of meteorological observations are collated, which indicate that the mean temperature or the amount of rain had been subject to a slight variation depending on the age of the moon. But there was no reason to believe that these changes arose from any other cause than the accidental vicissitudes to which the weather is at all times subject. There is, perhaps, higher authority for the opinion that the rays of the full moon clear away clouds; but if we reflect that the effect of the sun itself in this respect is not very noticeable, and that the full moon gives only $\frac{1}{80000}$ of the heat of the sun, this opinion will appear extremely improbable.

§ 6. *The Planet Mars.*

The fourth planet in the order of distance from the sun, and the next one outside the orbit of the earth, is Mars. Its mean distance from the sun is about 141 millions of miles. The eccentricity of its orbit is such that at perihelion it is only 128 millions of miles from the sun, while in aphelion it is 154 millions distant. It is, next to Mercury, the smallest of the primary planets, its diameter being little more than 4000 miles. It makes one revolution in its orbit in less than two years (more nearly in 687 days, or $43\frac{1}{2}$ days short of two Julian years). If the period were exactly two years, it would make one revolution while the earth made two, and the oppositions would occur at intervals of two years. But, going a little faster than this, it takes the earth, on the average, fifty days over the two years to catch up to it. The times of opposition are shown in the following table:

1873......................April 27th.	1877................September 5th.
1875......................June 20th.	1879................November 12th.

The times of several subsequent oppositions may be found with sufficient exactness for the identification of the planet by adding two years and two months for every opposition, except during the spring months, when only one month is to be added. Oppositions will occur in January, 1882, and February, 1884. At the times of opposition Mars rises when the sun sets, and may be seen during the entire night.

Aspect of Mars.—Mars is easily recognized with the naked eye when near its opposition by its fiery-red light. It is much more brilliant at some oppositions than at others, but always exceeds an ordinary star of the first magnitude. The variations of its brilliancy arise from the eccentricity of its orbit, and the consequent variations of its distance from the earth and the sun. The perihelion of Mars is in the same longitude in which the earth is on August 27th; and when an opposition occurs near that date, the planet is only 35 millions of miles from the earth. This is about the closest approach which the two planets can ever make. When an opposition occurs in February or March the planet is near its aphelion—154 millions of miles from the sun and 62 millions from the earth. The result of these variations of distance is that Mars is more than four times brighter when an opposition occurs in August or September than when it occurs in February or March. The opposition of 1877 (September 5th) is quite remarkable in this respect, as it occurs only nine days after the planet has passed its perihelion. At that time Mars will form a conspicuous object in the south-eastern sky during the early evening.

Mars has been an interesting object of telescopic research from the fact that it is the planet which exhibits the greatest analogy with our earth. The equatorial regions, even with a small telescope, can be distinctly seen to be divided into light and dark portions, which some observers suppose to be continents and oceans. Around each pole is a region of brilliant white, which the same class of astronomers suppose to be due to a deposit of snow. The outlines of the dark and light portions are sometimes so hard to trace as to give rise to the suspicion of clouds in a Martial atmosphere. At the same time, a single look at Mars through a large telescope would convince most observers that these resemblances to our earth have a very small foundation in observation, the evidence being negative rather than positive. It must be said in their favor that if our earth were viewed at the distance at which we view Mars, and with the same optical power, it would present a similar telescopic aspect. But it is also possible that if the optical power of our telescopes were so increased that we could see Mars as from a distance of a thousand miles, the resemblances would all vanish as completely as they did in the case of the moon.

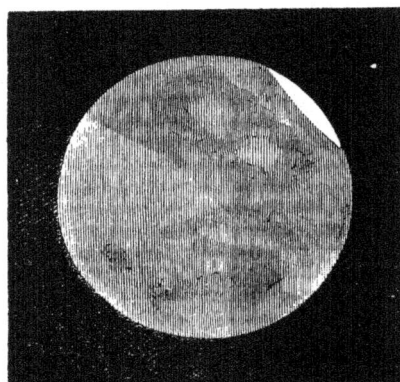

Fig. 82.—The planet Mars on June 23d, 1875, at 10 hours 45 minutes, as seen by Professor Holden with the great Washington telescope.

So many drawings of Mars in various positions have been made by the numerous observers who have studied it, that it has become possible to construct tolerably accurate maps of the surface of the planet. We give a copy of one of these sets of maps by Kaiser, the late Leyden astronomer. Kaiser does not pretend to call the different regions continents and oceans, but merely designates them as light and dark portions.

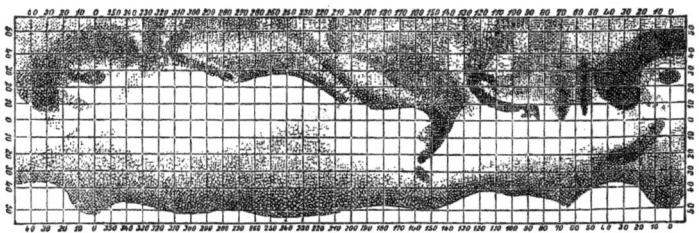

Fig. 83.—Map of Mars, after Kaiser, on Mercator's projection.

Rotation of Mars.—Mars is the only planet besides the earth of which we can be sure that the time of axial rotation admits of being determined with entire precision. Drawings by Hooke, two centuries ago, exhibit markings which can still be recognized, and from a comparison of them with recent ones Mr. Proctor has found for the period of rotation 24 hours 37

THE SMALL PLANETS. 323

minutes 22.73 seconds, which he considers correct within three or four hundredths of a second. The equator of Mars is inclined to the plane of its orbit about 27°, so that the vicissitudes of the seasons are greater on Mars than on the earth in the proportion of 27° to 23½°. Owing to this great obliquity, we can sometimes see one pole of the planet, and sometimes the other, from the earth. When in longitude 350°, that is, in the same

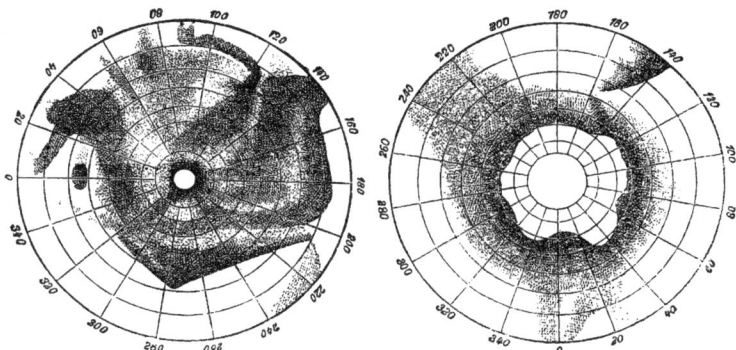

Fig. 84.—Northern hemisphere of Mars.   Fig. 85.—Southern hemisphere of Mars.

direction from the sun in which the earth is situated on September 10th, the south pole of the planet is inclined towards the sun; and if the planet is then in opposition, it will be inclined towards the earth also, so that we can see the region of the planet to a distance of 27° beyond the pole. At an opposition in March the north pole of the planet is inclined towards the sun, and towards the earth also. We have just seen that Mars is much farther at the latter oppositions than at the former, so that we can get much better views of the south pole of the planet than of the north pole.

§ 7. *The Small Planets.*

It was impossible to study the solar system, as it was known to modern astronomy before the beginning of the present century, without being struck by the great gap which existed between Mars and Jupiter. Except this gap, all the planets then known succeeded each other according to a tolerably regular law, and by interpolating a single planet at nearly double the distance of Mars the order of distances would be complete. The idea that an unknown planet might really exist in this region was entertained from the time of Kepler. So sure were some astronomers of this that, in 1800, an association of twenty-four observers was formed, having for its object a systematic search for the planet. The zodiac was divided into twenty-four parts, one of which was to be searched through by each observer. But by one of those curious coincidences which have so frequently occurred in the history of science, the planet was accidentally discovered by an outside astronomer before the society could get fairly to work. On January 1st, 1801, Piazzi, of Palermo, found a star in the constellation Taurus which did not belong there, and on observing it the night after, he found that it had changed its position among the surrounding stars, and must, therefore, be a planet. He followed it for a period of about six weeks, after which it was lost in the rays of the sun without any one else seeing it. When it was time to emerge again in the following autumn, its rediscovery became a difficult problem. But the skill of the great mathematician Gauss came to the rescue with a method by which the orbit of any planetary body could be completely and easily determined from three or four observations. He was thus able to tell observers where their telescopes must be pointed to rediscover the planet, and it was found without difficulty before the end of the year. Piazzi gave it the name *Ceres*. The orbit found by Gauss showed it to revolve between Mars and Jupiter at a little less than double the distance of the former, and therefore to be the long-thought-of planet. But the discovery had a sequel which no one anticipated, and of which we have not yet seen the end. In March, 1802, Olbers discovered a second planet, which was also found to be revolving between Mars and Jupiter, and to which he gave the name *Pallas*. The most extraordinary feature of its orbit was its great inclination, which exceeded 34°. Olbers thereupon suggested his celebrated hypothesis that the two bodies might be fragments of a single planet which had been shattered by some explosion. If such were the case, the orbits of all the fragments would at first intersect each other at the point where the explosion occurred. He therefore thought it likely that other fragments would be found, especially if a search were kept up near the point of intersection of the orbits of Ceres and Pallas. Acting on this idea, Harding, of Lilienthal, found a third planet in 1804, while Olbers found a fourth one in 1807. These were called *Juno* and *Vesta*. The former came quite near to Olbers's theory that the orbits should all pass near the same point, but the latter did not. Olbers continued a search for additional planets of this group for a number of years, but at length gave it up, and died without the knowledge of any but these four.

In December, 1845, thirty-eight years after the discovery of Vesta, Hencke, of Driesen, being engaged in the preparation of star-charts, found a fifth planet of the group, and thus recommenced a series of discoveries which have continued till the present time. No less than three were discovered in 1847, and at least one has been found every year since. To show the rate at which discovery has gone on, we divide the time since 1845 into periods of five years each, and give the number found during each period:

In 1846–50.......... 8 were discovered.	In 1861–65.......... 23 were discovered.
" 1851–55.......... 24 " "	" 1866–70.......... 27 " "
" 1856–60.......... 25 " "	" 1871–75.......... 45 " "

In 1876, 12 were discovered, and three additional ones have been found during the first five months of 1877, making a total of 172 known at the present time (May, 1877).

It will be seen that the rate of discovery has been pretty steadily increasing during thirty years. This is not because the number of those visible, but not yet found, is so great that it is as easy as ever to find one, but because they are now sought after with more skill and more system than formerly.*

---

* In illustration of this the writer has been informed by Professor Peters that in searching for these bodies he falls upon several already known for every new one that he finds. Consequently, were they all lost, he alone could now rediscover them at a more rapid rate than they actually have been discovered by the efforts of all the observers engaged in the search.

Of those discovered during the last ten years, nearly half have been found by two American observers, Professors Peters and Watson. American discoveries of these bodies were commenced by Mr. James Ferguson, who discovered Euphrosyne at Washington on September 1st, 1854.

All the planets of this group are remarkable for their minuteness. The disks are all so small as to defy exact measurement, presenting the appearance of mere stars. A rough estimate of their diameters can, however, be made from the amount of light which they reflect; and although, in the absence of exact knowledge of their reflecting power, the results of this method are not very certain, they are the best we can obtain. It is thus found that Ceres and Vesta are the largest of the group, their diameters lying somewhere between 200 and 400 miles; while, if we omit some very lately discovered, the smallest are Atalanta, Maja, and Sappho, of which the diameters may be between 20 and 40 miles. We may safely say that it would take several thousand of the largest of these small planets to make one as large as the earth.

It has sometimes been said that some of these bodies are of irregular shape, and thus favor Olbers's hypothesis that they are fragments of an exploded planet. But this opinion has no other foundation than a suspected variability of their light, which may be an illusion, and which, if it exists, might result from one side of the planet being darker in color than the other. The latter supposition is not at all improbable, as many of the satellites are known to be variable from this or some analogous cause. As the supposed irregularities of form have never been seen, and are not necessary to account for the variations of brilliancy, there is no sufficient reason for believing in their existence.

Olbers's Hypothesis. — The question whether these bodies could ever have formed a single one has now become one of cosmogony rather than of astronomy. If a planet were shattered, the orbit of each fragment would, at first, pass through the point at which the explosion occurred, however widely they might be separated through the rest of their course. But owing to the secular changes produced by the attractions of the other planets, this coincidence would not continue. The orbits would slowly move away, and after the lapse of a few thousand years no trace of a common intersection would be seen. It is, therefore, curious that Olbers and his contemporaries should have expected to find such a region of intersection, as it implied that the explosion had occurred within a few thousand years. The fact that the required conditions were not fulfilled was no argument against the hypothesis, because the explosion might have occurred millions of years ago, and in the mean time the perihelion and node of each orbit would have made many entire revolutions; so that the orbits would have been completely mixed up.

Desirous of seeing whether the orbits passed nearer a common point of intersection in times past than at present, Encke computed their secular variations. The result seemed to be adverse to Olbers's hypothesis, as it showed that the orbits were farther from having a common point in ages past than at present. But this result was not conclusive either, because he only determined the rates at which the orbits are now changing, whereas, as previously explained, the orbits of all the planets really go through periodic oscillations; and it is only by calculating these oscillations that their positions can be determined for very remote epochs. They have since been determined for some of the planets in question, and the result seems to show that the orbits could never have intersected unless some of them have, in the mean time, been altered by the attraction of the small planets on each other. Such an action is not impossible; but it is impossible to determine it, owing to the great number of these bodies, and our ignorance of their masses. We can, however, say that if the explosion ever did occur, an immense interval, probably millions of years, must have elapsed in the mean time. A different explanation of the group is given by the nebular hypothesis, of which we shall hereafter speak, so that Olbers's hypothesis is no longer considered by astronomers.

The planets in question are distinguished from the others,

not only by their small size, but by the great eccentricities and inclinations of their orbits. If we except Mercury, none of the larger planets has an eccentricity amounting to one-tenth the diameter of its orbit, nor is any orbit inclined more than two or three degrees to the ecliptic. But the inclinations of many of the small planets exceed ten degrees, and the eccentricities frequently amount to a fourth of the radii of their orbits. The result is that the same small planet is at very different distances from the sun in various points of its orbit. Add to this the fact that the mean distances of these bodies from the sun have a pretty wide range, and we shall find that they extend through a quite broad zone. The inside edge of this zone seems pretty well marked, its distance being about 180 millions of miles from the sun, or between 30 and 40 millions beyond the orbit of Mars. On the outside, it terminates more gradually, but nowhere extends within 50 millions of miles of the orbit of Jupiter. If any of the small planets ever ranged outside of certain limits, the attraction of Mars or Jupiter was so great as to completely derange their orbits, so that we have a physical law which sets a limit to the zone; but whether the limit thus set would coincide with the actual limit we cannot at present say.

There are also within the limits of the group certain positions, in which, if the orbits were placed, they would be greatly changed by the action of Jupiter. These positions are those in which the time of revolution would be some simple exact fraction of that of Jupiter, as $\frac{1}{2}$, $\frac{1}{3}$, $\frac{2}{3}$, $\frac{3}{7}$, etc. Professor Daniel Kirkwood has pointed out the curious fact that there are gaps in the series of small planets corresponding to these periodic times. Whether these gaps are really due to the relations of the periodic times, or are simply the result of chance, cannot yet be settled. The fact that quite a number of the small planets have a period very nearly three-eighths that of Jupiter, may lead us to wait for further evidence before concluding that we have to deal with a real law of nature in the cases pointed out by Professor Kirkwood.

Number and Total Mass of the Small Planets.—At present it

THE SMALL PLANETS.

is not possible to set any certain limits to the probable number of the small planets. Although a hundred and seventy-two are now known, there is as yet no sensible diminution in the rate at which they are being discovered. The question of their total number depends very largely on whether there is any limit to their minuteness. If there is no such limit, then there may be an indefinite number of them, too small to be found with the telescopes now engaged in searching for them; and the larger the telescopes engaged in the search, the more will be found. On the other hand, if they stop at a certain limit—say twenty miles in diameter—we may say with considerable confidence that their total number is also limited, and that by far the largest part of them will be discovered by the present generation of astronomers.

So far as we can now see, the preponderance of evidence is on the side of the number and magnitude being limited. The indications in this direction are that the newly discovered ones are not generally the smallest objects which could be seen with the telescopes which have made the discovery, and do not seem, on the average, to be materially smaller than those which were discovered ten years ago. It is not likely that the number of this average magnitude which still remain undiscovered can be very great, and new ones will probably be found to grow decidedly rare before another hundred are discovered. Then it will be necessary to employ greater optical power in the search. If this results in finding a number of new ones too small to be found with the former telescopes, we shall have to regard the group as unlimited in number. But if no such new ones are thus found, it will show that the end has been nearly reached.

In gravitational astronomy, the question of the total mass of the small planets is more important than that of their total number, because on this mass depends their effect in altering the motions of the large planets. Any individual small planet is so minute that its attraction on the other planets is entirely insensible. But it is not impossible that the whole group might, by their combined action, produce a secular variation in the form of the orbits of Mars and Jupiter which, in the course of years, will be clearly shown by the observations. But, although accurate observations of these planets have been made for more than a century, no such effect has yet been noticed. The sum total of their masses must, therefore, be much less than that of an average planet, though we cannot say precisely what the limit is. The apparent magnitude of those which have been discovered is entirely accordant with the opinion that the mass of the entire group is so small that it cannot make itself felt by its attraction on the other planets for many years to come. In fact, if their diameters be estimated from their brightness, in the manner already indicated, we shall find that if all that are yet known were made into a single planet the diameter would be less than 400 miles; and if a thousand more, of the average size of those discovered since 1850 should exist, their addition to the consolidated planet would not increase its diameter to 500 miles. Such a planet would be only $\frac{1}{4000}$ of the bulk of the earth, and, unless we supposed it to possess an extraordinary specific gravity, could not much exceed $\frac{1}{4000}$ of the mass of the earth, or $\frac{1}{20}$ of the mass of Mercury. We may fairly conclude that unless the group of small planets actually consists of tens of thousands of minute bodies, of which only a few of the brightest have yet been discovered, their total volume and mass are far less than those of any one of the major planets.

The number of these bodies now known is so great that the mere labor of keeping the run of their motions, so that they shall not be lost, is out of proportion to the value of its results. It is mainly through the assiduity of German students that most of them are kept from being lost. Should many more be found, it may be necessary to adopt the suggestion of an eminent German astronomer, and let such of them as seem unimportant go again, and pursue their orbit undisturbed by telescope or computer.

CHAPTER IV.

THE OUTER GROUP OF PLANETS.

§ 1. *The Planet Jupiter.*

JUPITER is the "giant planet" of our system, his mass largely exceeding that of all the other planets combined. His mean diameter is about 85,000 miles; but owing to his rapid rotation on his axis, his equatorial exceeds his polar diameter

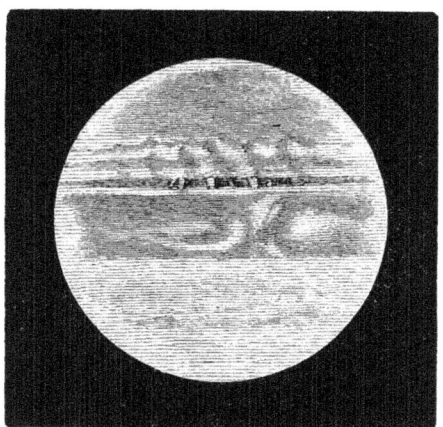

FIG. 86.—Jupiter as seen with the great Washington telescope, March 21st, 1876, 15 hours 38 minutes mean time. Drawn by Professor Holden.

by 5000 miles. In volume he exceeds our earth about 1300 times, while in mass he exceeds it about 213 times. His specific gravity is, therefore, far less than that of the earth, and even less than that of water. His mean distance from the sun is 480 millions of miles, but, owing to the eccentricity of his orbit, his actual distance ranges between 457 and 503 millions. His time of revolution is fifty days less than twelve years.

Jupiter is easily recognized by his brilliant white light, with which he outshines every other planet except Venus. To facilitate his recognition, we give the dates of opposition during a few years.

1877	June 19th.	1879	August 31st.
1878	July 25th.	1880	October 7th.

During the four years following 1880 he will be in opposition, on the average, about a month and seven days later each year; namely, in the middle of November, 1881; towards the latter part of December, 1882, and so on. A month or two before opposition he can be seen rising late in the evening, while during the three months following opposition he will always be seen in the early evening somewhere between south-east and south-west.

The Surface of Jupiter.—Except the sun and moon, there is no object of our system which has during the last few years been the subject of more careful examination than this planet. Unlike Mars, there are no really permanent markings on his surface, and a map of Jupiter is therefore impossible. But this surface always presents a very diversified appearance. The earlier telescopic observers described light and dark belts as extending across it. Until a quite recent period, it has been customary to describe these belts as two in number, one north of the equator, and the other south of it. Commonly, they are seen as dark bands on the bright disk of the planet; but it is curious that Huyghens represents them as brighter than the rest of the surface. As telescopic power was increased, it was seen that these so-called bands were of a far more complex structure than had been supposed, and consisted of great numbers of stratified, cloud-like appearances of the most variegated forms. These forms change so rapidly that the face of the planet hardly ever presents the same appearance on two successive nights. They are most strongly marked at some distance on each side of the Jovian equator, and thus give rise to the appearance of two belts when a very small or imperfect telescope is used.

Both the outlines of these belts and the color of some parts of the planet, seem subject to considerable changes. The equatorial regions, and indeed the spaces between the belts generally, are often of a rosy tinge. This coloring is sometimes so strongly marked as to be evident to the most superficial observer, while at other times hardly a trace of it can be seen.

Spots which are much more permanent than the ordinary markings on the belt are sometimes visible. By watching these spots from day to day, and measuring their distance from the apparent disk, the time of rotation of Jupiter on his axis has been determined. Commonly the spots are dark; but on some rather rare occasions the planet is seen with a number of small, round, bright spots like satellites. Of these bright spots no explanation has been given.

FIG. 87.—View of Jupiter, as seen in Lord Rosse's great telescope on February 27th, 1861, at 12 hours 30 minutes.

From the changeability of the belts, and indeed of nearly all the visible features on the surface of Jupiter, it is clear that what we see on that planet is not the surface of a solid nucleus, but vaporous or cloud-like formations which cover the entire surface and extend to a great depth below. To all appearance, the planet is covered with a deep and dense atmosphere, through which light cannot penetrate on account of thick masses of clouds and vapor. In the arrangements of these clouds in streaks parallel to the equator, and in the change of their forms with the latitude, there may be something analogous to the zones of clouds and rain on the earth. But of late years it has been noticed that the physical constitution of Jupiter seems to offer more analogies to that of the sun than to that of the earth. Like the sun, he is brighter in the centre than near the edges. This is shown in the most striking manner in the transits of his satellites over his disk. When the satellite first enters on the disk, it commonly seems like a bright spot on a dark background; but as it approaches the centre, it appears like a dark spot on the bright background of the planet. The brightness of the centre is probably two or three times greater than that of the limb. This diminution of light towards the edge may arise, as in the case of the sun, from the light near the edge passing through a greater depth of atmosphere, and thus becoming fainter by absorption.

A still more remarkable resemblance to the sun has sometimes been suspected—nothing less, in fact, than that Jupiter shines partly by his own light. It was at one time supposed that he actually emitted more light than fell upon him from the sun; and if this were proved, it would show conclusively that he was self-luminous. If all the light which the sun shed upon the planet were equally reflected in every direction, we might speak with some certainty on this question; but in the actual state of our knowledge we cannot. Zöllner has found that the brightness of Jupiter may be accounted for by supposing him to reflect 62 per cent. of the sunlight which he receives. But if this is his average reflecting power, the reflecting power of his brighter portions must be much greater; in fact, they are so bright that they must shine partly by their own light, unless they reflect a disproportionate share of the sunlight back in the direction of the earth and sun. Clouds would not be likely to do this. On the other hand, if we assume that the planet emits any great amount of light, we are

THE PLANET JUPITER. 335

met by the fact that, if this were the case, the satellites would shine by this light when they were in the shadow of the planet. As these bodies totally disappear in this position, the quantity of light emitted by Jupiter must be quite small. On the whole, there is a small probability that the brighter spots of this planet are from time to time slightly self-luminous.

Again, the interior of Jupiter seems to be the seat of an activity so enormous that we can attribute it only to a very high temperature, like that of the sun. This is shown by the rapid movements always going on in his visible surface, which frequently changes its aspect in a few hours. Such a powerful effect could hardly be produced by the rays of the sun, because, owing to the great distance of the planet, he receives only between one-twenty-fifth and one-thirtieth of the light and heat which we do. It is therefore probable that Jupiter is not yet covered by a solid crust, as our earth is, but that his white-hot interior, whether liquid or gaseous, has nothing to cover it but the dense vapors to which that heat gives rise. In this case the vapors may be self-luminous when they have freshly arisen from the interior, and may rapidly cool off after reaching the upper limit to which they ascend.

Rotation of Jupiter.—Owing to the physical condition of Jupiter, no precisely determinate time of rotation can be assigned him, as in the case of Mars. Without a solid crust which we can see from time to time, the observed times of rotation will be those of liquid or vaporous formations, which may have a proper motion of their own. A spot has, however, on some occasions been observed for several months, and it has thus been pretty certainly determined that the time of rotation is about 9 hours $55\frac{1}{2}$ minutes. The first observation of a spot of this kind was made by Cassini, who found the time of rotation to be 9 hours 55 minutes 58 seconds. No further exact observations were made until the time of Schröter, who observed a number of transient spots during 1785 and 1786. The times of rotation varied from 9 hours 55 minutes to 9 hours 56 minutes, from which he concluded that heavy storms raged on the surface of the planet, and gave the cloudy masses which formed the spots a motion of their own. In November, 1834, a remarkable spot was observed by Mädler, of Dorpat, which lasted until the following April, from which the time of rotation came out 9 hours 55 minutes 30 seconds; but the observations showed that the spot did not move uniformly. Professor Airy, who observed the same spot at Cambridge, found the period to be 9 hours 55 minutes 21.3 seconds.

Recent observations and researches indicate that the equatorial regions of Jupiter rotate in less time, and with more irregularity, than the others, thus showing still another analogy between that planet and the sun. Thus, in 1871, Dr. Lohse, of Bothkamp, observed a spot near Jupiter's equator, which during several days performed its revolution in a period of 9 hours 51 minutes 47 seconds. Other equatorial spots had a very irregular motion, but their period was generally less than that found by Mädler and Airy.

§ 2. *The Satellites of Jupiter.*

One of the earliest telescopic discoveries by Galileo was that Jupiter was accompanied by four satellites, which revolved round him as a centre, thus forming a miniature copy of the solar system. As in the case of spots on the sun, Galileo's announcement of this discovery was received with incredulity by those philosophers of the day who believed that everything in nature was described in the writings of Aristotle. One eminent astronomer — Clavius — said that to see the satellites one must have a telescope which would produce them; but he changed his mind as soon as he saw them himself. Another philosopher, more prudent, refused to put his eye to the telescope lest he should see them and be convinced. He died shortly afterwards. "I hope," said the caustic Galileo, "that he saw them while on his way to heaven."

A very small telescope, or even a good opera-glass, is sufficient to show these bodies. Indeed, very strong evidence is on record that they have been seen with the naked eye. That they could be seen by any good eye, if the planet were out of the way, there is no doubt, the difficulty in seeing them arising from the glare of the planet on the eye. If the lenses of the eye are so transparent and pure that there is no such glare, it is quite possible that the two outer satellites might be seen, especially if they should happen to be close together.

According to the best determinations, which are, however, by no means certain, the diameters of the satellites of Jupiter range between 2200 and 3700 miles, the third from the planet being the largest, and the second the smallest. The volume of the smallest is, therefore, very near that of our moon.

The light of these satellites varies to an extent which it is difficult to account for, except by supposing very violent changes constantly going on on their surfaces. It has sometimes been supposed that some of them, like our moon, always present the same face to Jupiter, and that the changes in their brilliancy are due to differences in the color of the parts of the satellites which are successively turned towards us during one revolution round the planet. But the careful measures of their light made by Auwers, of Berlin, and Engelmann, of Leipsic, show that this hypothesis does not account for the changes of brilliancy, which are sometimes sudden in a surprising degree. The satellites are so distant as to elude telescopic examination of their surfaces. We cannot, therefore, hope to give any certain explanation of these changes.

The satellites of Jupiter offer problems of great difficulty to the mathematician who attempts to calculate the effect of their mutual attractions. The secular variations of their orbits are so rapid that the methods applied in the case of the planets cannot be applied here without material alterations. The most curious and interesting effect of their mutual attraction is that there is a connection between the motions of the three inner satellites such as exists nowhere else in the solar system. The connection is shown by these two laws:

1. *That the mean motion of the first satellite added to twice the mean motion of the third is exactly equal to three times the mean motion of the second.*

2. *That if to the mean longitude of the first satellite we add*

338   THE SOLAR SYSTEM.

twice the mean longitude of the third, and subtract three times the mean longitude of the second, the difference is always 180°.

The first of these relations is shown in the following table of the mean daily motions of the satellites:

Satellite I. in one day moves	203°.4890
" II. " "	101°.3748
" III. " "	50°.3177
" IV. " "	21°.5711
Motion of Satellite I.	203°.4890
Twice that of Satellite III.	100°.6354
Sum	304°.1244
Three times motion of Satellite II.	304°.1244

It was first found from observations that the three satellites moved together so nearly according to this law that no certain deviation could be detected. But it was not known whether this was a mere chance coincidence, or an actual law of nature, till Laplace showed that, if they moved so nearly in this way as observations had shown them to, there would be an extremely minute force arising from their mutual gravitation, sufficient to keep them in this relative position forever. There is, in this case, some analogy to the rotation of the moon, which, being once started presenting the same face to the earth, is always held in that position by a minute residual of the earth's attraction.

We have already spoken of the discovery of the progressive motion of light from the eclipses of these satellites, and of the uses of these eclipses for the rough determination of longitudes. Both the eclipses, and the transits of their bodies over the face of Jupiter afford interesting subjects of observation with a telescope of sufficient power, say four inches aperture or upwards. To facilitate such observations the times of these phenomena are predicted in both the American and British Nautical Almanacs.

§ 3. *Saturn and its System, Physical Aspect, Belts, Rotation.*

Saturn is the sixth of the major planets in the order of distance from the sun, around which it revolves in $29\frac{1}{2}$ years at a mean distance of about 880 millions of miles. In mass and size it stands next to Jupiter. To show the disparity in the masses of the planets we may refer to the table already given, showing that although Saturn is not one-third the mass of Jupiter, it has about three times the mass of the six planets, which are smaller than itself put together. Its surroundings are such as to make it the most magnificent object in the solar system. While no other planet is known to have more than

FIG. 88.—View of Saturn and his rings.

four satellites, Saturn has no less than eight. It is also surrounded by a pair of rings, the interior diameter of which is about 100,000 miles. The aspect of these rings is subject to great variations, for reasons which will soon appear. The great distance of the planet renders the study of its details difficult unless the highest telescopic power is applied. The whole combination of Saturn, his rings, and his satellites is often called the *Saturnian System.*

The planet Saturn generally shines with the brilliancy of a

moderate first-magnitude star, and with a dingy, reddish light, as if seen through a smoky atmosphere. Its apparent brightness is, however, different at different times: during the years 1876–1879 it is fainter than the average, owing to its ring being seen nearly edgewise. From 1878 till 1885 it will constantly grow brighter, on account both of the opening out of the ring and the approach of the planet to its perihelion. The times of opposition are as follow:

| 1877.............September 9th. | 1879.................October 5th. |
| 1878.............September 22d. | 1880.................October 18th. |

In subsequent years opposition will occur about thirteen days later every year, so that by adding this amount to the date for each year the oppositions can be found until the end of the century without an error of more than a few days.

The physical constitution of Saturn seems to bear a great resemblance to that of Jupiter; but, being twice as far away, it cannot be so well studied. The farther an object is from the sun, the less brightly it is illuminated; and the farther from the earth, the smaller it looks, so that there is a double difficulty in getting the finest views of the more distant planets. When examined under favorable circumstances, the surface of Saturn is seen to be diversified with very faint markings; and if high telescopic powers are used, two or more very faint streaks or belts may be seen parallel to its equator, the strongest ones lying on, or very near, the equator. As in the case of Jupiter, these belts change their aspect from time to time, but they are so faint that the changes cannot be easily followed. It is therefore, in general, difficult to say with certainty whether we do or do not see the same face of Saturn on different nights; and, consequently, it is only on extraordinary occasions that the time of rotation can be determined.

The first occasion on which a well-defined spot was known to remain long enough on Saturn to determine the period of its rotation was in the time of Sir W. Herschel, who, from observations extending over several weeks, found the time of rotation to be 10 hours 16 minutes.* No further opportunity for determining this period seems to have offered itself until 1876, when an appearance altogether new suddenly showed itself on the globe of this planet. On the evening of December 7th, 1876, Professor Hall, who had been engaged in measures of the satellites of Saturn with the great Washington telescope, saw a brilliant white spot near the equator of the planet. It seemed as if an immense eruption of white-hot matter had suddenly burst up from the interior. The spot gradually spread itself out in the direction which would be east on the planet, so as to assume the form of a long light streak, of which the brightest point was near the following end. It continued visible until January, when it became faint and ill-defined, and the planet was lost in the rays of the sun.

Immediately upon the discovery of this remarkable phenomenon, messages were sent to other observers in various parts of the country, and on the 10th it was seen by several observers, who noted the time at which it crossed the centre of the disk in consequence of the rotation of the planet. From all the observations of this kind, Professor Hall found the period of Saturn to be 10 hours 14 minutes, taking the brightest part of the streak, which, as we have said, was near one end. Had the middle of the streak been taken, the time would have been less, because the bright matter seemed to be carried along in the direction of the planet's rotation. Attributing this to a wind, the velocity of the latter would have been between 50 and 100 miles an hour.

§ 4. *The Rings of Saturn.*

The most extraordinary feature of Saturn is the magnificent system of rings by which he is surrounded. To the early telescopists, who could not command sufficient optical power to see exactly what it was, this feature was a source of great

---

* It is very curious that nearly all modern writers give about 10 hours 29 minutes as the time of rotation of Saturn which Herschel finally deduced. I can find no such result in Herschel's papers. A suspicious coincidence is that this period agrees with that assigned for the time of rotation of the ring.

perplexity and difference of opinion. To Galileo it made the planet appear triform—a large globe with two small ones affixed to it, one on each side. After he had observed it for a year or two, he was greatly perplexed to find that the appendages had entirely disappeared, leaving Saturn a single round globe, like the other planets. His chagrin was heightened by the fear, not unnatural under the circumstances, that the curious form he had before seen might be due to some optical illusion connected with his telescope. It is said (I do not know on what authority) that his annoyance at the supposed deception into which he had fallen was so great that he never again looked at Saturn.

A very few years sufficed to show other observers, who had command of more powerful telescopes, that the singularity of form was no illusion, but that it varied from time to time. We give several pictures from Huyghens's *Systema Saturnium*, showing how it was represented by various observers during the first forty years of the telescope. If the reader will compare these with the picture of Saturn and his rings as they actually are, he will see how near many of the observers came to a representation of the proper apparent form, though none divined to what sort of an appendage the appearance was due.

The man who at last solved the riddle was Huyghens, of whose long telescopes we have already spoken. Examining Saturn in March and April, 1655, he saw that instead of the appendages presenting the appearance of curved handles, as in previous years, a long narrow arm extended straight out on each side of the planet. The spring following, this arm had disappeared, and the planet appeared perfectly round as Galileo had seen it in 1612. In October, 1655, the handles had reappeared, much as he had seen them a year and a half before. To his remarkably acute mathematical and mechanical mind this mode of disappearance of the handles sufficed to suggest the cause which led to their apparent form. Waiting for entire confirmation by future observations, he communicated his theory to his fellow-astronomers in the following com-

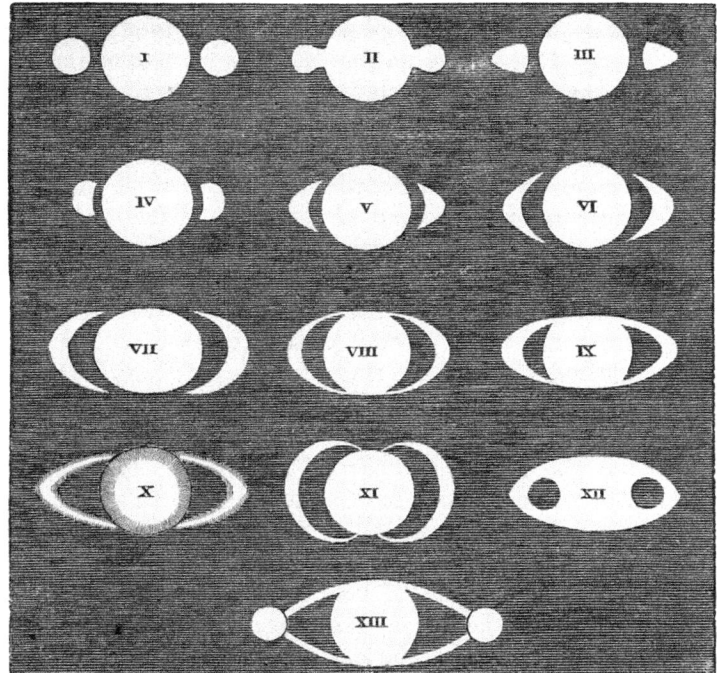

FIG. 89.—Specimens of drawings of Saturn by various observers before the rings were recognized as such: I. Form as given by Galileo in 1610; II. Drawing by Scheiner, in 1614, "showing ears to Saturn;" III. Drawing by Ricciolus, in 1640 and 1643; IV.,V., VI., and VII. are by Hevelius, and show the changes due to the different angles under which the rings were seen; VIII. and IX. are by Ricciolus, between 1648 and 1650, when the ring was seen at the greatest angle; X. is by a Jesuit who passed under the pseudonym of *Eustachius de Divinis;* XI. is by Fontana; XII. by Gassendi and Blancanus, and XIII. by Ricciolus.

bination of letters, printed without explanation at the end of a little pamphlet on his discovery of the satellite of Saturn:

aaaaaaa ccccc d eeeee g h iiiiiii llll mm nnnnnnnnn oooo pp q rr s ttttt uuuuu,

which, properly arranged, read—

"*Annulo cingitur, tenui, plano, nusquam cohærente, ad eclipticam inclinato*" (It is girdled by a thin plane ring, nowhere touching, inclined to the ecliptic).

This description is remarkably complete and accurate; and enabled Huyghens to give a satisfactory explanation of the various phases which the ring had assumed as seen from the earth. Owing to the extreme thinness and flatness of the object, it was completely invisible in the telescopes of that time when its edge was presented towards the observer or towards the sun. This happens twice in each revolution of Saturn, in much the same way that the earth's equator is twice directed towards the sun in the course of the year. The ring is inclined to the plane of the planet's orbit by 27°, corresponding to the angle of $23\frac{1}{2}°$ between the earth's equator and the ecliptic. The general aspect from the earth is very near the same as from the sun. As the planet revolves around the sun, the axis and plane of the ring preserve the same absolute direction in space, just as the axis of the earth and the plane of the equator do.

When the planet is in one part of its orbit, an observer at the sun or on the earth will see the upper or northern side of the ring at an inclination of 27°. This is the greatest angle at which the ring can ever be seen, the position occurring when the planet is in 262° of longitude, in the constellation Sagittarius. When the planet has moved through a quarter of a revolution, the edge of the ring is turned towards the sun, and, owing to its extreme thinness, it is visible only in the most powerful telescopes as an exceedingly fine line of light, stretching out on each side of the planet. In this position the planet is in longitude 352°, in the constellation Pisces. When the planet has moved 90° farther, an observer on the sun or earth again sees the ring at an angle of 27°; but now it is the lower or southern side which is visible. The planet is now in longitude 82°, between the constellations Taurus and Gemini. When it has moved 90° farther, to longitude 172°, in the constellation Leo, the edge of the ring is again turned towards the earth and sun.

Thus there are a pair of opposite points of the orbit of Saturn in which the rings are turned edgewise to us, and another pair half-way between the first in which the ring is seen at its maximum inclination of about 27°. Since the planet performs a revolution in $29\frac{1}{2}$ years, these phases occur at average intervals of about seven years and four months. The following are some of the times of their occurrence:

1870. The planet being between Scorpio and Sagittarius, the ring was seen open to its greatest breadth, the north side being visible. The same phase recurs at the end of 1899.

1878 (February 7th). The edge of the ring is turned towards the sun, so that only a thin line of light will be visible. The planet is then between Aquarius and Pisces.

1885. The planet being in Taurus (the Bull) the south side of the rings will be seen at the greatest elevation.

1892. The edge of the ring is again turned towards the sun, the planet being in Leo (the Lion).

Owing to the motion of the earth, the times when the edge of the ring is turned towards it do not accurately correspond to those when it is turned towards the sun, and the points of Saturn's orbit in which this may occur range over a space of several degrees. The most interesting times for viewing the rings with powerful telescopes are on those rare occasions when the sun shines on one side of the ring, while the dark side is directed towards the earth. On these occasions the plane of the ring, if extended out far enough, would pass between the sun and the earth. This will be the case between February 9th and March 1st, 1878; but, unfortunately, at that time the earth and Saturn are on opposite sides of the sun, so that the planet is nearly lost in the sun's rays, and can be observed only low down in the west just after sunset. In 1891 the position of Saturn will be almost equally unfavorable for the observation in question, as it can be made only in the early mornings of the latter part of October of that year, just after Saturn has risen. In fact, a good opportunity will not occur till 1907. In northern latitudes the finest telescopic views of Saturn and his ring may be obtained between 1881 and 1889, because during that interval Saturn passes his perihelion, and also the point of greatest northern declination, while the ring is opened out to its widest extent. In fact, these three most favorable conditions all fall nearly together during the years 1881–'85.

After Huyghens, the next step forward in discoveries on Saturn's ring was made by an English observer, named Ball, otherwise unknown in astronomy, who found that there were really two rings, divided by a narrow dark line. The breadth of the rings is very unequal, the inner ring being several times broader than the outer one. A moderate-sized telescope is sufficient to show this division near the extreme points of the ring if the atmosphere is steady; but it requires both a large telescope and fine seeing to trace it all the way across that part of the ring which is between the observer and the ball of the planet. Other divisions, especially in the outer ring, have at times been suspected by various observers, but if they really existed, they must have been only temporary, forming and closing up again.

In December, 1850, the astronomical world was surprised by the announcement that Professor Bond, of Cambridge, had discovered a third ring to Saturn. It lay between the rings already known and the planet, being joined to the inner edge of the inner ring. It had the appearance of a ring of crape, being so dark and obscure that it might easily have been overlooked in smaller telescopes. It was seen in England by Messrs. Lassell and Dawes before it was formally announced by the Bonds. Something of the kind had been seen by Dr. Galle, at Berlin, as far back as 1838; but the paper on the subject by Encke, the director of the observatory, did not describe the appearance very clearly. Indeed, on examining the descriptions of observers in the early part of the eighteenth century, some reason is found for suspecting that they saw this dusky ring; but none of the descriptions are sufficiently definite to establish the fact, though it is strange if an object so plain as this ring now is should have been overlooked by all the older observers.

The question whether changes of various sorts are going on in the rings of Saturn is one which is still unsettled. There is some reason to believe that the supposed additional divisions noticed in the rings from time to time are only errors of vision, due partly to the shading which is known to exist on

THE RINGS OF SATURN. 347

various parts of the ring. By reference to the diagram of Saturn, it will be seen that the outer ring has a shaded line extending around it about two-thirds of the way from its inner to its outer edge. This line, however, is not fine and sharp, like the known division, but seems to shade off gradually towards each edge. As observers who have supposed themselves to see a division in this ring saw it where this shaded line is, and do not speak of the latter as anything distinct from the former, there is reason to believe that they mistook this permanent shading for a new division. The inner ring is brightest near its outer edge, and shades off gradually towards its inner edge. Here the dusky ring joins itself to it, and extends about half-way in to the planet.

As seen with the great Washington equatorial in the autumn of 1874, there was no great or sudden contrast between the inner or dark edge of the bright ring and the outer edge of the dusky ring. There was some suspicion that the one shaded into the other by insensible gradations. No one could for a moment suppose, as some observers have, that there was a separation between these two rings. All these considerations give rise to the question whether the dusky ring may not be growing at the expense of the inner bright ring.

A most startling theory of changes in the rings of Saturn was propounded by Struve, in 1851. This was nothing less than that the inner edge of the ring was gradually approaching the planet in consequence of the whole ring spreading inwards, and the central opening thus becoming smaller. The data on which this theory was founded were the descriptions and drawings of the rings by the astronomers of the seventeenth century, especially Huyghens, and the measures executed by later astronomers up to the time at which Struve wrote. The rate at which the space between the ring and the planet was diminishing seemed to be about $1''.3$ per century. The following are the numbers used by Struve, which are deduced from the descriptions by the ancient observers, and the measures by the modern ones:

	Year.	Distance between Ring and Planet.	Breadth of 1 i .g.
Huyghens........................	1657	6.5	4.6
Huyghens and Cassini.......	1695	6.0	5.1
Bradley...........................	1719	5.4	5.7
Herschel..........................	1799	5.12	5.98
W. Struve.......................	1826	4.36	6.74
Encke and Galle................	1838	4.04	7.06
Otto Struve.....................	1851	3.67	7.43

If these estimates and measures were certainly accurate, they would place the fact of a progressive approach of the rings to the ball beyond doubt, an approach which, if it continued at the same rate, would bring the inner edge of the ring into contact with the planet about the year 2150. But in measuring such an object as the inner edge of the ring of Saturn, which, as we have just said, seems to fade gradually into the obscure ring, different observers will always obtain different results, and the differences among the four observers commencing with W. Struve are no greater than are often seen in measuring an object of such uncertain outline. Hence, considering the great improbability of so stupendous a cosmical change going on with so much rapidity, Struve's theory has always been viewed with doubt by other astronomers.

At the same time, it is impossible to reconcile the descriptions by the early observers with the obvious aspect of the ring as seen now without supposing some change of the kind. The most casual observer who now looks at Saturn will see that the breadth of the two bright rings together is at least half as great again, if not twice as great, as that of the dark space between the inner edge of the bright ring and the planet. But Huyghens describes the dark space as about equal to the breadth of the ring, or a little greater. Supposing the ring the same then as now, could this error have arisen from the imperfection of his telescope? No; because the effect of the imperfection would have been directly the opposite. The old telescopes all represented planets and other bright objects too large, and therefore would show dark spaces too small, owing to the irradiation produced by their imperfect glasses. A strong confirmation of Struve's view is found in the old pictures given in Fig. 89 by those observers who could not clearly make out the ring. In nearly all cases the dark spaces were more conspicuous than the edges of the ring. But if we now look at Saturn through a very bad atmosphere, though the elliptical outline of the ring may be clearly made out, the dark space will be almost obliterated by the encroachment of the light of the planet and ring upon it. The question is, therefore, one of those the complete solution of which must be left to future observers.

§ 5. *Constitution of the Ring.*

The difficulties which investigators have met with in accounting for the rings of Saturn are of the same nature as those we have described as arising from spectroscopic discoveries respecting the envelopes of the sun. They illustrate the philosophic maxim that surprise—in which term we may include all difficulty and perplexity which men meet with in seeking to account for the phenomena of nature—is a result of partial knowledge, and cannot exist either with entire ignorance or complete knowledge. Those who are perfectly ignorant are surprised at nothing, because they expect nothing, while perfect knowledge of what is to happen also precludes the same feeling. The astronomers of two centuries ago saw nothing surprising in the fact of a pair of rings surrounding a planet, and accompanying it in its orbit, because they were not acquainted with the effects of gravitation on such bodies as the rings seemed to be. But when Laplace investigated the subject, he found that a homogeneous and uniform ring surrounding a planet could not be in a state of stable equilibrium. Let it be balanced ever so nicely, the slightest external force, the attraction of a satellite or of a distant planet, would destroy the equilibrium, and the ring would soon be precipitated upon the planet. He therefore remarked that the rings must have irregularities in their form, such as Herschel supposed he had seen; but he did not investigate the question whether with those irregularities the equilibrium would really be stable.

The question was next taken up in this country by Professors Peirce and Bond. The latter started from the supposed result of observations—that new divisions show themselves from time to time in the ring, and then close up again. He thence inferred that the rings must be fluid, and, to confirm this view, he showed the impossibility of even an irregular solid pair of rings fulfilling all the necessary conditions of stability and freedom of motion. Professor Peirce, taking up the same subject from a mathematical point of view, found that no conceivable form of irregular solid ring would be in a state of stable equilibrium; he therefore adopted Bond's view that the rings were fluid. Following up the investigation, he found that even a fluid ring would not be entirely stable without some external support, and he attributed that support to the attractions of the satellites. But as Laplace did not demonstrate that irregularities would make the ring stable, so Peirce merely fell back upon the attraction of the satellites as a sort of forlorn hope, but did not demonstrate that the fluid ring would really be stable under the influence of their attraction. Indeed, it now seems very doubtful whether this attraction would have the effect supposed by Peirce.

The next, and, we may say, the last, important step was taken by Professor J. Clerk Maxwell, of England, in the Adams prize essay for 1856. He brought forward objections which seem unanswerable against both the solid and the fluid ring, and revived a theory propounded by Cassini about the beginning of the last century.* This astronomer considered the ring to be formed by a cloud of satellites, too small to be separately seen in the telescope, and too close together to admit of the intervals between them being visible. This is the view of the constitution of the rings of Saturn now most generally adopted. The reason why the ring looks solid and continuous is that the satellites are too small and too numerous to be seen singly. They are like the separate little drops of

---

* See Memoirs of the French Academy of Sciences for 1715, p. 47; or Cassini's "Élémens d'Astronomie," p. 338, Paris, 1740.

water of which clouds and fog are composed, which, to our eyes, seem like solid masses. In the dusky ring the particles may be so scattered that we can see through the cloud, the reason that it looks dusky being simply the comparatively small number of the particles, so that to the distant eye they appear like the faint stippling of an engraving.

The question arises whether the comparative darkness of some portions of the bright ring may not be due to the paucity of the particles, which allows the dark background of the sky to be seen through. This question cannot be positively answered until further observations are made; but the preponderance of evidence favors the view that the entire bright ring is opaque, and that the dark shading is due entirely to a darker color of that part of the ring. Indeed, for anything we certainly know, the whole ring may be continuous and opaque, the darker shade of some parts arising solely from the particles being there black in color. The only way to settle conclusively the questions whether these parts of the ring look black, owing to the sky beyond showing through openings, as it were, or from a black color of the ring, is to find whether a star or other object can be seen through the dark spaces. But an opportunity for seeing a bright star through the ring has never yet presented itself. The most obvious way of settling the question in respect to the dusky ring is to notice whether the planet itself can be seen through it; but this is much more difficult than might be supposed, owing to the ill-defined aspect of the ring. The testimony of both Lassell and Trouvelot is in favor of the view that this ring is partially transparent; but their observations will need to be repeated when the ring is opened out to our sight after 1882.

§ 6. *The Satellites of Saturn.*

When Huyghens commenced his observations of Saturn in 1655, he saw a star near the planet which a few days' observation enabled him to recognize as a satellite revolving round it in about fifteen days. In his "*Systema Saturnium*," he ventured to express the opinion that this discovery completed the solar system, which now comprised six planets (Saturn being then the outermost known planet) and six satellites (one of the earth, four of Jupiter, and this one of Saturn), making the perfect number of twelve. He was, therefore, confident that no more satellites were left to discover, and through failing to search for others, he probably lost the honor of additional discoveries.

Twelve years after this prediction, Cassini discovered a second satellite outside that found by Huyghens, and within a few years more he found three others inside of it. The discovery of four satellites by one astronomer was so brilliant a result of French science that the Government of France struck a medal in commemoration of it, bearing the inscription *Saturni Satellites primum cogniti*. These five satellites completed the number known for more than a century. In 1789 Herschel discovered two new ones still nearer the ring than those found by Cassini. The space between the ring and the inner one is so small that the satellite is generally invisible, even in the most powerful telescopes. Finally, in September, 1848, the Messrs. Bond, at the Observatory of Harvard College, found an eighth satellite, while examining the ring of Saturn. By a singular coincidence, this satellite was found by Mr. Lassell, of England, only a couple of nights after it was detected by the Bonds. The names which have been given to these bodies are shown in the following list, in which the satellites are arranged in the order of their distance from the planet. The distances are given in semidiameters of Saturn. More exact elements will be found in the Appendix to this volume.

No.	Name.	Distance from Planet.	Discoverer.	Date.
1	Mimas.....	3.3	Herschel..	1789, September 17th.
2	Enceladus.	4.3	Herschel..	1789, August 28th.
3	Tethys.....	5.3	Cassini....	1684, March.
4	Dione......	6.8	Cassini....	1684, March.
5	Rhea.......	9.5	Cassini....	1672, December 23d.
6	Titan.......	20.7	Huyghens.	1655, March 25th.
7	Hyperion .	26.8	Bond.......	1848, September 16th.
8	Japetus....	64.4	Cassini....	1671, October.

The brightness, or rather, the visibility, of these satellites follows the same order as their discovery. The smallest telescope will show Titan, and one of very moderate size will show Japetus in the western part of its orbit. Four or five inches aperture will show Rhea, and perhaps Tethys and Dione, while seven or eight inches are required for Enceladus, and even with that aperture it will probably be seen only near its greatest elongation from the planet. Mimas can be seen only near the same position, unless the ring is seen edgewise, and will then require a large telescope, probably twelve inches or upwards. Finally, Hyperion can be recognized only with the most powerful telescopes; not only on account of its faintness, but of the difficulty of distinguishing it from minute stars.

All these satellites, except Japetus, revolve very nearly in the plane of the ring. Consequently, when the edge of the ring is turned towards the earth, the satellites seem to swing from one side of the planet to the other in a straight line, running along the thin edge of the ring, like beads on a string. This phase affords the best opportunity of seeing the inner satellites Mimas and Enceladus, because they are no longer obscured by the brilliancy of the ring.

Japetus, the outer satellite of all, exhibits this remarkable peculiarity, that while in one part of its orbit it is the brightest of the satellites, except Titan, in the opposite part it is almost as faint as Hyperion, and can be seen only in large telescopes. When west of the planet, it is bright; when east of it, faint. This peculiarity has been accounted for only by supposing that the satellite, like our moon, always presents the same face to the planet, and that one side of it is white and the other intensely black. The only difficulty in the way of this explanation is that it is doubtful whether any known substance is so black as one side of the satellite must be to account for such great changes of brilliancy.

§ 7. *Uranus and its Satellites.*

Uranus, the next planet beyond Saturn, is at a mean distance from the sun of about 1770 millions of miles, and performs a revolution in 84 years. It shines as a star of the sixth magnitude, and can therefore be seen with the naked eye, if one knows exactly where to look for it. It was in opposition February 11th, 1877, and the time of opposition during the remainder of the present century may be found by adding $4\frac{1}{2}$ days for every year subsequent to 1877. To find it readily, either with a telescope or the naked eye, recourse must be had to the *Nautical Almanac*, where the position (right ascension and declination) is given for each day in the year.

Of course the smallest telescopes will show this planet as a star, but to recognize its disk a magnifying power of at least 100 should be used, and 200 will be necessary to any one who is not a practised observer. As seen in a large telescope, the planet has a decided sea-green color. No markings have ever been certainly seen on the disk, and therefore no changes which could be due to an axial rotation have ever been established; but it may be regarded as certain that it does rotate in the same plane in which the satellites revolve around it.

Discovery of Uranus.— This planet was discovered by Sir William Herschel, in March, 1781. Perceiving by its disk that it was not a star, and by its motion that it was not a nebula, he took it for a comet. The possibility of its being a new planet did not at first occur to him; and he therefore communicated his discovery to the Royal Society as being one of a new comet. Various computing astronomers thereupon attempted to find the orbit of the supposed comet, from the observations of Herschel and others, assuming it to move in a parabola, like other comets. But the actual motion of the body constantly deviated from the orbits thus computed to such an extent that new calculations had to be repeatedly made. After a few weeks it was found that if it moved in a parabola, the nearest distance to the sun must be at least fourteen times that of the earth from the sun, a perihelion distance many times greater than that of any known comet. This announcement gave the hint that some other hypothesis must be resorted to, and it was then found that all the observations could be well represented by a circular orbit, with a radius nineteen times that of the earth's orbit. The object was, therefore, a planet moving at double the distance of Saturn.

With a commendable feeling of gratitude towards the royal patron who had afforded him the means of making his discoveries, Herschel proposed to call the new planet *Georgium Sidus* (the Star of the Georges). This name, contracted to "the Georgian," was employed in England until 1850, but never came into use on the Continent. Lalande thought the most appropriate name of the planet was that of its discoverer, and therefore proposed to call it Herschel. But this name met with no more favor than the other. Several other names were proposed, but that of Uranus at length met with universal adoption. It was proposed by Bode as the most appropriate, on the ground that the most distant body of our system might be properly named after the oldest of the gods.

After the elliptic orbit of the planet had been accurately computed, and its path mapped out in the heavens, it was found that it had been seen a surprising number of times as a star without the observers having entertained any suspicion of its planetary nature. It had passed through the field of their telescopes, and they had noted the time of its transit, or its declination, or both, but had entered it in their journals simply as an unnamed star of the constellation in which it happened to be at the time. It had been thus seen five times by Flamsteed, the first observation being in 1690, nearly a century before the discovery by Herschel. What is most extraordinary, it had been observed eight times in rapid succession by Le Monnier, of Paris, in December, 1768, and January, 1769. Had that astronomer merely taken the trouble to reduce and compare his observations, he would have anticipated Herschel by twelve years. Indeed, considering how easily the planet can be seen with the naked eye, it is illustrative of the small amount of care devoted to cataloguing the stars that it was not discovered without a telescope.

Satellites of Uranus. — In January and February, 1787, Herschel found that Uranus was accompanied by two satellites, of which the inner performed a revolution in a little less than nine days, and the outer in thirteen days and a half. The existence of these two satellites was well authenticated by his observations, and they have been frequently observed in recent times. They can be seen with a telescope of one-foot aperture or upwards. Afterwards Herschel made a very assiduous search for other satellites. He encountered many difficulties, not only from the extreme faintness of the objects, but from the difficulty of deciding whether any object he might see was a satellite, or a small star which happened to be in the neighborhood. He at length announced the probable existence of four additional satellites, the orbit of one being inside of those of the two certain ones, one between them, and two outside them. This made an entire number of six; and though the evidence adduced by Herschel in favor of the existence of the four additional ones was entirely insufficient, and their existence has been completely disproved, they figure in some of our books on astronomy to this day.

For half a century no telescope more powerful than that of Herschel was turned upon Uranus, and no additional light was thrown upon the question of the existence or non-existence of the questionable objects. At length, about 1846, Mr. William Lassell, of England, constructed a reflector of two feet aperture, of which we have already spoken, and of very excellent definition, which in optical power exceeded any of the older instruments. With this he succeeded in discovering two new satellites inside the orbits of the two brighter ones,* but found no trace of any of the additional satellites of Herschel. In the climate of England, he could make only very imperfect observations of these bodies; but in 1852 he moved his telescope temporarily to Malta, to take advantage of the purer sky of that latitude, and there he succeeded in determining their orbits with considerable accuracy. Their times of revolution are about $2\frac{1}{2}$ and 4 days respectively. They may fairly be

---

* These difficult objects were also sought for by Otto Struve with the fifteen-inch telescope of the Pulkowa Observatory, and occasional glimpses of them were, he believed, attained before they were certainly found by Mr. Lassell, but he was not able to follow them so continuously as to fix upon their times of revolution.

regarded as the most difficult known objects in the planetary system; indeed, it is only with a few of the most powerful telescopes in existence that they have certainly been seen.

The non-existence of Herschel's suspected satellites is proved by the fact that they have been sought for in vain, both with Mr. Lassell's great reflectors and with the Washington twenty-six-inch refractor, all of which are optically more powerful than the telescopes of Herschel. There may be additional satellites which have not yet been discovered; but if so, they must be too faint to have been recognized by Herschel. Professor Holden, of the Naval Observatory, has sought to show that some of Herschel's observations of his supposed inner satellites were really glimpses of the objects afterwards discovered by Mr. Lassell. This he has done by calculating the positions of these inner satellites from tables for the date of each of Herschel's observations, and comparing them with the position of the object noted by Herschel. In four cases, the agreement is sufficiently close to warrant the belief that Herschel actually saw the real satellites; but Mr. Lassell attributes these coincidences to chance, and contests Professor Holden's views.

The most remarkable peculiarity of the satellites of Uranus is the great inclination of their orbits to the ecliptic. Instead of being inclined to it at small angles, like the orbits of all the other planets and satellites, they are nearly perpendicular to it; indeed, in a geometrical sense, they are more than perpendicular, because the direction of the motion of the satellites in their orbits is retrograde. To change the position of the orbit of an ordinary satellite into that of the orbits of these satellites, it would have to be tipped over 100°; so that, supposing the orbit a horizontal plane, the point corresponding to the zenith would be 10° below the horizon, and the upper surface would be inclined beyond the perpendicular, so as to be the lower of the two surfaces.

Observations of the satellites afford the only accurate way of determining the mass of Uranus; because, of the adjoining planets, Saturn and Neptune, the observations of the first are too uncertain and those of the last too recent to give any certain result. Measures made with the great Washington telescope show this mass to be $\frac{1}{22600}$; a result which is probably correct within $\frac{1}{200}$ part of its whole amount.*

§ 8. *Neptune and its Satellite.*

The discovery of this planet is due to one of the boldest and most brilliant conceptions of modern astronomy. The planet was felt, as it were, by its attraction upon Uranus; and its direction was thus calculated by the theory of gravitation before it had been recognized by the telescope. An observer was told that if he pointed his telescope towards a certain point in the heavens, he would see a new planet. He looked, and there was the planet, within a degree of the calculated place. It is difficult to imagine a more striking illustration of the certainty of that branch of astronomy which treats of the motions of the heavenly bodies and is founded on the theory of gravitation.

To describe the researches which led to this result, we shall have to go back to 1820. In that year, Bouvard, of Paris, prepared improved tables of Jupiter, Saturn, and Uranus, which, although now very imperfect, have formed the basis of most of the calculations since made on the motions of those bodies. He found that while the motions of Jupiter and Saturn were fairly in accord with the theory of gravitation, it was not so with those of Uranus. After allowing for the perturbations produced by the known planets, it was impossible to find any orbit which would satisfy both the ancient and the recent observations of Uranus. By the ancient observations we mean those accidental ones made by Flamsteed, Le Monnier, and others, before the planetary character of the object was suspected; and by the recent ones, those made after the discovery of the planet by Herschel, in 1781. Bouvard, therefore, rejected the older observations, founding his tables on the modern ones alone; and leaving to future investigators the

---

* Washington Observations for 1873: Appendix.

question whether the difficulty of reconciling the two systems arose from the inaccuracy of the ancient observations, or from the action of some extraneous influence upon the planet.

Only a few years elapsed, when the planet began to deviate from the tables of Bouvard. In 1830 the error amounted to 20''; in 1840, to 90''; in 1844, to 2'. From a non-astronomical point of view, these deviations were very minute. Had two stars moved in the heavens, the one in the place of the real planet, the other in that of the calculated planet, it would have been an eye of wonderful keenness which could have distinguished the two from a single star, even in 1844. But, magnified by the telescope, it is a large and easily measurable quantity, not for a moment to be neglected. The probable cause of the deviation was sometimes a subject of discussion among astronomers, but no very definite views respecting it seem to have been entertained, nor did any one express the decided opinion that it was to be attributed to a trans-Uranian planet, natural as it seems to us such an opinion would have been.

In 1845, Arago advised his then young and unknown friend Leverrier, whom he knew to be an able mathematician and an expert computer, to investigate the subject of the motions of Uranus. Leverrier at once set about the task in the most systematic manner. The first step was to make sure that the deviations did not arise from errors in Bouvard's theory and tables; he therefore commenced with a careful recomputation of the perturbations of Uranus produced by Jupiter and Saturn, and a critical examination of the tables. The result was the discovery of many small errors in the tables, which, however, were not of a character to give rise to the observed deviations.

The next question was whether any orbit could be assigned which, after making allowance for the action of Jupiter and Saturn, would represent the modern observations. The answer was in the negative, the best orbit deviating, first on one side and then on the other, by amounts too great to be attributed to errors of observation. Supposing the deviations to be due to the attraction of some unknown planet, Leverrier next inquired where this planet must be situated. Its orbit could not lie between those of Saturn and Uranus, because then it would disturb the motions of Saturn as well as those of Uranus. Outside of Uranus, therefore, the planet must be looked for, and probably at not far from double the distance of that body; this being the distance indicated by the law of Titius. Complete elements of the orbit of the unseen planet were finally deduced, making its longitude 325° as seen from the earth at the beginning of 1847. This conclusion was reached in the summer of 1846.

Leverrier was not alone in reaching this result. In 1843, Mr. John C. Adams, then a student at Cambridge University, England, having learned of the discordances in the theory of Uranus from a report of Professor Airy, attacked the same problem which Leverrier took hold of two years later. In October, 1845, he communicated to Professor Airy elements of the planet so near the truth that, if a search had been made with a large telescope in the direction indicated, the planet could hardly have failed to be found. The Astronomer Royal was, however, somewhat incredulous, and deferred his search for further explanations from Mr. Adams, which, from some unexplained cause, he did not receive. Meanwhile the planet, which had been in opposition about the middle of August, was lost in the rays of the sun, and could not be seen before the following summer. A most extraordinary circumstance was that nothing was immediately published on the subject of Mr. Adams's labors, and no effort made to secure his right to priority, although in reality his researches preceded those of Leverrier by nearly a year.

In the summer of 1846, M. Leverrier's elements appeared, and the coincidence of his results with those of Mr. Adams was so striking, that Professor Challis, of the Cambridge Observatory, commenced a vigorous search for the planet. Unfortunately, he adopted a mode of search which, although it made the discovery of the planet certain, was extremely laborious. Instead of endeavoring to recognize it by its disk, he sought to detect it by its motion among the stars — a course which required all the stars in the neighborhood to have their positions repeatedly determined, so as to find which of them had changed its position. Observations of the planet as a star were actually made on August 4th, 1846, and again on August 12th; but these observations, owing to Mr. Challis's other engagements, were not reduced, and so the fact that the planet was observed did not appear. His mode of proceeding was much like that of a man who, knowing that a diamond had dropped near a certain spot on the sea-beach, should remove all the sand in the neighborhood to a convenient place for the purpose of sifting it at his leisure, and should thus have the diamond actually in his possession without being able to recognize it.

Early in September, 1846, while Professor Challis was still working away at his observations, entirely unconscious that the great object of search was securely imprisoned in the pencilled figures of his note-book, Leverrier wrote to Dr. Galle, at Berlin, suggesting that he should try to find the planet. It happened that a map of the stars in the region occupied by the planet was just completed, and on pointing the telescope of the Berlin Observatory, Galle soon found an object which had a planetary disk, and was not on the star map. Its position was carefully determined, and on the night following it was re-examined, and found to have changed its place among the stars. No further doubt could exist that the long-sought-for planet was found. The date of the optical discovery was September 23d, 1846. The news reached Professor Challis October 1st, and, looking into his note-book, he found his own observations of the planet, made nearly two months before.

As between Leverrier and Adams, the technical right of priority in this wonderful investigation lay with Leverrier, although Adams had preceded him by nearly a year, for the double reason that the latter did not publish his results before the discovery of the planet, and that it was by the directions of Leverrier to Dr. Galle that the actual discovery was made. But this does not diminish the credit due to Mr. Adams for his boldness in attacking, and his skill in successfully solving, so noble a problem. The spirit of true science is advancing to a stage in which contests about priority are looked upon as below its dignity. Discoveries are made for the benefit of mankind; and if made independently by several persons, it is fitting that each should receive all the credit due to success in making it. We should consider Mr. Adams as entitled to the same unqualified admiration which is due to a sole discoverer; and whatever claims to priority he may have lost by the more fortunate Leverrier will be compensated by the sympathy which must ever be felt towards the talented young student in his failure to secure for his work that immediate publicity which was due to its interest and importance.

The discovery of Neptune gave rise to a series of researches, in which American astronomers took a distinguished part. One of the first questions to be considered was whether the planet had, like Uranus, been observed as a star by some previous astronomer. This question was taken up by Mr. Sears C. Walker, of the Naval Observatory. A few months' observation sufficed to show that the distance of the planet from the sun was not far from 30 (the distance of the earth being, as usual, unity), and, assuming a circular orbit, he computed the approximate place of the planet in past years. He traced its course back from year to year in order to find whether at any time it passed through a region which was at the same time being swept by the telescopes of observers engaged in preparing catalogues of stars. He was not successful till he reached the year 1795. On the 8th and 10th of May of that year, Lalande, of Paris, had swept over the place of the planet. It must now be decided whether any of the stars observed on those nights could have been Neptune. Although the exact place of the planet could not yet be fixed for an epoch so remote, it was easy to mark out the apparent position of its orbit as a line among the stars, and it must then have been somewhere on that line. After taking out the stars which were too far from the line, and those which had been seen by subsequent observers, there remained one, observed on May

10th, which was very near the computed orbit. Walker at once ventured on the bold prediction that if this region of the heavens were examined with a telescope, *that star* would be found missing. He communicated this opinion officially to Lieutenant Maury and other scientific men in Washington, and asked that the search might be made. On the first clear evening the examination was made by Professor Hubbard, and, surely enough, the star was not there.

There was, however, one weak point in the conclusion that this was really the planet Neptune. Lalande had marked his observation of the missing star with a colon, to indicate that there was a doubt of its accuracy: therefore it was possible that the record of the supposed star might have been the simple result of some error of observation. Happily, the original manuscripts of Lalande were carefully preserved at the Paris Observatory; and as soon as the news of Walker's researches reached that city an examination of the observations of May 8th and 10th, 1795, was entered upon. The extraordinary discovery was made that there was no mark of uncertainty in the original record, but that Lalande had observed the planet both on the 8th and 10th of May. The object having moved slightly during the two days' interval, the observations did not agree; and Lalande supposed that one of them must be wrong, entirely unconscious that in that little discrepancy lay a discovery which would have made his name immortal. Without further examination, he had rejected the first observation, and copied the second as doubtful on account of the discrepancy, and thus the pearl of great price was dropped, not to be found again till a half-century had elapsed.

For several years the investigation of the motion of the new planet was left in the hands of Mr. Walker and Professor Peirce. The latter was the first one to compute the perturbations of Neptune produced by the action of the other planets. The results of these computations, together with Mr. Walker's elements, are given in the Proceedings of the American Academy of Arts and Sciences.

Physical Aspect of Neptune.—On the physical appearance of

this planet very little can be said. In the largest telescopes and through the finest atmosphere, it presents the appearance of a perfectly round disk about 3″ in diameter, of a pale-blue color. No markings have been seen upon it. When first seen by Mr. Lassell, he suspected a ring, or some such appendage; but future observations under more favorable circumstances showed this suspicion to be without foundation. To recognize the disk of Neptune with ease, a magnifying power of 300 or upwards must be employed.

Satellite of Neptune.—Soon after the discovery of Neptune, Mr. Lassell, scrutinizing it with his two-foot reflector, saw on various occasions a point of light in the neighborhood. During the following year it proved to be a satellite, having a period of revolution of about 5 days 21 hours. During 1847 and 1848 the satellite was observed, both at Cambridge by the Messrs. Bond, and at Pulkowa by Struve. These observations showed that its orbit was inclined about 30° to the ecliptic, but it was impossible to decide in which direction it was moving, since there were two positions of the orbit, and two directions of motion, in which the apparent motion, as seen from the earth, would be the same. After a few years the change in the direction of the planet enabled this question to be decided, and showed that the motion was retrograde. The case was more extraordinary than that of the satellites of Uranus, since, to represent both the position of the orbit and the direction of motion in the usual way, the orbit would have to be tipped over 150°; it is, in fact, nearly upside down. The determinations of the elements of the satellite have been extremely discordant, a circumstance which we must attribute to its extreme faintness. It is a minute object, even in the most powerful telescopes.

Measures of the distance of the satellite from the planet, made with the great Washington telescope, show the mass of Neptune to be $\frac{1}{19380}$. The mass deduced from the perturbations of Uranus is $\frac{1}{19700}$, an agreement as good as could be expected in a quantity so difficult to determine.

CHAPTER V.

COMETS AND METEORS.

§ 1. *Aspects and Forms of Comets.*

THE celestial motions which we have hitherto described take place with a majestic uniformity which has always impressed the minds of men with a sense of the unchangeableness of the heavens. But this uniformity is on some occasions broken by the apparition of objects of an extraordinary aspect, which hover in the heavens for a few days or weeks, like some supernatural visitor, and then disappear. We refer to comets, bodies which have been known from the earliest times, but of which the nature is not yet deprived of mystery.

Comets bright enough to be noticed with the naked eye consist of three parts, which, however, are not completely distinct, but run into each other by insensible degrees. These are the *nucleus*, the *coma*, and the *tail*.

The *nucleus* is the bright centre which to the eye presents the appearance of an ordinary star or planet. It would hardly excite remark but for the coma and tail by which it is accompanied.

The *coma* (which is Latin for *hair*) is a mass of cloudy or vaporous appearance, which surrounds the nucleus on all sides. Next to the nucleus, it is so bright as to be hardly distinguishable from it, but it gradually shades off in every direction. Nucleus and coma combined present the appearance of a star, more or less bright, shining through a small patch of fog, and are together called the *head* of the comet.

The *tail* is a continuation of the coma, and consists of a stream of milky light, growing wider and fainter as it recedes from the comet, until the eye can no longer trace it. A curious feature, noticed from the earliest times, is that the tail is always turned from the sun. The extent of the tail is very different in different comets, that appendage being brighter and longer the more brilliant the comet. Sometimes it might almost escape notice, while in many great comets recorded in history it has extended half-way across the heavens. The actual length, when one is seen at all, is nearly always many millions of miles. Sometimes, though rarely, the tail of the comet is split up into several branches, extending out in slightly different directions.

Such is the general appearance of a comet visible to the naked eye. When the heavens were carefully swept with telescopes, it was found that comets thus visible formed but a small fraction of the whole number. If a diligent search is kept up, as many comets are sometimes found with the telescope in a single year as would be seen in a lifetime with the unaided eye. These "telescopic comets" do not always present the same aspect as those seen with the naked eye. The coma, or foggy light, generally seems to be developed at the expense of the nucleus and the tail. Sometimes either no nucleus at all can be seen with the telescope, or it is so faint and ill-defined as to be hardly distinguishable. In the cases of such comets, it is generally impossible to distinguish the coma from the tail, the latter being either entirely invisible, or only an elongation of the coma. Many well-known comets consist of hardly anything but a patch of foggy light of more or less irregular form.

Notwithstanding these great apparent differences between the large comets and the telescopic ones, yet, when we closely watch their respective modes of development, we find them all to belong to one class. The differences are like those between some animals, which, to the ordinary looker-on, have nothing in common, but in which the zoologist sees that every part of the one has its counterpart in the other—indeed, the analogy between what the astronomer sees in the growth of comets and the zoologist in the growth of animals is quite worthy of remark. As a general rule, all comets look nearly

ASPECTS AND FORMS OF COMETS.

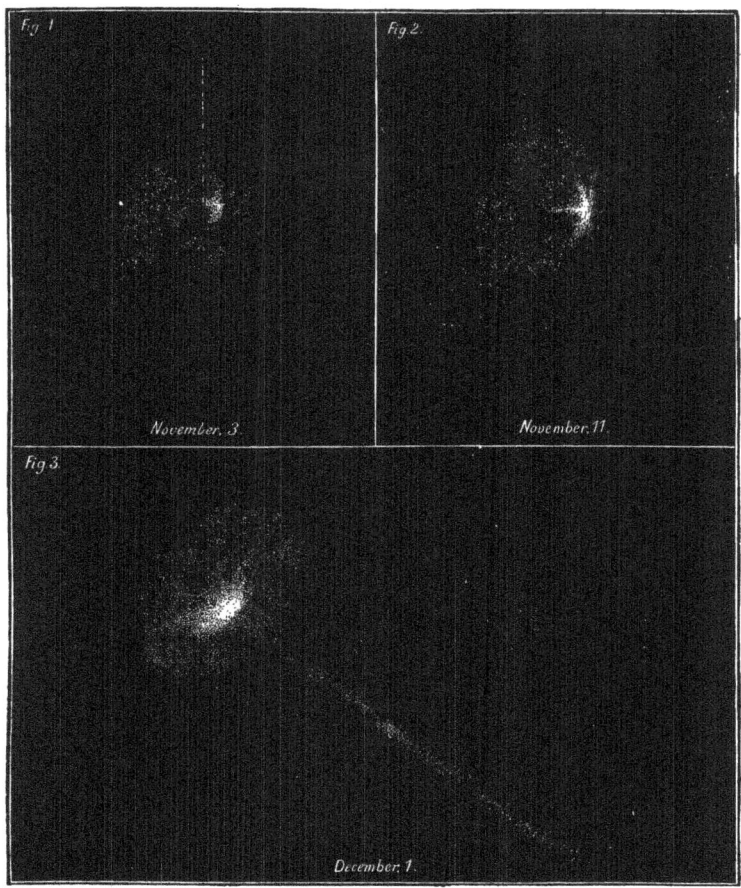

Fig. 90.—Views of Encke's comet in 1871, by Dr. Vogel.

alike when they first come within reach of the telescope, the subsequent diversities arising from the different developments of corresponding parts. The first appearance is that of a little foggy patch without any tail, and very often without any visible nucleus. Thus, in the case of Donati's comet of 1858, one of the most splendid on record, it was more than two months after the first discovery before there was any appearance of a tail. To enable the reader to see the relation of this to a very diffused telescopic comet, we present a telescopic view of the head of this great comet when near its brightest, and three drawings of Encke's comet, made by Dr. Vogel, in November and December, 1871.

When the nucleus of a telescopic comet begins to show itself, it is commonly on the side farthest from the sun. Several little branches will then be seen stretched out in the direction of the sun, so that it will appear as if the comet had a small fan-shaped tail directed towards the sun, instead of from it, as is usual. Thus, in the pictures of Encke's comet in Figs. 1 and 2, the sun is towards the left, and we see what

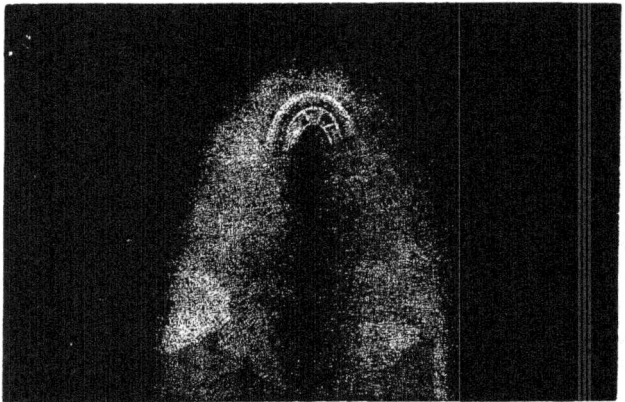

FIG. 91.—Head of Donati's great comet of 1858, after Bond.

looks like three little tails, the middle one pointed towards the sun. But if we look at the view of Donati's comet, Fig. 91, we see several little lines branching upwards from the centre of the head, and it is to these, and not to the tail, that the little tails in the figures of Encke's comet correspond. In fact, the general rule is that the heads of comets have a fan-shaped structure, the handle of the fan being in the nucleus, and the middle arm pointing towards the sun; and it is this appendage which first shows itself.

In the larger comets, this fan is surrounded by one or more

semicircular arches, or envelopes, the inner one forming its curved border; but this arch does not show itself in very faint comets. The true tail of the comet, when it appears, is always directed from the sun, and therefore away from the fan. In Fig. 90, No. 3, a very faint true tail will be seen extending out towards the lower right-hand corner of the picture, which was opposite to the direction of the sun. On the other hand, though the branches turned towards the sun have disappeared, the fan-like form can still be traced in the head. In Fig. 91, the true tail is turned downwards: owing to the large scale of the picture, only the commencement of it can be seen. The central line of the tail, it will be remarked, is comparatively dark. This is very generally the case with bright comets.

§ 2. *Motions, Origin, and Number of Comets.*

When it was found by Kepler that all the planets moved around the sun in conic sections, and when Newton showed that this motion was the necessary result of the gravitation of the planets towards the sun, the question naturally arose whether comets moved according to the same law. It was found by Newton that the comet of 1680 actually did move in such an orbit, but instead of being, like the planetary orbits, nearly circular, it was very eccentric, being to all appearance a parabola.

A parabola being one of the orbits which gravitation would cause to be described, it was thus made certain that comets gravitated towards the sun, like planets. It was, however, impossible to say whether the orbit was really a parabola or a very elongated ellipse. The reason of this difficulty is that comets are visible in only a very small portion of their orbits, quite close to the sun, and in this portion the forms of a parabola and of a very eccentric ellipse are so nearly the same, that they cannot always be distinguished.

There is this very important difference between an elliptical and a parabolic orbit—that the former is closed up, and a comet moving in it must come back some time, whereas the two branches of the latter extend out into infinite space without ever meeting. A comet moving in a parabolic orbit will, therefore, never return, but, after once sweeping past the sun, will continue to recede into infinite space forever. The same thing will happen if the comet moves in an hyperbola, which is

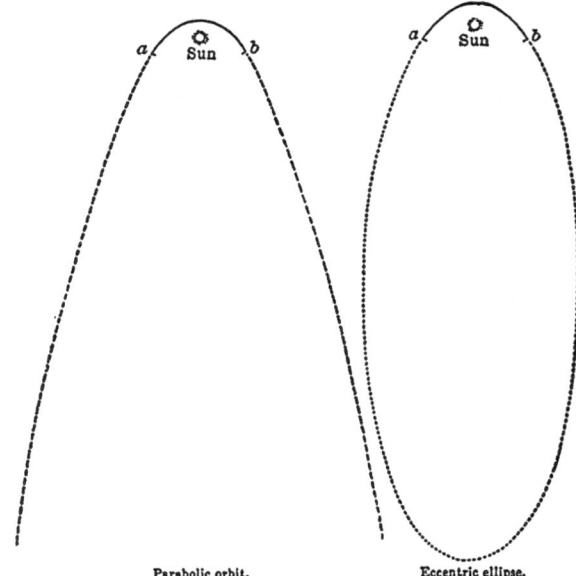

Fig. 92.—Parabolic and elliptic orbit of a comet. The comet is invisible in the dotted part of the orbits, and the forms of the visible parts, *a*, *b*, cannot be distinguished in the two orbits. But the ellipse forms a closed curve, while the two branches of the parabola continue forever without meeting.

the third class of orbit that may be described under the influence of gravitation. In a parabola, the slightest retardation of a comet would change the orbit into an ellipse, the velocity being barely sufficient to carry the comet off forever, whereas in an hyperbola there is more or less velocity to spare. Thus the parabola is a sort of dividing curve between the hyperbola and the ellipse.

The astronomer, knowing the position of an orbit, can tell exactly what velocity is necessary at any point of it in order that a body moving in it may go off, never to return. A body thrown from the earth's surface with a velocity of seven miles a second, and not retarded by the atmosphere, would never return to the earth, but would describe some sort of an orbit round the sun. It would, in fact, be a little planet. If the earth were out of the way, a body moving past the earth's orbit at the rate of twenty-six miles a second would have just the velocity necessary to describe a parabola. If the velocity of a comet exceeds this limit at that point of its orbit which is $92\frac{1}{2}$ millions of miles from the sun, then the comet must go off into infinite space, never to return to our system. But with a less velocity the comet must be brought back by the sun's attraction at some future time, the time being longer the more nearly the velocity reaches twenty-six miles per second. It is by the velocity that the astronomer must, in general, determine the form of the orbit. If it corresponds exactly to the calculated limit, the orbit is a parabola; if it exceeds this limit, it is an hyperbola; if it falls short of it, it is an ellipse.

Now, in the large majority of comets the velocity is so near the parabolic limit that it is not possible to decide, from observations, whether it falls short of it or exceeds it. In the case of a few comets the observations indicate an excess of velocity, but an excess is so minute that its reality cannot be confidently asserted. It cannot, therefore, be said with certainty that any known comet revolves in a hyperbolic orbit, and thus it is possible that all comets belong to our system, and will ultimately return to it. It is, however, certain that in the majority of cases the return will be delayed many centuries, nay, perhaps many thousand years. There are quite a number of comets which are known to be periodic, returning to the sun at regular intervals in elliptic orbits. Some of these have been observed at several returns, so that their exact period has been determined with great certainty: in the case of others, the periodicity has been inferred only from the fact that the velocity fell so far short of the parabolic limit that there could be no doubt of the fact that the comet moved in an ellipse.

In this question of cometary orbits is involved the very interesting one, whether comets should be considered as belonging to our system, or as mere visitors from the stellar spaces. We may conceive of them as stray fragments of original nebulous matter scattered through the great wilderness of space around us, drawn towards our sun one by one as the long ages elapse. If no planets surrounded the sun, or if, surrounding it, they were immovable, a comet thus drawn in would whirl around the sun in a parabolic orbit, and leave it again, not to return until millions of years had elapsed, because the velocity it would acquire by falling towards the sun would be just sufficient to carry it back into the infinite void from which it came. But owing to the motions of the several planets in their orbits, the comet would have its velocity changed in passing each of them, the change being an acceleration or a retardation, according to the way in which it passed. If the total accelerations produced by all the planets exceeded the retardations, the comet would leave our system with more than the parabolic velocity, and would certainly never return. If the retarding forces chanced to be in excess, the orbit would be changed into an ellipse more or less elongated, according to the amount of this excess. In the large majority of cases, the retardation would be so slight that the most delicate observations could not show it, and it could be known only by calculation, or by the return of the comet after tens or hundreds of thousands of years. But should the comet chance to pass very near a planet, especially a large planet like Jupiter, the retardation might be so great as to make the comet revolve in an orbit of quite short period, and thus become a seemingly permanent member of our system. So near an approach of a comet to a planet would not be likely to occur more than once in a number of centuries, but every time it did occur there would be an even chance for an additional comet of short period, the orbit of which would, at first, almost intersect that of the planet which had deranged it. It might not, however, be a known comet, because the orbit might be wholly beyond the reach of our vision.

It is impossible, in the present state of science, to say with certainty whether the periodic comets were thus brought into our system; but it seems probable that they were, from the fact that many, if not all, of the orbits of these comets pass near the orbits of some of the planets. That the planetary and cometary orbits in such a case should intersect now is not to be expected, because both would change by the secular variations resulting from the action of the planets. Future research will probably throw more light on this question.

Number of Comets.—It was the opinion of Kepler that the celestial spaces were as full of comets as the sea of fish, only a small proportion of them coming within the range of our telescopes. That only an insignificant fraction of all existing comets have ever been observed, we may regard as certain. Owing to their extremely elongated orbits, they can be seen only when near their perihelion, and as it is probable that the period of revolution of the large majority of those which have been observed is counted by thousands of years—if, indeed, they ever return at all—our observations must be continued for many thousand years before we have seen all which come within range of our telescopes. It is also probable that all which can ever be seen will be but a small fraction of the number which exist, because a comet can seldom be seen unless its perihelion is either inside the orbit of the earth, or but little outside of it. There are a few exceptions to the rule that only such comets are seen, the most notable one being that of the comet of 1729, which, at perihelion, was more than four times the earth's distance from the sun. This comet must have been one of extraordinary magnitude, as almost every other known comet would have disappeared entirely from the most powerful telescopes of that time, if placed at the distance at which it was observed.

The actual number of comets recorded as visible to the naked eye since the Christian era is given in the table on the following page.*

---

* This table is taken at second-hand, principally from Arago ("Astronomie Populaire," livre xvii., chap. xv.). Arago mentions but *eight* as visible during the eighteenth century. I have considered the number thirty-six, given by Klein, as more probable.

Years of our Era.	Number of Comets.	Years of our Era.	Number of Comets.
From 0 to 100	22	From 1001 to 1100	36
" 101 " 200	23	" 1101 " 1200	26
" 201 " 300	44	" 1201 " 1300	26
" 301 " 400	27	" 1301 " 1400	29
" 401 " 500	16	" 1401 " 1500	27
" 501 " 600	25	" 1501 " 1600	31
" 601 " 700	22	" 1601 " 1700	12
" 701 " 800	16	" 1701 " 1800	36
" 801 " 900	42	" 1801 " 1875	16
" 901 " 1000	26		

In round numbers, about five hundred comets visible to the naked eye have been recorded since our era, making a general average of one every four years. Besides these, nearly two hundred telescopic comets have been observed since the invention of the telescope; so that the total number of these bodies observed during the period in question does not fall far short of seven hundred. Several new telescopic comets are now discovered nearly every year, the number sometimes ranging up to six or eight. It is probable that the annual number of this class discovered depends very largely on the skill, assiduity, and good-fortune of the astronomers who chance to be engaged in searching for them.

§ 3. *Remarkable Comets.*

In unenlightened ages comets were looked on with terror, as portending pestilence, war, the death of kings, or other calamitous or remarkable events. Hence it happens that in the earlier descriptions of these bodies, they are generally associated with some contemporaneous event. The descriptions of the comets themselves are, however, so vague and indefinite as to be entirely devoid of either instruction or interest, as it often happens that not even their course in the heavens is stated.

The great comet of 1680 is, as already said, remarkable for being not only a brilliant comet, but the one by which Newton proved that comets move under the influence of the gravitation of the sun. It first appeared in the autumn of 1680, and continued visible most of the time till the following spring.

It fell down almost in a direct line to the sun, passing nearer to that luminary than any comet before known. It passed its perihelion on December 18th, and, sweeping round a large arc, went back in a direction not very different from that from which it came. The observations have been calculated and the orbit investigated by many astronomers, beginning with Newton; but the results show no certain deviation from a parabolic orbit. Hence, if the comet ever returns, it is only at very long intervals. Halley, however, suspected, with some plausibility, that the period might be 575 years, from the fact that great comets had been recorded as appearing at that interval. The first of these appearances was in the month of September, after Julius Cæsar was killed; the second, in the year 531; the third, in February, 1106; while that of 1680 made the fourth. If, as seems not impossible, these were four returns of one and the same comet, a fifth return will be seen by our posterity about the year 2255. Until that time the exact period must remain doubtful, because observations made two centuries ago do not possess the exactitude which will decide so delicate a point.

Halley's Comet.—Two years after the comet last described, one appeared which has since become the most celebrated of modern times. It was first seen on August 19th, 1682, and observed about a month, when it disappeared. Halley computed the position of the orbit, and, comparing it with previous orbits, found that it coincided so exactly with that of a comet observed by Kepler in 1607, that there could be no doubt of the identity of the two orbits. So close were they together that, if drawn on the heavens, the naked eye would almost see them joined into a single line. The chances against two separate comets moving in the same orbit were so great that Halley could not doubt that the comet of 1682 was the same that had appeared in 1607, and that it therefore revolved in a very elliptic orbit, returning about every seventy-five years. His conclusion was confirmed by the fact that a comet was observed in 1531, which moved in apparently the same orbit. Again subtracting the period of seventy-five years, it was found that the comet had appeared in 1456, when it spread such terror throughout Christendom that Pope Calixtus ordered prayers to be offered for protection against the Turks and the comet. This is supposed to be the circumstance which gave rise to the popular myth of the Pope's Bull against the Comet.

This is the earliest occasion on which observations of the course of the comet were made with such accuracy that its orbit could be determined. If we keep subtracting $75\frac{1}{2}$ years, we shall find that we sometimes fall on dates when the apparition of a comet was recorded; but without any knowledge of the orbits of these bodies, it cannot be said with certainty that they are identical. However, in the returns of 1456, 1531, 1607, and 1682, at nearly equal intervals, Halley had good reason for predicting that the comet would return again about 1758. This gave the mathematicians time to investigate its motions; and the establishment, in the mean time, of the theory of gravitation showed them how to set about the work. It was necessary to calculate the effect of the attraction of the planets on the motion of the comet during the entire seventy-six years. This immense labor was performed by Clairaut, who found that, in consequence of the attractions of Jupiter and Saturn, the return of the comet would be delayed 618 days, so that it would not reach its perihelion until the middle of April, 1759. Not having time to finish his calculations in the best way, he considered that this result was uncertain by one month. The comet actually did pass its perihelion at midnight on March 12th, 1759.

Seventy-six years more were to elapse, and the comet would again appear about 1835. Meanwhile, great improvements were made in the methods of computing the effects of planetary attraction on the motions of a comet, so that mathematicians, without expending more labor than Clairaut did, were enabled to obtain much more accurate results. The French were still the leading nation of the world in this sort of investigation, and the computation of the return of the comet was undertaken independently by two of their leading astronomers,

De Damoiseau and De Pontécoulant. Of these, the first announced that it would reach its perihelion on November 4th, 1835; while De Pontécoulant, after revising his computations with more exact determinations of the masses of the planets, assigned November 13th, at 2 A.M., as the date. The expected comet was, of course, looked for with the greatest assiduity, and was first seen on August 5th. Approaching the sun, it passed its perihelion on November 16th, at eleven o'clock in the morning, only three days after the time predicted by De Pontécoulant.

This was the last return of the celebrated comet of Halley. It was followed until May 17th, 1836, when it disappeared from the sight of the most powerful telescopes of the time, and has not been seen since. But the astronomer can follow it with the eye of science with almost as much certainty as if he had it in the field of view of his telescope. We cannot yet fix the time of its return with certainty; but we know that it reached the farthest limit of its course, which extends some distance beyond the orbit of Neptune, about 1873, and that it is now on its return journey. We present a diagram of its orbit, showing its position in 1874. Its velocity will constantly increase from year to year, and we may expect it to reach perihelion about the year 1911.

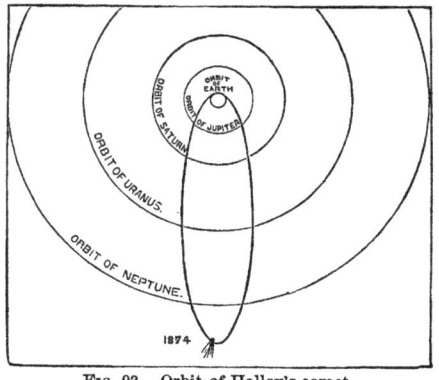

FIG. 93.—Orbit of Halley's comet.

The exact date cannot be fixed until the effect of the action of all the planets is computed, and this will be a greater labor than before, not only because greater accuracy will be aimed at, but because the action of more planets must be taken into account. When Clairaut computed the return of 1759, Saturn was the outermost known planet. When the return of 1835 was computed,

Uranus had been added to the list, and its action had to be taken into account. Since that time Neptune has been discovered; and the astronomer who computes the return of 1911 must add its action to that of the other planets. By doing so, we may hope that the time of reaching perihelion will be predicted within one or two days.

The Lost Biela's Comet.—Nothing could more strikingly illustrate the difference between comets and other heavenly bodies than the fact of the total dissolution of one of the former. In 1826, a comet was discovered by an Austrian named Biela, which was found to be periodic, and to have been observed in 1772, and again in 1805. The time of revolution was found to be six years and eight months. In the next two returns, the earth was not in the right part of its orbit to admit of observing the comet; the latter was therefore not seen again till 1845. In November and December of that year it was observed as usual, without anything remarkable being noticed. But in January following, the astronomers of the Naval Observatory found it to have suffered an accident never before known to happen to a heavenly body, and of which no explanation has ever been given. The comet had separated into two distinct parts, of quite unequal brightness, so that there were two apparently complete comets, instead of one. During the month following, the lesser of the two continually increased, until it became equal to its companion. Then it grew smaller, and in March vanished entirely, though its companion was still plainly seen for a month longer. The distance apart of the two portions, according to the computations of Professor Hubbard, was about 200,000 miles.

The next return of the comet took place in 1852, and was, of course, looked for with great interest. It was found still divided, and the two parts were far more widely separated than in 1846, their distance having increased to about a million and a half of miles. Sometimes one part was the brighter, and sometimes the other, so that it was impossible to decide which ought to be regarded as representing the principal comet. The pair passed out of view about the end of September, 1852, and have not been seen since. They would, since then, have made three complete revolutions, returning in 1859, 1865, and 1872. At the first of these returns, the relative positions of the comet and the earth were so unfavorable that there was no hope of seeing the former. In 1865, it could not be found; but it was thought that this might be due to the great distance of the comet from us. In 1872, the relative positions were extremely favorable, yet not a trace of the object could be seen.* It had seemingly vanished, not into thin air, but into something of a tenuity compared with which the thinnest air was as a solid millstone. Some invisible fragments were, however, passing along the comet's orbit, and produced a small meteoric shower, as will be explained in a later section.

The Great Comet of 1843.—This remarkable comet burst suddenly into view in the neighborhood of the sun about the end of February, 1843. It was visible in full daylight, so that some observers actually measured the angular distance between the comet and the sun. It was followed until the middle of April. The most remarkable feature of the orbit of this comet has been already mentioned: it passed nearer the sun than any other known body—so near it, in fact, that, with a very slight change in the direction of its original motion, it would actually have struck it. Its orbit did not certainly deviate from a parabola. The most careful investigation of it—that of Professor Hubbard, of Washington—indicated a period of 530 years; but the velocity which would produce this period is so near the parabolic limit that the difference does not exceed the uncertainty of the observations.

Donati's Comet of 1858.—This great comet, one of the most magnificent of modern times, which hung in the western sky during the autumn of 1858, will be well remembered by all who were then old enough to notice it. It was first seen at

---

* Just after the meteoric shower, Mr. Pogson, of Madras, obtained observations of an object which, it was supposed, might have been a fragment of this comet. But the object was some two months behind the computed position of the comet, so that the identity of the two has never been accepted by astronomers.

380 THE SOLAR SYSTEM.

Florence, on June 2d, 1858, by Donati, who described it as a very faint nebulosity, about 3' in diameter. About the end of the month it was discovered independently by three American observers: H. P. Tuttle, at Cambridge; H. M. Parkhurst, at Perth Amboy, New Jersey; and Miss Maria Mitchel, at Nantucket. During the first three months of its visibility it gave no indications of its future grandeur. No tail was noticed until the middle of August, and at the end of that month it was only half a degree in length, while the comet itself was barely visible to the naked eye. It continued to approach the sun till the end of September, and during this

FIG. 94.—Great comet of 1858.

month developed with great rapidity, attaining its greatest brilliancy about the first half of October. Its tail was then 40° in length, and 10° in breadth at its outer end, and of a curious feather-like form. About October 20th it passed so far south as to be no longer visible in northern latitudes; but it was followed in the southern hemisphere until March following.

Observations of the position of this comet soon showed its orbit to be decidedly elliptic, with a period of about 2000 years or less. A careful investigation of all the observations was made by Mr. G. W. Hill, who found a period of 1950 years. If this period is correct, the comet must have appeared about ninety-two years before our era, and must appear again about the year 3808; but the uncertainty arising from the imperfections of the observations may amount to fifty years.

§ 4. *Encke's Comet, and the Resisting Medium.*

The comet which in recent times has most excited the attention of astronomers is that known as Encke's, from the astronomer who first carefully investigated its motion. It was first seen in January, 1786, but the observations only continued through two days, and were insufficient to determine the orbit. In 1795, a comet was found by Miss Caroline Herschel, on which observations were continued about three weeks; but no very accurate orbit was derived from these observations. In 1805, the same comet returned again to perihelion, but its identity again failed to be recognized. As in the previous returns, the observations continued through less than a month. It was found, for the fourth time, by Pons, of Marseilles, in 1818. When its orbit was calculated, it was seen to coincide so closely with that of the comet of 1805 as to leave no doubt that the two were really the same body. But the first astronomers who noticed this were unable to decide whether this was its first return since 1805, or whether it had in the mean time made several revolutions.

The motions of the comet were now taken up by Encke, of Berlin, and investigated with a thoroughness before unknown. He found the period to be about 1200 days, four complete revolutions having been made between 1805 and 1818. Knowing this, there was no longer any difficulty in identifying the comet of 1795 as also being the same, three complete revolutions having been made between that date and 1805. In the intermediate returns to perihelion, its position had been so unfavorable that it had not been observed at all. This result was received by astronomers with the greatest interest, because it was the first known case of a comet of short period. Its return in 1822 was duly predicted, but it was found that when near its greatest brilliancy it would be visible only in the southern hemisphere. Happily, Sir Thomas Brisbane had an observatory at Paramatta, New South Wales, and his assistant, Rumker, was so fortunate as to find the comet. It was so near the position predicted by Encke that, by constantly pointing the telescope in the direction predicted by that astronomer, the comet was in the field of view during its whole course.

Encke continued to investigate the course of the comet during each revolution up to the time of his death, in 1865. At some returns it could not be seen, owing to its distance from the earth, or the otherwise unfavorable position of our planet; but generally very accurate observations of its course were made. By a comparison of its motions with those which would result from the gravitation of the sun and planets, he found that the periodic time was constantly diminishing, and was thus led to adopt the famous hypothesis of Olbers, that the comet met with a resisting medium in space. The diminution of the period was about two hours and a half in each revolution. The conclusion of Encke and Olbers was that the planetary spaces are filled with a very rare medium—so rare that it does not produce the slightest effect on the motion of such massive bodies as the planets. The comet being a body of extreme tenuity, probably far lighter than air, it might be affected by such a medium. The existence of this medium cannot, however, be considered as established by Encke's researches. In the first place, if we grant the fact that the time of revolution is continually diminishing, as maintained by the great German astronomer, it does not follow that a resisting medium is the only cause to which we can attribute it. But the main point is, that the computations on which Encke founded his hypothesis are of such intricacy as to be always liable to small errors, and their results cannot be received with entire confidence until some one else has examined the subject by new and improved methods.

Such an examination is now being made by Dr. Von Asten, of Pulkowa; and, although it is still unfinished, it seems likely, in the end, to confirm Encke's results, at least in part. Dr. Von Asten commenced by calculating the motion of the comet from the theory of gravitation during the period from 1865 to 1871, within which the comet made two entire revolutions, and was surprised to find that during this time there was no deviation from the computed positions which could be attributed to the action of a resisting medium. But on carrying the calculation back to 1861, he found that between that epoch and 1865 there must have been a retarding action like that supposed by Encke. Carrying his work forward to 1875, he found that between 1871 and 1875 there was once more evidence of a retardation about two-thirds as great as that found by Encke. The absence of such an action between 1865 and 1871, therefore, seems quite exceptional, and difficult of explanation.

To judge whether the deviations in the motion of Encke's comet are really due to a resisting medium, we should know whether the motions of other comets exhibit similar anomalies. So far as is yet known, no other one does. There is at least one which has returned a sufficient number of times, and of which the motions have been computed with sufficient care, to lead to an entirely definite conclusion on this point, namely, the periodic comet of Faye, which has been investigated by Möller.* This comet was discovered in 1843 by the astronomer whose name it bears, and was soon found to move in an elliptic orbit, with a period of a little more than seven years. As it has been observed at several returns since, Möller investigated its motions with a view of finding whether its period was affected by any resisting medium. At first he thought there was such an effect, his general result being of the same nature with that reached by Encke. But on repeating his calculations with the improved data afforded by a first calculation, he found that the result arose from the imperfection of the latter, and that the comet really showed no sign of a change in its mean motion. It therefore seems certain that, if there is a resisting medium, it does not extend out far enough from the sun to meet the orbit of Faye's comet. But

---

* Professor Axel Möller, director of the observatory at Lund, Sweden.

this orbit lies wholly outside the orbit of Mars; so that if the sun were surrounded by an atmosphere extending out to Mars, and no farther, the comet would never enter it. On the other hand, Encke's comet, when in perihelion, is nearer the sun than Mercury is, and might there meet a resisting medium which did not extend so far out as the orbit of Mars. We must therefore adopt one of two conclusions: either the cause which is supposed to affect the motion of Encke's comet is not a resisting medium, or, if it is such, it is confined to the neighborhood of the sun. Considering the improbability of the sun having any atmosphere which can extend to such a distance, the former should be deemed the more probable alternative. We can accept it the more readily, from the fact that comets in general exhibit deviations from their calculated orbits many times larger than those of the planets, so that an exact agreement between theory and observations can never be expected in the case of those bodies.

The next subject to which we would ask the attention of the reader is that of the physical constitution of comets. But this subject can be discussed only in connection with another, to which, at first sight, it seems to have no relation, though so curious a relation has really been discovered as greatly to modify our views of what a comet probably is. We refer to the phenomena of meteors, meteoric showers, and shooting-stars, which next claim our attention.

§ 5. *Meteors and Shooting-stars.*

If we carefully watch the heavens on a cloudless night, we shall frequently see an appearance as of a star rapidly shooting through a short space in the sky, and then suddenly disappearing. Three or four such shooting-stars may generally be seen in the course of an hour. Generally they are visible only for a second or two, but sometimes move slowly, and are seen much longer. Occasionally they are so brilliant as to illuminate the whole heavens, and they are then known as meteors—a term which is equally applicable to the ordinary shooting-stars. In general, they are seen only one at a time,

METEORS AND SHOOTING-STARS. 385

and are so minute as hardly to attract attention. But they have on some occasions shown themselves in such numbers as to fill the beholders with terror, lest the end of the world had come. The Chinese, Arabian, and other historians have handed down to us many accounts of such showers of meteors, which have been brought to light by the researches of Edward Biot, Quetelet, Professor H. A. Newton, and others. As an example of these accounts, we give one from an Arabian writer:

"In the year 599, on the last day of Moharrem, stars shot hither and thither, and flew against each other like a swarm of locusts; this phenomenon lasted until daybreak; people were thrown into consternation, and made supplication to the Most High: there was never the like seen except on the coming of the messenger of God, on whom be benediction and peace."

In 1799, on the night of November 12th, a remarkable shower was seen by Humboldt and Bonpland, who were then on the Andes. Humboldt described the shower as commencing a little before two o'clock, and the meteors as rising above the horizon between east and north-east, and moving over towards the south. From not continuing his observations long enough, or from some other cause, he failed to notice that the lines in which the meteors moved all seemed to converge towards the same point of the heavens, and thus missed the discovery of the real cause of the phenomenon.

The next great shower was seen in this country in 1833. All through the Southern States, the negroes, like the Arabs of a previous century, thought the end of the world had come at last. The phenomenon was observed very carefully at New Haven by Professor Olmsted, who worked out a theory of its cause. Although his ideas are in many respects erroneous, they were the means of suggesting the true theory to others. The recurrence of the shower at this time suggested to the astronomer Olbers the idea of a thirty-four-year period, and led him to predict a return of the shower in 1867. A few years before the expected time, the subject was taken up by

Professor Newton, of Yale College, to whose researches our knowledge of the true cause of the phenomenon is very largely due.

The phenomena of shooting-stars branch out in yet another direction. As we have described them, they are seen only in the higher and rarer regions of the atmosphere, far above the clouds: no sound is heard from them, nor does anything reach the surface of the earth from which the nature of the object can be inferred. But on rare occasions meteors of extreme brilliancy are followed by a loud sound, like the discharge of heavy artillery; while on yet rarer occasions large masses of metallic or stony substances fall to the earth. These aërolites were the puzzle of philosophers. Sometimes there was much scepticism as to the reality of the phenomenon itself, it appearing to the doubters more likely that those who described such things were mistaken than that heavy metallic masses should fall from the air. When their reality was placed beyond doubt, many theories were propounded to account for them, the most noteworthy of which was that they were thrown from volcanoes in the moon. The problem of the motion of a body projected from the moon was investigated by several great mathematicians, the result being that such a body could not reach the earth unless projected with a velocity far exceeding anything seen on our planet.

When aërolites were examined by chemists and mineralogists, it was found that although they contained no new chemical elements, yet the combinations of these elements were quite unlike any found on the earth, so that they must have originated outside the earth. Moreover, these combinations exhibited certain characteristics peculiar to aërolites, so that the mineralogist, from a simple examination and analysis of a substance, could detect it as part of such a body, though it had not been seen to fall. Great masses of matter thus known to be of meteoric origin have been found in various parts of the earth, especially in Northern Mexico, where, at some unknown period, an immense shower of these bodies seems to have fallen.

Cause of Shooting-stars.—It is now universally conceded that the celestial spaces are crowded with innumerable minute bodies moving around the sun in every possible kind of orbit When we say crowded, we use the word in a relative sense; they may not average more than one in a million of cubic miles, and yet their total number exceeds all calculation. Of the nature of the minuter bodies of this class nothing is certainly known. But whatever they may be, the earth is constantly encountering them in its motion around the sun. They are burned by passing through the upper regions of our atmosphere, and the shooting-star is simply the light of that burning. We shall follow Professor Newton in calling these invisible bodies *meteoroids*.

The question which may be asked at this stage is, Why are these bodies burned? Especially, how can they burn so suddenly, and with so intense a light, as to be visible hundreds of miles away? These questions were the stumbling-block of investigators until they were answered, clearly and conclusively, by the discovery of the mechanical theory of heat. It is now established that heat is only a certain form of motion; that hot air differs from cold air only in a more rapid vibration of its molecules, and that it communicates its heat to other bodies simply by striking them with its molecules, and thus setting their molecules in vibration. Consequently, if a body moves rapidly through the air, the impact of the air upon it ought to heat it just as warm air would, even though the air itself were cold. This result of theory has been experimentally proved by Sir William Thomson, who found that a thermometer placed in front of a rapidly moving body rose one degree when the body moved through the air at the rate of 125 feet per second. With higher velocities, the increase of temperature was proportional to the square of the velocity, being 4 degrees with a velocity of 250 feet, 16 degrees with one of 500 feet per second, and so on. This result is in exact accordance with the mechanical theory of heat. To find the effective temperature to which a meteoroid is exposed in moving through our atmosphere, we divide its velocity in feet per second by 125; the square of the quotient will give the temperature in degrees.

Let us apply this principle to the case of the meteoroids. The earth moves in its orbit at the rate of 98,000 feet per second; and if it met a meteoroid at rest, our atmosphere would strike it with this velocity. By the rule we have given for the rise of temperature $(98{,}000 \div 125)^2 = 784^2 = 600{,}000$ degrees, nearly. This is many times any temperature ever produced by artificial means. If, as will commonly be the case, the meteoroid is moving to meet the earth, the velocity, and therefore the potential temperature, will be higher. We know that the meteoroids which produce the November showers already described move in a direction nearly opposite that of the earth with a velocity of 26 miles per second, so that the relative velocity with which the meteoroids meet our atmosphere is 44 miles per second. By the rule we have given, this velocity corresponds to a temperature of between three and four million degrees. We do not mean that the meteoroids are actually heated up to this temperature, but that the air acts upon them as if it were heated up to the point mentioned; that is, it burns or volatilizes them in less than a second with an enormous evolution of light and heat, just as a furnace would if heated to a temperature of three million degrees. It is not at all necessary that the body should be combustible; the light and heat of ordinary burning are nothing at all compared with the deflagration which such a temperature would cause by acting on the hardest known body. A few grains of platinum or iron striking the atmosphere with the velocity of the celestial motions might evolve as much light and heat as are emitted by the burning of a pint of coal-oil or several pounds of gunpowder; and as the whole operation is over in a second, we may imagine how intense the light must be.

The varied phenomena of aërolites, meteors, shooting-stars, and meteoric showers depend solely on the number and nature of the meteoroids which give rise to them. If one of these bodies is so large and firm as to pass through the atmosphere and reach the earth without being destroyed by the potential heat, we have an aërolite. As this passage only occupies a few seconds, the heat has not time to penetrate far into the interior of the body, but expends itself in melting and volatilizing the outer portions. When the body first strikes the denser portion of the atmosphere, the resistance becomes so enormous that the aërolite is frequently broken to pieces with such violence that it seems to explode. Further color is given to the idea of an explosion by the loud detonation which follows, so that the explosion is frequently spoken of as a fact, and as the cause of the detonation. Really, there is good reason to believe that both of these phenomena are due to the body striking the air with a velocity of ten, twenty, or thirty miles a second.

If, on the other hand, the meteoroid is so small or so fusible as to be dissipated in the upper regions of the atmosphere, we have a common shooting-star, or a meteor of greater or less brilliancy. Very careful observations have been made from time to time, with a view of finding the height of these bodies above the earth at their appearance and disappearance. An attempt of this kind was made by the Naval Observatory on the occasion of the meteoric shower of November 13th, 1867, when Professor Harkness was sent to Richmond to map the paths of the brighter meteors as seen from that point. By comparing these paths with those mapped at Washington, the parallaxes, and thence the altitudes, of these bodies were determined. The lightning-like rapidity with which the meteors darted through their course rendered it impossible to observe them with astronomical precision; but the general result was that they were first seen at an average height of 75 miles, and disappeared at a height of 55 miles. There was no positive evidence that any meteor commenced at a height much greater than 100 miles. It is remarkable that this corresponds very nearly to the greatest height at which the most brilliant meteors are ever certainly seen. These phenomena seem to indicate that our atmosphere, instead of terminating at a height of 45 miles, as was formerly supposed, really extends to a height of between 100 and 110 miles.

390   THE SOLAR SYSTEM.

The ordinary meteors, which we may see on every clear evening, move in every direction, thus showing that their orbits lie in all possible positions, and are seemingly scattered entirely at random. But the case is quite different with those meteoroids which give rise to meteoric showers. Here we have a swarm of these bodies, all moving in the same direction in parallel lines. If we mark, on a celestial globe, the

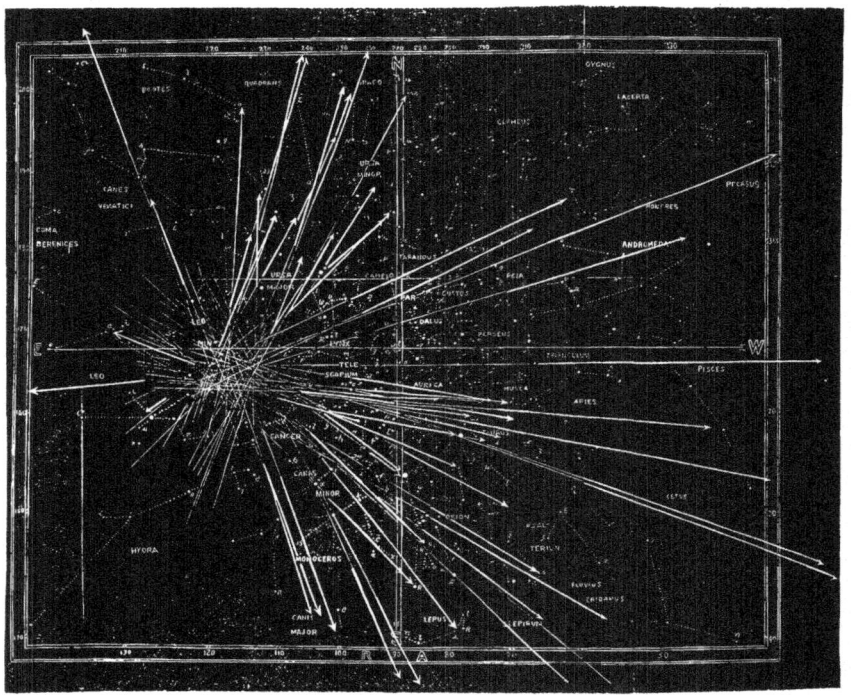

Fig. 95.—Meteor paths, illustrating the radiant point.

apparent paths of the meteors which fall during a shower, or if we suppose them marked on the celestial sphere, and then continue them backwards, we shall find them all to meet in the same point of the heavens. This is called the *radiant point*. It always appears in the same position, wherever the observer is situated, and does not partake of the diurnal motion of the earth; that is, as the stars seem to move towards the west in their diurnal course, the radiant point moves with them. The point in question is purely an effect of perspective, being the "vanishing point" of the parallel lines in which the meteors really move. These lines do not appear in their real direction in space, but are seen as projected on the celestial sphere. A good visible illustration of the effect in question may be afforded by looking upwards and watching falling snow during a calm. The flakes which are falling directly towards the observer do not seem to move at all, while the surrounding flakes seem to separate from them on all sides. So with the meteoric showers. A meteor coming directly towards the observer does not seem to move at all, and marks the radiant point from which all the others seem to diverge. The great importance of the determination of the radiant point arises from the fact that it marks the direction in which the meteors are moving relatively to the earth, and thus affords some data for determining their orbits.

§ 6. *Relations of Comets and Meteoroids.*

We have now to mention a series of investigations which led to the discovery of a curious connection between meteoroids and comets. These investigations were commenced by Professor Newton on the November meteoric showers. Tracing back the historical accounts of these showers to which we have already alluded, he found that the thirty-three-year period, which had been suspected by Olbers, was confirmed by records reaching back a thousand years. Moreover, the showers in question occurred only at a certain time of the year: in 1799 and 1833, it was on November 12th or November 13th. In other words, the shower occurred only as the earth passed a certain point of its orbit. But this point was found not to be always the same, the showers being found to occur about a couple of days earlier every century as they were traced back. The principal conclusions to which these facts led were as follows:

1. That the swarm of meteoroids which cause the Novem-

ber showers revolve around the sun in a definite orbit, which intersects the orbit of the earth at the point which the latter now passes on November 13th.

2. The point of intersection of the two orbits moves forwards about 52″ per annum, or nearly a degree and a half a century, owing to a change in the position of the meteoric orbit.

3. The swarm of meteoroids is not equally scattered all around their orbit, but the thickest portion extends along about one-fifteenth of the orbit.

4. The earth meets this swarm, on the average, once in 33.25 years. At other times the swarm has not arrived at the point of crossing, or has already passed it, and a meteoric shower cannot occur unless the earth and the swarm cross at the same time.

Professor Newton did not definitely determine the time of revolution of the meteors in their orbit, but showed that it must have one of five values. The greatest of these values, and the one which it seems most natural to select, is that of the mean interval between the showers, or $33\frac{1}{4}$ years. Adopting this period, it would follow that between 1799, when Humboldt saw the meteoric shower, and 1833, when it was seen throughout the United States, the swarm of meteoroids had been flying out as far as the planet Uranus in a very elliptical orbit, and returning again. But the periodic time might also be one year and about eleven days. Then the group which Humboldt saw on November 12th, 1799, would not reach the same point of its orbit until November 23d, 1800, when the earth would have passed by. Passing 11 days later every year, it would make about 33 revolutions in 34 years, and thus would pass about the middle of November once more, and another shower would occur. In a word, giving exact numbers, we might suppose that in the period of $33\frac{1}{4}$ years the meteoroids made one revolution, or $32\frac{1}{4}$, $34\frac{1}{4}$, $65\frac{1}{2}$, or $67\frac{1}{2}$ revolutions, and the conditions of the problem would be equally satisfied.

At the same time, Professor Newton gave a test by which

the true time could be determined. As we have said, he showed that the node of the orbit changed its position 52" a century, and there could be no doubt that this change was due to the attraction of the planets. If, then, the effect of this attraction was calculated for each of the five orbits, it would be seen which of them would give the required change. This was done by Professor Adams, of England, and the result was that the thirty-three-year period, and that alone, was admissible.

These researches of Professor Newton were published in 1864, and ended with a prediction of the return of the shower on November 13th of one or more of the three following years—probably 1866. This prediction was verified by a remarkable meteoric shower seen in Europe on that very day, which, however, was nearly over before it could become visible in this country. On the same date of the year following, a shower was visible in this country, and excited great public interest. From the data derived from the first of these showers, Schiaparelli, an Italian astronomer, was led to the discovery of a remarkable relation between meteoric and cometary orbits. Assuming the period of the November meteoroids to be $33\frac{1}{4}$ years, he computed the elements of their orbit from the observed position of the radiant point. A similar computation was made by Leverrier, and the results were presented to the French Academy of Sciences on January 21st, 1867.

The exact orbit which these bodies followed through space, crossing the earth's orbit at one point, and extending out beyond the planet Uranus at another, was thus ascertained. But, as these bodies were absolutely invisible, no great interest seemed to attach to their orbit until it was found that a comet was moving in that very orbit. This was a faint telescopic comet discovered by Tempel, at Marseilles, in December, 1865. It was afterwards independently discovered by Mr. H. P. Tuttle, at the Naval Observatory, Washington. It passed its perihelion in January, and, receding from the sun, vanished from sight in March. It was soon found to move in an elliptic orbit, but, owing to the uncertainty of observa-

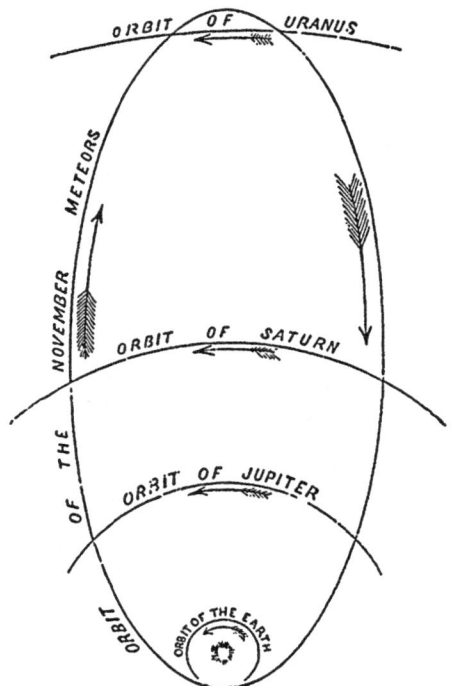

Fig. 96.—Orbit of November meteors and the comet of 1861.

tions on such a body, there was at first some disagreement as to the exact periodic time. The subject was taken up by Dr. Oppolzer, of Vienna, who, in January, 1867, was able to present a definitive orbit of the comet, which was published in the *Astronomische Nachrichten* on the 28th of that month. We now present the orbit of the comet, as found by Oppolzer, and that of the meteors, as found by Leverrier, premising that these orbits were computed and published within a few days of each other, without any knowledge on the part of either astronomer of the results obtained by the other:

	The Comet.	Meteoroids.
Period of revolution..................	33.18 yrs.	33.25 yrs.
Eccentricity...........................	0.9054	0.9044
Perihelion distance...................	0.9765	0.9890
Inclination of orbit...................	162° 42'	165° 19'
Longitude of the node...............	51° 26'	51° 18'
Longitude of perihelion.............	42° 24'	Near node.

The similarity of these orbits is too striking to be the result of chance. The only element of which the values differ materially is the inclination, and this difference proceeds from Leverrier not having used a very exact position of the radiant point in making his computations. Professor Adams found by a similar calculation that the inclination of the orbit of the meteoroids was 163° 14′, only half a degree different from that of the orbit of Tempel's comet. The result of these investigations was as follows:

The November meteoric showers arise from the earth encountering a swarm of particles following Tempel's comet in its orbit.

When this fact came out, Schiaparelli had been working on the same subject, and had come to a similar conclusion with regard to another group of meteors. It had long been known that about August 9th of every year an unusual number of meteors shoot forth from the constellation Perseus. At times these showers have been inferior only to those of November. Thus, on August 9th, 1798, they succeeded each other so rapidly as to keep the eye of the observer almost constantly engaged, and several hundred may nearly always be counted on the nights of the 9th, 10th, and 11th. These August meteors are remarkable in that they leave trails of luminous vapor which often last several seconds. Assuming the orbit of this group to be a parabola, it was calculated by Schiaparelli, and is substantially the same with that of a comet observed in 1862. The following are the elements of the orbits of the two bodies:

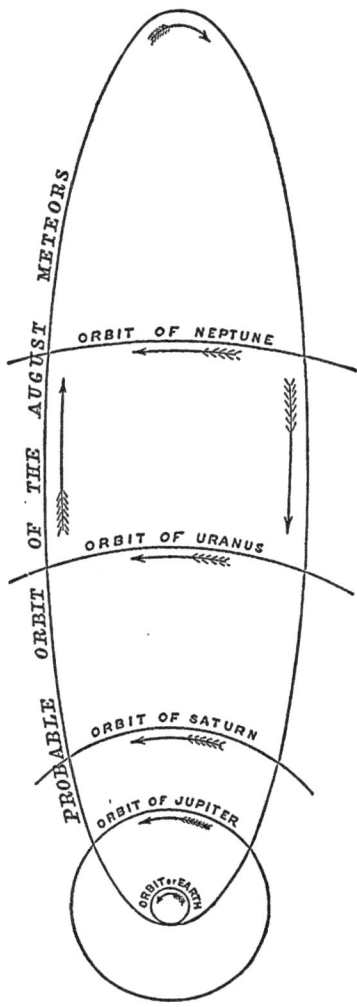

FIG. 97.—Orbit of the third comet of 1862.

	Comet II., 1862.	August Meteoroids.
Perihelion distance..................	0.9626	0.9643
Inclination of orbit..................	113° 35'	115° 57'
Longitude of the node..............	137° 27'	138° 16'
Longitude of the perihelion........	344° 41'	343° 28'

It appears that the August meteors are caused by a long stream of bodies following the second comet of 1862 in its orbit, or, rather, moving in the same orbit with it. The orbit of this comet is decidedly elliptic; the difference from the parabola is, however, too small to be determined with great precision. According to Oppolzer, the period derived from the observations would be 124 years, which, however, may be ten years or more in error.

A third striking case of the connection between comets and meteors which we are showing is afforded by the actual prediction of a meteoric shower on the night of November 27th, 1872. I have already described Biela's comet as first breaking into two pieces and then entirely disappearing, as though its parts had become completely scattered. This is one of the few comets which may come very near the earth, the latter passing the orbit of the comet on November 27th of each year. By calculation, the comet should have passed the point of crossing early in September, 1872, while the earth reached the same point between two and three months later. Judging from analogy, there was every reason to believe that the earth would encounter a stream of meteoroids consisting of the remains of the lost comet, and that a small meteoric shower would be the result. Moreover, it was shown that the meteors would all diverge from a certain point in the constellation Andromeda, as the radiant point, because that would be the direction from which a body moving in the orbit of the comet would seem to come. The prediction was fully verified in every respect. The meteors did not compare, either in numbers or brilliancy, with the great displays of November; but, though faint, they succeeded each other so rapidly that the most casual observer could not fail to notice them, and they all moved in the predicted direction.

That the meteoroids in these cases originally belonged to the comet, few will dispute. Accepting this, the phenomena of the November showers lead to the conclusion that the comet of 1866, with which they are associated, was not an original member of our system, but has been added to it within a time which, astronomically speaking, is still recent. The separate meteoroids which form the stream will necessarily have slightly different periodic times. Such being the case, they will, in the course of many revolutions, gradually scatter themselves around their entire orbit; and then we shall have an equal meteoric shower on every 13th of November. This complete scattering seems to have actually taken place in the case of the August meteoroids, since we have nearly the same sort of shower on every 9th or 10th of August. But in the case of the November meteors, the stream is not yet scattered over one-tenth of the orbit. If we suppose that the motions of the slowest and the swiftest bodies of the stream only differ by a thousandth part of their whole amount—which is not an unreasonable supposition—it would follow that the stream had only made about 100 revolutions around the sun, and had therefore been revolving only about 3300 years. Though this number is purely hypothetical, we may say with confidence that the stream has not been in existence many thousand years.

This opinion is strongly supported by the fact that the orbit of this meteoric comet passes very near that of Uranus as well as that of the earth, so that there is reason to believe that it was introduced into our system by the attraction of one of these planets, probably of Uranus. If the comet is seen on its next return, in 1899, we may hope that its periodic time will be determined with sufficient accuracy to enable us to fix with some probability the exact date at which Uranus brought it into our system. Indeed, Leverrier has attempted to do this already, having fixed upon the year 126 of our era as the probable date of this event; but, unfortunately, neither the position of the orbit nor the time of revolution is yet known with such accuracy as to inspire confidence in this result.

The idea that this November group is something comparatively new is strengthened by a comparison with that which produces the August meteors, where we find a decided mark of antiquity. Here the swiftest of the group has, in the course of numerous revolutions, overtaken the slowest, so that the group is now spread almost equally around the entire orbit. The time of revolution being, in this case, more than a century, this equal distribution would take a much longer time than in the other case, where the period is only thirty-three years; so that we can say, with considerable probability, that the August group has been in our system at least twenty times as long as the November group.

§ 7. *The Physical Constitution of Comets.*

A theory of the physical constitution of comets, to be both complete and satisfactory, must be founded on the properties of matter as made known to us here at the surface of the earth. That is, we must show what forms and what combinations of known substances would, if projected into the celestial spaces, present the appearance of a comet. Now, this has never yet been completely done. Theories without number have been propounded, but they fail to explain some of the phenomena, or explain them in a manner not consistent with the known laws of matter or force. We cannot stop even to mention most of these theories, and shall therefore confine our attention to those propositions which are to some extent sustained by facts, and which, on the whole, seem to have most probability in their favor.

The simplest form of these bodies is seen in the telescopic comets, which consist of minute particles of a cloudy or vaporous appearance. Now, we know that masses which present this appearance at the surface of the earth, where we can examine them, are composed of detached particles of solid or liquid matter. Clouds and vapor, for instance, are composed of minute drops of water, and smoke of very minute particles of carbon. Analogy would lead us to suppose that the telescopic comets are of the same constitution. They are generally tens of thousands of miles in diameter, and yet of such tenuity that the smallest stars are seen through them. The strongest evidence of this constitution is, however, afforded by the phenomena of meteoric showers described in the last section. We have seen that these are caused by our atmosphere encountering the débris of comets, and this débris presents itself in the form of detached meteoroids, of very small magnitude, but hundreds of miles apart.

The only alternative to this theory is that the comet is a mass of true gas, continuous throughout its whole extent. This gaseous theory derives its main support from the spectroscope, which shows the spectrum of the telescopic comets to consist of bright bands, the mark of an incandescent gas. Moreover, the resemblance of these bands to those produced by the vapor of carbon is so striking that it is quite common among spectroscopists to speak of a comet as consisting of the gas of some of the compounds of carbon. But there are several difficulties which look insuperable in the way of the theory that a comet is nothing but a mass of gas. In the first place, the elastic force of such a mass would cause it to expand beyond all limits when placed in a position where there is absolutely no pressure to confine it, as in the celestial spaces. Again, a gas cannot, so far as experiment has ever gone, shine by its own light until it is heated to a high temperature, far above any that can possibly exist at distances from the sun so great as those at which comets have been situated when under examination with the spectroscope. Finally, in the event of a purely gaseous comet being broken up and dissipated, as in the case of Biela's comet, it is hardly possible to suppose that it would separate into innumerable widely detached pieces, as this comet did. The gaseous theory can, therefore, not be regarded as satisfactory. It may be that comets will hereafter be found to consist of some combination of solid and gaseous matter, the exact nature of which is not yet determined; or it may be that this matter is of a nature or in a form wholly unlike anything that we are acquainted with or can produce here on the earth. As the case now stands, we must regard the spectrum of a comet as something not yet satisfactorily accounted for.

When we turn from telescopic comets to those brilliant ones which exhibit a nucleus and a tail, we can trace certain operations which are not seen in the case of the others. What the nucleus is—whether it is a solid body several hundred miles in diameter, or a dense mass of the same materials which compose a telescopic comet—we are quite unable to say. But there can hardly be any reasonable doubt that it is composed of some substance which is vaporized by the heat of the solar rays. The head of such a comet, when carefully examined with the telescope, is found to be composed of successive envelopes or layers of vapor; and when these envelopes are watched from night to night, they are found to be gradually rising upwards, growing fainter and more indistinct in outline as they attain a greater elevation, until they are lost in the outlying parts of the coma. These rising masses form the fan-shaped appendage described in a preceding section.

The strongest proof that some evaporating process is going on from the nucleus of the comet is afforded by the movements of the tail. It has long been evident that the tail could not be an appendage which the comet carried along with it, and this for two reasons: first, it is impossible that there could be any cohesion in a mass of matter of such tenuity that the smallest stars could be seen through a million of miles of it, and which, besides, constantly changes its form; secondly, as a comet flies around the sun in its immediate neighborhood, the tail appears to move from one side of the sun to another with a rapidity which would tear it to pieces, and send the separate parts flying off in hyperbolic orbits, if the movement were real. The inevitable conclusion is that the tail is not a fixed appendage of the comet, which the latter carries with it, but a stream of vapor rising from it, like smoke from a chimney. As the line of smoke which we now see coming from the chimney is not the same which we saw a minute ago, because the latter has been blown away and dissipated, so we do not see the same tail of a comet all the time, because the matter which makes up the tail is constantly streaming outwards, and constantly being replaced by new vapor rising from the nucleus. The evaporation is, no doubt, due to the heat of the sun, for there can be no evaporation without heat, and the tails of comets increase enormously as they approach the sun. Altogether, a good idea of the operations going on in a comet will be obtained if we conceive the nucleus to be composed of water or other volatile fluid which is boiling away under the heat of the sun, while the tail is a column of steam rising from it.

We now meet a question to which science has not yet been able to return a conclusive answer. Why does this mass of vapor always fly away from the sun? That the matter of the comet should be vaporized by the sun's rays, and that the nucleus should thus be enveloped in a cloud of vapor, is perfectly natural, and entirely in accord with the properties of matter which we observe around us. But, according to all known laws of matter, this vapor should remain around the head, except that the outer portions would be gradually detached and thrown off into separate orbits. There is no known tendency of vapor, as seen on the earth, to recede from the sun, and no known reason why it should so recede in the celestial spaces. Various theories have been propounded to account for it; but as they do not rest on causes which we have verified in other cases, they must be regarded as purely hypothetical.

The first of these explanations, in the order of time, is due to Kepler, who conceived the matter of the tail to be driven off by the impulsion of the solar rays, which thus bleached the comet as they bleach cloths here. If light were an emission of material particles, as Newton supposed it to be, this view would have some plausibility. But light is now conceived to consist of vibrations in an ethereal medium; and there is no known way in which they could exert any propelling force on matter. Two or three years ago, it was for a while supposed that the "radiometer" of Mr. Crookes might really indicate such an action of the solar rays upon matter in a vacuum, but it is now found that the action exhibited is really due to a minute quantity of air left in the instrument. Had Mr. Crookes shown that the motion of his radiometer was really due to the impulsion of the solar rays, we might be led to the remarkable conclusion that Kepler's theory, though rejected for more than two centuries, was, after all, quite near the truth.

Sir Isaac Newton, being the author of the emission theory of light, could not dispute the possibility of Kepler's views being correct, but nevertheless gave the preference to another hypothesis. He conceived the celestial spaces to be filled with a very rare medium, through which the sun's rays passed without heating it, as they pass through cold air. But the comet being warmed up by the rays, the medium surrounding it is warmed up by contact, and thus a warm current is sent out from the comet, just as a current of warm air rises from a heated body on the surface of the earth. This current carries the vapor of the comet with it, and thus gives rise to the tail in the same way that the current of warm air rising from a chimney carries up a column of smoke. It has long been established that there is no medium in the planetary spaces in which such an effect as this is possible: Newton's theory is, therefore, no longer considered.

In recent times, Zöllner has endeavored to account for the tail of the comet by an electrical action between the sun and the vapor rising from the nucleus of the comet. The various papers in which he has elaborated his views of the constitution of comets are marked by profound research; and we must regard his theories as those which, on the whole, most completely explain all the phenomena. But they still lack the one thing needful to secure their reception: there is no evidence that the sun acts as an electrified body; and until such evidence is adduced by experiment, or by observation on other bodies than comets, the electric theory of the comet's tail can only be regarded as a more or less probable hypothesis. Indeed, some physicists claim that any such electric action in the planetary spaces is impossible. Before any theory can be definitely settled upon, accurate observations must be made upon the tails of comets with a view of learning the law according to which the vapor is repelled from the sun. Such observations were made by Bessel on Halley's comet in 1835, and by various observers on the great comet of 1858. The former were investigated by Bessel himself, and the latter by several mathematicians, among them Professor Peirce, whose results are found in a paper communicated to the American Academy in 1859. He found the repulsive force of the sun upon the particles which form the front edge of the tail to be $1\frac{1}{2}$ times its attractive force upon ordinary bodies at the same distance. It seemed constantly to diminish as the back edge of the tail was approached; but, owing to the poor definition of this edge, and the uncertainty whether it was composed of a continuous stream of particles, the amount of the diminution could not be accurately fixed. The successive envelopes were found to ascend uniformly towards the sun at the rate of about thirty-five miles an hour. Bond, from a careful examination of all the observations, was led to the result that the rate of ascent diminished as the height became greater.

An apparently necessary conclusion from this constant evaporation and expulsion of vapor from comets with tails is, that such bodies are constantly wasting away when in the neighborhood of the sun. This conclusion is strengthened by the fact that not a single comet of very short period has a considerable tail, the probability being that all the volatile matter which once went to form the tail has been evaporated. Indeed, from the descriptions of the old chroniclers, it has been supposed that Halley's comet had a much more conspicuous tail at the time of its earliest recorded apparitions than it has exhibited at its last few returns. There is, however, no necessity for supposing the diminution so rapid as this, for the amount of matter really necessary to make the most splendid tail is so extremely small that a comet might lose it a hundred times over without becoming perceptibly smaller. This constant loss of matter through the tail affords an additional ground for the view that comets in general are visitors introduced into our system by the action of the planets. If, for instance, such a comet as Halley's had been a member of our system for millions of years, and had returned to perihelion a hundred thousand times, all its volatile matter must long ago have evaporated.

The question of the mass and density of comets is also one of those on which it is difficult to reach satisfactory conclusions. We cannot certainly decide from mere telescopic observation whether the nucleus is a single large body, like a planet or satellite, or whether it is merely the densest part of an immense cloud of meteoroids. The mass of nebulous matter which surrounds the nucleus increases so gradually as we approach the central parts, that it is hardly possible to decide where the nucleus begins: the more powerful the telescope, the smaller the nucleus generally appears. Moreover, in the same comet, the apparent magnitude of the nucleus is subject to immense variations, thus showing that it cannot be a solid body out to its apparent limits. If we considered only this circumstance, and the general analogy with telescopic comets, we should say that even the densest part of the comet was nothing but a cloud of solid or liquid particles so thick that it looked solid, as a cloud does in our sky. But if this was the case, as Professor Peirce showed in his investigations of the comet of 1858, the comets of 1680 and of 1843 must have been completely pulled apart by the enormous tidal forces generated by their near approach to the sun. In the opinion of this investigator, the fact that they went through such an ordeal shows them to be of metallic density.

The question is frequently asked, What would be the effect if a comet should strike the earth? This would depend upon what sort of a comet it was, and what part of the comet came in contact with our planet. The latter might pass through the tail of the largest comet without the slightest effect being produced, the tail being so thin and airy that a million miles thickness of it looks only like gauze in the sunlight. It is not at all unlikely that such a thing may have happened without ever being noticed. A passage through a telescopic comet would be accompanied by a brilliant meteoric shower, probably a far more brilliant one than has ever been recorded. No more serious danger would be encountered than that arising from a possible fall of meteorites. But a collision between the nucleus of a large comet and the earth might be a serious matter. If, as Professor Peirce supposes, the nucleus is a solid body of metallic density, many miles in diameter, the effect where the comet struck would be terrific beyond conception. At the first contact in the upper regions of the atmosphere, the whole heavens would be illuminated with a resplendence beyond that of a thousand suns, the sky radiating a light which would blind every eye that beheld it, and a heat which would melt the hardest rocks. A few seconds of this, while the huge body was passing through the atmosphere, and the collision at the earth's surface would in an instant reduce everything there existing to fiery vapor, and bury it miles deep in the solid earth. Happily, the chances of such a calamity are so minute that they need not cause the slightest uneasiness. There is hardly a possible form of death which is not a thousand times more probable than this. So small is the earth in comparison with the celestial spaces, that if one should shut his eyes and fire a gun at random in the air, the chance of bringing down a bird would be better than that of a comet of any kind striking the earth.

§ 8. *The Zodiacal Light.*

This object consists of a very soft, faint column of light, which may be seen rising from the western horizon after twilight on any clear winter or spring evening: it may also be seen rising from the eastern horizon just before daybreak in the summer or autumn. It really extends out on each side of the sun, and lies nearly in the plane of the ecliptic. The reason it cannot be well seen in the summer and autumn evenings is, that in our latitudes the course of the ecliptic in the south-west is, during those seasons, so near the horizon that the light in question is extinguished by the great thickness of atmosphere through which it has to pass. Near the equator, where the ecliptic always rises high above the horizon, the light can be seen about equally well all the year round. It grows fainter the farther it is from the sun, and can generally be traced to about 90° from that luminary, when it gradually fades away. But in a very clear atmosphere, between the tropics, it has been traced all the way across the heavens, from east to west, thus forming a complete ring.

Such is the zodiacal light as it appears to the eye. Putting its appearances all together, we may see that it is due to a lens-shaped appendage of some sort surrounding the sun, and extending out a little beyond the earth's orbit. It lies very nearly in the plane of the ecliptic, but its exact position is difficult to determine, not only owing to its indistinct outline, but because in northern latitudes the southern edge will be dimmed by the greater thickness of atmosphere through which it is seen, and thus the light will look farther north than it really is. The nature of the substance from which this light emanates is entirely unknown. Its spectrum has been examined by several observers, some of whom have reported it as consisting of a single yellow line, and therefore arising from an incandescent gas. This would indicate a lenticular-shaped atmosphere of inconceivable rarity surrounding the sun, and extending out near the plane of the ecliptic beyond the orbit of the earth. But Professor Wright, of Yale College, who has made the most careful observations of this spectrum, finds it to be continuous. For several reasons, too minute to enter into now, this observation seems to the writer more likely to be correct. Accepting it, we should be led to the conclusion that the phenomenon in question is due to reflected sunlight, probably from an immense cloud of meteoroids filling up the space between the earth and sun. But further researches must be made before a conclusive result can be reached.

PART IV.—THE STELLAR UNIVERSE.

INTRODUCTORY REMARKS.

HITHERTO our attention has been principally occupied with the bodies which surround our sun and make up the solar system. Notwithstanding the immense distances at which these bodies are found, we may regard them, in comparison with the fixed stars, as an isolated family immediately surrounding us, since a sphere as large as the whole solar system would only appear as a point to the vision if viewed from the nearest star. The space which separates the orbit of Neptune from the fixed stars and the fixed stars from each other is, so far as we can learn, entirely void of all visible matter, except occasional waste nebulous fragments of a meteoric or cometary nature which are now and then drawn in by the attraction of our sun.

The widest question which the study of the stars presents to us may be approached in this way: We have seen, in our system of sun, planets, and satellites, a very orderly and beautiful structure, every body being kept in its own orbit through endless revolutions by a constant balancing of gravitating and centrifugal forces. Do the millions of suns and clusters scattered through space, and brought into view by the telescope, constitute a greater system of equally orderly structure? and, if so, what is that structure? If we measure the importance of a question, not by its relations to our interests and our welfare, but by the intrinsic greatness of the subject to which it relates, then we must regard this question as one of the noblest with which the human mind has ever been occupied. In piercing the mystery of the solar system, and showing that the earth on which we dwell was only one of the smaller of eight planets which move around the sun, we made a great step in the way of enlarging our ideas of the immensity of creation and of the comparative insignificance of our sublunary interests. But when, on extending our view, we find our sun to be but one out of unnumbered millions, we see that our whole system is but an insignificant part of creation, and that we have an immensely greater fabric to study. When we have bound all the stars, nebulæ, and clusters which our telescopes reveal into a single system, and shown in what manner each stands related to all the others, we shall have solved the problem of the material universe, considered, not in its details, but in its widest scope.

From the time that Copernicus showed the stars to be self-luminous bodies, situated far outside of our solar system, the question thus presented has occupied the attention of the philosophical class of astronomers. The original view, which has been the starting-point of all speculation on the subject, we have described in the Introduction as that of a spherical universe. The apparent sphericity of the vault of heaven, the uniformity of the diurnal revolution, and the invariability of the relative positions of the stars, all combined to strengthen the idea that the latter were set on the interior surface of a hollow sphere, having the earth or the sun in its centre. This sphere constituted the firmament of the ancients, outside of which was situated the empyrean, or kingdom of fire. Copernicus made no advance whatever on this idea. Galileo and Kepler seem to have made the first real advance—the former by resolving the Milky Way into stars with his telescope, the latter by suggesting that our sun might be simply one of numerous stars scattered through space, looking so bright only on account of our proximity to it. In the problem of the stellar system this conception held the same important place which that of the earth as a planet did in the problem of the solar system. But Kepler was less fortunate than Copernicus in that he failed to commend his idea, even to his own judgment. It was by affording a starting-point for the researches of Kant and Herschel that Kepler's suggestion really bore fruit.

Notwithstanding the amount of careful research which Herschel and his successors have devoted to it, we are still very far from having reached even an approximate solution of the problem of which we speak. In whatever direction we pursue it, we soon find ourselves brought face to face with the infinite in space and time. Especially is this the case when we seek to know, not simply what the universe is to-day, but what causes are modifying it from age to age. All the knowledge that man has yet gathered is then found to amount to nothing but some faint glimmers of light shining here and there through the seemingly boundless darkness. The glimmer is a little brighter for each successive generation, but many centuries must elapse before we can do much more than tell how the nearer stars are situated in space. Indeed, we see as yet but little hope that an inhabitant of this planet will ever, from his own observations and those of his predecessors, be able to completely penetrate the mystery in which the structure and destiny of the cosmos are now enshrouded. However this may be in the future, all we can do at present is to form more or less probable conjectures, founded on all we know of the general character of natural law. In a strictly scientific treatise, such conjectures would find no place; and if we had to grope in absolute darkness, they would be entirely inappropriate in any but a poetical or religious production. But the subject is too fascinating to permit us to neglect the faintest light by the aid of which we may penetrate the mystery; we shall therefore briefly set forth both what men of the past have thought on the subject, what the science of to day enables us to assert with some degree of probability, and what knowledge it wholly denies us. To proceed in scientific order, we must commence by laying a wide foundation of facts. Our first step will therefore be to describe the heavens as they appear to the naked eye, and as they are seen in the telescope.

CHAPTER I.

THE STARS AS THEY ARE SEEN.

§ 1. *Number and Orders of Stars and Nebulæ.*

THE total number of stars in the celestial sphere visible with the average naked eye may be estimated, in round numbers, as 5000. The number varies so much with the perfection and training of the eye, and with the atmospheric conditions, that it cannot be stated very definitely. When the telescope is pointed at the heavens, it is found that for every star visible to the naked eye there are hundreds, or even thousands, too minute to be seen without artificial aid. From the counts of stars made by Herschel, Struve has estimated that the total number of stars visible with Herschel's twenty-foot telescope was about 20,000,000. The great telescopes of modern times would, no doubt, show a yet larger number; but a reliable estimate has not been made. The number is probably somewhere between 30,000,000 and 50,000,000.

At a very early age, the stars were classified according to their apparent brightness or magnitude. The fifteen brightest ones were said to be of the first magnitude; the fifty next in order were termed of the second magnitude, and so on to the sixth, which comprised the faintest stars visible to the naked eye. The number of stars of each order of magnitude between the north pole and the circle 35° south of the equator is about as follows:

Of magnitude 1 there are about	14	stars.
" 2 "	48	"
" 3 "	152	"
" 4 "	313	"
" 5 "	854	"
" 6 "	2010	"
Total visible to naked eye	3391	"

This limit includes all the stars which, in the Middle States, culminate at a greater altitude than 15°. The number of the sixth magnitude which can be seen depends very much upon the eye of the observer and the state of the sky. The foregoing list includes all that can be seen by an ordinary good eye in a clear sky when there is no moonlight; but the German astronomer Heis, from whom these numbers are taken, gives a list of 1964 more which he believes he can see without a glass.

The system of expressing the brightness of the stars by a series of numbers is continued to the telescopic stars. The smallest star visible with a six-inch telescope under ordinary circumstances is commonly rated as of the thirteenth magnitude. On the same scale, the smallest stars visible with the largest telescopes of the world would be of about the sixteenth magnitude, but no exact scale for these very faint stars has been arranged.

Measures of the relative brilliancy of the stars indicate that, as we descend in the scale of magnitude, the quantity of light emitted diminishes in a geometrical ratio, the stars of each order being, in general, between two-fifths and one-third as bright as those of the order next above them. This order of diminution is not, however, exact, because the arrangement of magnitudes has been made by mere estimation of individual observers who may have hit on different and varying ratios; but it is a sufficient approach to the truth for common purposes. From the second to the fifth magnitude the diminution is probably one-third in each magnitude, after that about two-fifths. Supposing the ratio two-fifths to be exact, we find that it would take about

$2\frac{1}{2}$ stars of the second magnitude to make one of the first.
6 " third " " "
16 " fourth " " "
40 " fifth " " "
100 " sixth " " "
10,000 " eleventh " " "
1,000,000 " sixteenth " " "

The number of stars of the several scales of magnitude vary in a ratio not far different from the inverse of that of their brightness, the ratio being a little greater in the case of the higher magnitudes, and probably a little less in the case of the lower ones. Thus, we see that there are about three times as many stars of the second magnitude as of the first, three times as many of the third as of the second, and after that something less than three times as many of each magnitude as of the magnitude next above. Comparing this with the table of relative brightness just given, we may conclude that if all the stars of each magnitude were condensed into a single one, the brightness of the combined stars thus formed would not vary extravagantly from one to another until we had passed beyond the ninth or tenth magnitude. But it is certain that the brightness would ultimately diminish, because otherwise there would be no limit to the total amount of light given by the stars, and the whole heavens would shine like the sun.

The reader will, of course, understand that this arrangement by magnitude is purely artificial. Really the stars are of every order of brightness, varying by gradations which are entirely insensible, so that it is impossible to distinguish between the brightest star of one magnitude and the faintest of the magnitude next above it. Hence, those astronomers who wish to express magnitudes with the greatest exactness, divide them into thirds or even tenths; so that, for instance, stars between the sixth and seventh magnitudes are called 6.1, 6.2, 6.3, and so on to 6.9, according to their brilliancy. Various attempts have been made to place the problem of the relative amounts of light emitted by the stars upon a more exact basis than this old one of magnitudes, but this is a very difficult thing to do, because there is no way of measuring light except by estimation with the eye. In order to measure the relative intensity of two lights, it is necessary to have some instrument by which the intensity of one or both the lights may be varied until the two appear to be equal. Instruments for this purpose are known as photometers, and are of various constructions. For comparing the light of different stars, the photometer most used at the present time is that of Zöllner. By this instrument the light of the stars, as seen through a small telescope, is compared both in color and intensity with that of an artificial star, the light of which can be varied at pleasure. A complete set of measures with this instrument, including most of the brighter stars, is one of the wants of astronomy which we may soon hope to see supplied. The most extended recent series of photometric estimates with which the writer is acquainted is that of Professor Seidel, of Munich, which includes 209 stars, the smallest of which are of the fifth magnitude. An interesting result of these estimates is that Sirius gives us four times as much light as any other star visible in our latitude.

Catalogues of Stars.—In nearly every age in which astronomy has flourished catalogues of stars have been made, giving their positions in the heavens, and the magnitude of each. The earliest catalogue which has come to us is found in the "Almagest" of Ptolemy, and is supposed to be that of Hipparchus, who flourished 150 years before the Christian era. It is said, but not on the best authority, that he constructed it in order that future generations might find whether any change had in the mean time taken place in the starry heavens. An examination of the catalogue shows that the constellations presented much the same aspect two thousand years ago that they do now. There are two or three stars of his catalogue which cannot now be certainly identified; but it is probable that the difficulty arises from the imperfection of the catalogue, and from the errors which may have crept into the numerous transcriptions of it during the sixteen centuries which elapsed before the art of printing was discovered. The catalogue of Hipparchus contains only about 1080 stars, so that he could not have given all that he was able to see. He probably omitted many stars of the smaller magnitudes. The actual number given in the "Almagest" is still less, being only 1030.

The next catalogue in the order of time is that of Ulugh Beigh, a son of the Tartar monarch Tamerlane, which dates from the fifteenth century. For the most part, the stars are the same as in the catalogue of Ptolemy, only the places were redetermined from the observations at Samarcand. It contains 1019 stars, eleven less than Ptolemy gives. Tycho Brahe, having made so great an improvement in the art of observation, very naturally recatalogued the stars, determining their positions with yet greater accuracy than his predecessors. His catalogue is the third and last important one formed before the invention of the telescope. It contains 1005 stars.

Our modern catalogues may be divided into two classes: those in which the position of each star in the celestial sphere (right ascension and declination) is given with all attainable precision, and those in which it is only given approximately, so as to identify the star, or distinguish it from others in its neighborhood. The catalogues of the former class are very numerous, but the more accurate ones are necessarily incomplete, owing to the great labor of making the most exact determination of the position of a star. There are, perhaps, between ten or twenty thousand stars the positions of which are catalogued with astronomical precision, and a hundred thousand more in which, though entire precision is aimed at, it is not attained. Of the merely approximate catalogues, the greatest one is the "Sternverzeichniss" of Argelander, which enumerates all the stars down to the ninth magnitude between the pole and two degrees south of the equator. The work fills three thin quarto volumes, and the entire number of stars catalogued in it exceeds three hundred thousand. This "star census" is being continued to the south pole at the observatory of Cordoba, South America, by Dr. Gould. Of the millions of stars of the tenth magnitude and upwards, hardly one in a thousand is, or can be, individually known or catalogued. Except as one or another may exhibit some remarkable peculiarity, they must pass unnoticed in the crowd.

Division into Constellations.—A single glance at the heavens shows that the stars are not equally scattered over the sky, but that great numbers of them, especially of the brighter ones, are collected into extremely irregular groups, known as constellations. At a very early age the heavens were represented as painted over with figures of men and animals, so arranged as to include the principal stars of each constellation. There is no historic record of the time when this was done, nor of the principles by which those who did it carried out their work; but many of the names indicate that it was during the heroic age. Some have sought to connect it with the Argonautic expedition, from the fact that several heroes of that expedition were among those thus translated to the heavens; but this is little more than conjecture. So little pains was taken to fit the figures to the constellations that we can hardly suppose them to have all been executed at one time, or on any well-defined plan. Quite likely, in the case of names of heroes, the original object was rather to do honor to the man than to serve any useful purpose in astronomy. Whatever their origin, these names have been retained to the present day, although the figures which they originally represented no longer serve any astronomical purpose. The constellation Hercules, for instance, still exists; but it no longer represents the figure of a man among the stars, but a somewhat irregular portion of the heavens, including the space in which the ancients placed that figure. In star-maps, designed for school instruction and for common use, it is still customary to give these figures, but they are not generally found on maps designed for the use of astronomers.

Naming the Stars.—The question how to name the individual stars in each constellation, so as to readily distinguish them, has always involved some difficulty. In the ancient catalogues they were distinguished by the part of the figure representing the constellation in which they were found; as, the eye of the Bull, the tail of the Great Bear, the right shoulder of Orion, and so on. The Arabs adopted the plan of giving special names to each of the brighter stars, or adopting such names from the Greeks. Thus, we have the well-known stars Sirius, Arcturus, Procyon, Aldebaran, and so on. Most of these names have dropped entirely out of astronomical use, though still found on some school maps of the stars. The system now most in use for the brighter stars was designed by Bayer, of Augsburg, Germany, about 1610. He published a set of star-maps, in which the individual stars of each constellation were designated by the letters of the Greek alphabet—α, β, γ, etc. The first letters were given to the brightest stars, the next ones to the next brightest, and so on. After the Greek letter is given the Latin name of the constellation in the genitive case. Thus, Alpha (α) Scorpii, or Alpha of the Scorpion, is the name of Arcturus, the brightest star in Scorpius; α Lyræ, of the brightest star in the Lyre; and so on. We have here a resemblance to our system of naming men, the Greek letter corresponding to the Christian name, and the constellation to the surname. When the Greek alphabet was exhausted, without including all the conspicuous stars, the Latin alphabet was drawn upon.

The Bayer system is still applied to all the stars named by him. Most of the other stars down to the fifth magnitude are designated by a system of numbers assigned by Flamsteed in his catalogue. Yet other stars are distinguished by their numbers in some well-known catalogue. When this method fails, owing to the star not being catalogued, the position in the heavens must be given.

The Milky Way, or Galaxy.—To the naked eye so much of the Galaxy as can be seen at one time presents the appearance of a white, cloud-like arch, resting on two opposite points of the horizon, and rising to a greater or less altitude, according to the position of the celestial sphere relative to the observer. Only half of the entire arch can be seen above the horizon at once, the other half being below it, and directly opposite the visible half. Indeed, there is a portion of it which can never be seen in our latitude, being so near the south pole that it is always below our horizon. If the earth were removed, or made transparent, so that we could see the whole celestial sphere at once, the Galaxy would appear as a complete belt extending around it. The telescope shows that the Galaxy arises from the light of countless stars, too minute to be separately visible with the naked eye. We find, then, that the telescopic stars, instead of being divided up into a limited number of constellations, are mostly condensed in the region of the Galaxy. They are least numerous in the regions most distant from the galactic belt, and grow thicker as we approach it. The more powerful the telescope, the more marked the condensation is. With the naked eye, the condensation is hardly noticeable, unless by actual count: a very small telescope will show a decided thickening of the stars in and near the Galaxy; while, if we employ the most powerful telescopes, a large majority of the stars they show are found to lie actually in the Galaxy. In other words, if we should blot out all the stars visible with a twelve-inch telescope, we should find that the greater part of the remaining stars were in the Galaxy. The structure of the universe which this fact seems to indicate will be explained in a subsequent section.

Clusters. — Besides this gradual and regular condensation towards the galactic belt, occasional condensations of stars into clusters may be seen. Indeed, some of these clusters are visible to the naked eye, sometimes as separate stars, like the Pleiades, but more commonly as milky patches of light, because the stars are too small to be seen separately. The number visible in powerful telescopes is, however, much greater. Sometimes there are hundreds, or even thousands, of stars visible in the field of the telescope at once; and sometimes the number is so great, and the individual stars so small, that they cannot be counted even in the most powerful telescopes ever made.

Nebulæ. — Another class of objects which are found in the celestial spaces are irregular masses of soft, cloudy light, which are hence termed nebulæ. Many objects which look like nebulæ in small telescopes are found by more powerful ones to be really star clusters. But, as we shall hereafter show, many of these objects are not composed of stars at all, but of immense masses of gaseous matter.

§ 2. *Description of the Principal Constellations.*

For the benefit of the reader who wishes to make himself acquainted with the constellations in detail, or to identify any bright star or constellation which he may see, we present a brief description of the principal objects which may be seen in the heavens at different seasons, illustrated by five maps, showing the stars to the fifth magnitude inclusive. The reader who does not wish to enter into these details can pass to the next section without any break of the continuity of thought.

For the purpose of learning the constellations, the star-maps will be a valuable auxiliary. It will be better to begin with the northern, or circumpolar, constellations, because these are nearly always visible in our latitude. The first one to be looked for is *Ursa Major* (the Great Bear, or the Dipper), from which the pole star can always be found by means of the pointers, as shown in Fig. 2, page 10. Supposing the observer to look for it at nine o'clock in the evening, he will see it in various positions, depending on the time of year, namely, in

April and May.....................north of the zenith.
July and August...................to the west of north, the pointers lowest.
October and November..........close to the north horizon.
January and February............to the east of north, the pointers highest.

These successive positions are in the same order with those which the constellation occupies in consequence of its diurnal motion around the pole. The pointers are in the body of the bear, while the row of stars on the other end of the constellation forms his tail.

Ursa Minor, or the Little Dipper, is the constellation to which the pole star belongs. It includes, besides the pole star, another star of the second magnitude, which lies nearly in the direction of the tail of Ursa Major.

Cassiopeia, or the Lady in the Chair, is on the opposite side of the pole from Ursa Major, at nearly the same distance. The constellation can be readily recognized from its three or four bright stars, disposed in a line broken into pieces at right angles to each other. In the ancient mythology, Cassiopeia is the queen of Cepheus; and in the constellation she is represented as seated in a large chair or throne, from which she is issuing her edicts.

Perseus is quite a brilliant constellation, situated in the

DESCRIPTION OF THE PRINCIPAL CONSTELLATIONS. 419

Milky Way, east* of Cassiopeia, and a little farther from the pole. It may be recognized by a row of conspicuous stars extending along the Milky Way, which passes directly through this constellation.

Other circumpolar constellations are Cepheus, the Camelopard, the Lynx, the Dragon (*Draco*), and the Lizard; but they do not contain any stars so bright as to attract especial attention. The reader who wishes to learn them can easily find them by comparing the star-maps with the heavens.

Owing to the annual motion of the sun among the stars, the constellations which are more distant from the pole cannot be seen at all times, but must be looked for at certain seasons, unless inconvenient hours of the night be chosen. We shall describe the more remarkable constellations as they are seen by an observer in middle north latitudes in four different positions of the starry sphere. The sphere takes all four of these positions every day, by its diurnal motion; but some of these positions will occur in the daytime, and others late at night or early in the morning.

First Position, Orion on the Meridian.—The constellations south of the zenith are those shown on Maps II. and III., the former being west of the meridian, the latter east. This position occurs on

> December 21st...at midnight.
> January 21st...at 10 o'clock P.M.
> February 20th..at 8 o'clock P.M.
> March 21st...at 6 o'clock P.M.

And so on through the year. In this position, Cassiopeia and Ursa Major are near the same altitude, the former high up in

---

* In the celestial sphere the points of the compass have, of necessity, a meaning which may seem different from that which we attribute to them on the earth. *North* always means towards the north pole; *south*, from it; *west*, in the direction of the diurnal motion; *east*, in the opposite direction. In Fig. 2, the arrows all point west, and by examining the figure it will be seen that below the pole north is upwards, and east is towards the west horizon. Really, these definitions hold equally true for the earth, the same differences being found between the points of the compass at different places on the earth—here and in China, for instance—that we see on the celestial sphere.

the north-west, the latter in the north-east. The Milky Way spans the heavens like an arch, resting on the horizon in the north-north-west and south-south-east. We shall first describe the constellations in its course.

Cygnus, the Swan, is sinking below the horizon, where the Milky Way rests upon it in the north-north-west, and only a few stars of it are visible. It will be better seen at another season.

Next in order come Cepheus, Cassiopeia, and Perseus, which we have already described as circumpolar constellations.

Above Perseus lies *Auriga*, the Charioteer, which may be readily recognized by a bright star of the first magnitude, called *Capella*, the Goat, now a few degrees north-west of the zenith. Auriga is represented as holding a goat in his arm, in the body of which this star is situated. About ten degrees east of Capella is the star β Aurigæ of the second magnitude; while still farther to the east is a group of small stars which also belongs to the same constellation. The latter extends some distance south of the zenith.

The Milky Way next passes between Taurus and Gemini, which we will describe presently, and then crosses the equator east of Orion, the most brilliant constellation in the heavens, having two stars of the first magnitude and four of the second. The former are Betelguese, or α Orionis, which is highest up, and may be recognized by its reddish color, and Rigel, or β Orionis, a sparkling white star, lower down, and a little to the west. The former is in the shoulder of the figure, the latter in the foot. Between the two, three stars of the second magnitude, in a row, form the belt of the warrior.

Canis Minor, the Little Dog, lies just across the Milky Way from Orion, and may be recognized by the bright star Procyon, of the first magnitude, due east from Betelguese.

Canis Major, the Great Dog, lies south-east of Orion, and is easily recognized by Sirius, the brightest fixed star in the heavens. A number of bright stars south and south-east of Sirius belong to this constellation, making it one of great brilliancy.

As the Milky Way approaches the south horizon, it passes

through *Argo Navis*, the Ship Argo, which is partly below the horizon. It contains Canopus, the next brightest star to Sirius; but this object is below the horizon, unless the observer is as far south as 35° of north latitude.

We can next trace such of the zodiacal constellations as are high enough above the horizon. In the west, one-third of the way from the horizon to the zenith, will be seen *Aries*, the Ram, which may be recognized by three stars of the second, third, and fourth magnitudes, respectively, forming an obtuse-angled triangle, the brightest star being the highest. The arrangement of these stars, and of some others of the fifth magnitude, may be seen by Map II.

Taurus, the Bull, is next above Aries, and may be recognized by the Pleiades, or "seven stars," as the group is commonly called. Really there are only six stars in the group clearly visible to ordinary eyes, and an eye which is good enough to see seven will be likely to see four others, or eleven in all. A telescopic view of this group will be given in connection with the subject of clusters of stars. Another group in this constellation is the Hyades, the principal stars of which are arranged in the form of the letter V, one extremity of the V being formed by Aldebaran, a red star ranked as of the first magnitude, but not so bright as α Orionis.

Gemini, the Twins, lies east of the Milky Way, and may be found on the left side of Map II. and the right of Map III. The brightest stars of this constellation are Castor and Pollux, or α and β, which lie twenty or thirty degrees south-east or east of the zenith, about one-fourth or one-third of the way to the horizon. They are almost due north from Procyon; that is, a line drawn from Procyon to the pole star passes between them. The constellation extends from Castor and Pollux some distance south and west to the borders of Orion.

Cancer, the Crab, lies east of Gemini, but contains no bright star. The most noteworthy object within its borders is Præsepe, a group of stars too small to be seen singly, which appears as a spot of milky light. To see it well, the night must be perfectly clear, and the moon not in the neighborhood.

Leo, the Lion, contains the bright star Regulus, about two hours above the eastern horizon. This star, with five or six smaller ones, forms a sickle, Regulus being the handle. The sickle is represented as in the breast, neck, and head of the lion, his tail extending nearly to the horizon, where it ends at the star Denebola, now just risen.

Such are the principal constellations visible in the supposed position of the celestial sphere. If the hour of observation is different from that supposed, the positions of the constellations will be different by the amount of diurnal rotation during the interval. For instance, if, in the middle of March, we study the heavens at eight o'clock instead of six, the western stars will be nearer the horizon, the southern ones farther west, and the eastern ones higher up than we have described them.

Second Position of the Celestial Sphere.— The meridian in twelve hours of right ascension, near the left-hand edge of Map III., and the right-hand edge of Map IV. The stars on Map III. are west of the meridian, those of Map IV. east of it. This position occurs on

> March 21st..at midnight.
> April 20th..at 10 o'clock.
> May 21st...at 8 o'clock.

In this position Ursa Major is near the zenith, and Cassiopeia in the north horizon. The Milky Way is too near the horizon to be visible; Orion has set in the west; and there are no very conspicuous constellations in the south. Castor and Pollux are visible in the north-west, at a considerable altitude, and Procyon in the west, about an hour and a half above the horizon. Leo is west of the meridian, extending nearly to it, while three new zodiacal constellations have come into sight in the east.

Virgo, the Virgin, has a single bright star—Spica—about the brilliancy of Regulus, now about one hour east of the meridian, and a little more than half-way from the zenith to the horizon.

Libra, the Balance, has no stars which will attract attention. The constellation may be recognized by its position between Virgo and Scorpius.

Scorpius, the Scorpion, is just rising in the south-east, and is not yet high enough to be well seen.

Among the constellations north of the zodiac we have:

Coma Berenices, the Hair of Berenice, now exactly on the meridian, and about ten degrees south of the zenith. It is a close, irregular group of very small stars, quite different from anything else in the heavens. In the ancient mythology, Berenice had vowed her hair to the goddess Venus; but Jupiter carried it away from the temple in which it was deposited, and made it into a constellation.

Boötes, the Bear-keeper, is a large constellation east of Coma. It is marked by Arcturus, a very bright but somewhat red star, an hour and a half east of Coma Berenices.

Canes Venatici, the Hunting Dogs, are north of Coma. They are held in a leash by Boötes, and are chasing Ursa Major round the pole.

Corona Borealis, the Northern Crown, lies next east of Boötes in the north-east. It is principally composed of a pretty semicircle of stars, supposed to form a chaplet, or crown.

Third Position of the Sphere.—The southern constellations are those shown on Maps IV. and V., those of Map IV. being west of the meridian, and those of Map V. east of it. This position occurs on

June 21st..at midnight.
July 21st ....at 10 o'clock.
August 21st...at 8 o'clock.
  etc.   .... etc.

In this position the Milky Way is once more in sight, and seems to span the heavens, but we do not see the same part of it which was visible in the first position. Cassiopeia is now in the north-east, and Ursa Major has passed over to the north-west. Arcturus is two or three hours high in the west, and Corona is above it, two or three hours west of the zenith. Commencing, as in the first position, with the constellations which lie along the Milky Way, we start upwards from Cassiopeia, pass Cepheus and Lacerta, neither of which contains any striking stars, and then reach

Cygnus, the Swan, now north-east from the zenith, which may be recognized by four or five stars forming a cross, directly in the Milky Way. The brightest of these stars somewhat exceeds the brightest ones of Cassiopeia.

Lyra, the Harp, is west and south-west of Cygnus, and near the zenith. It contains the bright star Vega, or α Lyræ, of the first magnitude, of a brilliant white color with a tinge of blue.

Passing south, over *Vulpecula*, the Little Fox, and *Sagitta*, the Arrow, the next striking constellation we reach is

Aquila, the Eagle, now midway between the zenith and the horizon, and two hours east of the meridian. It contains a bright star — Altair, or α Aquilæ — situated between two smaller ones, the row of three stars running nearly north and south.

We next pass west of the Milky Way, and direct our attention to a point two hours west of the meridian, and some distance towards the south horizon. Here we find

Scorpius, the Scorpion, a zodiacal constellation and a quite brilliant one, containing *Antares*, or α Scorpii, a reddish star of nearly the first magnitude, with a smaller star on each side of it, and a long curved row of stars to the west.

Sagittarius, the Archer, comprises a large collection of second-magnitude stars east of Scorpius, and in and east of the Milky Way, and now extending from the meridian to a point two hours east of it.

Capricornus, the Goat, another zodiacal constellation, is now in the south-east, but contains no striking stars. The same remark applies to *Aquarius*, the Water-bearer, which has just risen, and *Pisces*, the Fishes, partly below the eastern horizon.

Leaving the zodiac again, we find, north of Scorpius and west of the Milky Way, a very large pair of constellations, called *Ophiuchus*, the Serpent-bearer, and *Serpens*, the Serpent. Ophiuchus stands with one foot on Scorpius, while his head is marked by a star of the second magnitude twelve degrees north of the equator, and now on the meridian. It is, therefore, one-third or one-fourth of the way from the zenith to the

DESCRIPTION OF THE PRINCIPAL CONSTELLATIONS. 425

horizon. The Serpent, which he holds in his hands, lies with its tail in an opening of the Milky Way, south-west of Aquila, while its neck and head are formed by a collection of stars of the second, third, and fourth magnitudes some distance north of Scorpius, and extending up to the borders of Boötes.

Hercules is a very large constellation, bounded by Corona on the west, Lyra on the east, Ophiuchus on the south, and Draco on the north. It is now in the zenith, but contains no striking stars.

Draco, the Dragon, lies with his head just north of Hercules, while his body is marked by a long curved row of stars extending round the pole between the Great and the Little Bear. His head is readily recognized by a collection of stars of the second and third magnitudes which might well suggest such an object.

Fourth Position of the Sphere.—The southern constellations are now found on Maps V. and II.—those of Map V. west of the meridian, those of Map II. east of it. The times are:

September 21st..at midnight.
October 21st..at 10 o'clock.
November 20th...at 8 o'clock.
December 21st..at 6 o'clock.

In this position Cassiopeia is just north of the zenith, while Ursa Major is glimmering in the north horizon. Following the Milky Way from Cassiopeia towards the west, we shall cross Cepheus, Cygnus, Lyra, and Aquila, while towards the east we pass Perseus and Auriga, all of which have been described.

In the south, the principal constellation is *Pegasus*, the Flying Horse, distinguished by four stars of the second magnitude, which form a large square, each side of which is about fourteen degrees.

Andromeda, her hands in chains, is readily found by a row of three bright stars extending north-east from the north-east corner of Pegasus in the direction of Perseus.

Cetus, the Whale, is a large constellation in the south, extending from the meridian to a point three hours east of it.

Its brightest stars are β Ceti, now near the meridian, at an altitude of 20°, which stands by itself, and α Ceti, about 20° below Aries, which is now about 30° south-east from the zenith.

The reader who wishes to consult the constellations in greater detail can readily do so by means of the star-maps.

§ 3. *New and Variable Stars.*

The large majority of stars always appear to be of the same brightness, though it is quite possible that, if the quantity of light emitted by a star could be measured with entire precision, it would be found in all cases to vary slightly, from time to time. There are, however, quite a number of stars in which the variation is so decided that it has been detected by comparing their apparent brightness with that of other stars at different times. More than a hundred such stars are now known; but in a large majority of cases the variation is so slight that only careful observation with a practised eye can perceive it. There are, however, two stars in which it is so decided that the most casual observer has only to look at the proper times, in order to see it. These are β Persei and ο Ceti, or Algol and Mira, to which we might add η Argus, a star of the southern hemisphere, which exhibits variations of a very striking character.

Variations of Algol. — This star, marked β in the constellation Perseus, may be readily found on Maps I. and II., in right ascension 3 hours and declination 40° 23'. When once found, it is readily recognized by its position nearly in a line between two smaller stars. The most favorable seasons for seeing it in the early evening are the autumn, winter, and spring. In autumn it will, after sunset, generally be low down in the north-east; in winter, high up in the north, not far from the zenith; and in spring, low down in the north-west. Usually it shines as a faint second-magnitude star: on an accurate scale the magnitude is about $2\frac{1}{2}$. But at intervals of a little less than three days, it fades out to the fourth magnitude for a few hours, and then resumes its usual splendor once more. These changes were first noticed about two centuries ago, but it was not till 1782 that they were accurately observed. The period is now known to be 2 days, 20 hours, 49 minutes—that is, 3 hours 11 minutes less than three days. It takes about four hours and a half to fade away to its least brilliancy, and four hours more are spent in recovering its light; so that there are nine and a half hours during each period in which its light is below the average. But near the beginning and end of the variations, the change is very slow, so that there are not more than five or six hours during which the ordinary eye would see that the star was any smaller than usual.

The apparent regularity of this variation of light at first suggested, as an explanation of its cause, that a large dark planet was revolving round Algol, and passed over its face at every revolution, thus cutting off a portion of its light. This theory accounts very well for the salient features of the variation. But when the latter came to be studied more closely and carefully, it was found that there were small irregularities in the variation which the theory would not well account for. The period of the variation was found to change a little at different times, while the star does not lose and recover its light in the same time as it would if the passage of a dark body caused the changes.

Another remarkable variable star, but of an entirely different type, is o Ceti, or *Mira* (the Wonderful). It may be found on Map II., in right ascension 2 hours 12 minutes, declination 3° 39′ south. During most of the time this star is entirely invisible to the naked eye, but at intervals of about eleven months it shines forth with the brilliancy of a star of the second or third magnitude. It is, on the average, about forty days from the time it first becomes visible until it attains its greatest brightness, and it then requires about two months to become invisible; so that it comes into sight more rapidly than it fades away. It is expected to attain its greatest brilliancy in November, 1877; in October, 1878, and so on, about a month earlier each year; but the period is quite irregular, ranging from ten to twelve months, so that the times of its appearance cannot be predicted with certainty. Its maximum brilliancy is also variable, being sometimes of the second magnitude, and at others only of the third or fourth.

η Argus.—Perhaps the most extraordinary known variable star in the heavens is η Argus, of the southern hemisphere, of which the position is, right ascension, 10 hours 40 minutes; declination, 59° 1' south. Being so far south of the equator, it cannot be seen in our latitudes, and the discovery and observations of the variations of its light have been generally made by astronomers who have visited the southern hemisphere. In 1677, Halley, while at St. Helena, found it to be of the fourth magnitude. In 1751, Lacaille found that it had increased to the second magnitude. From 1828 to 1838 it ranged between the first and second magnitudes. The first careful observations of its variability were made by Sir John Herschel while at the Cape of Good Hope. He says: "It was on the 16th December, 1837, that, resuming the photometrical comparisons, my astonishment was excited by the appearance of a new candidate for distinction among the very brightest stars of the first magnitude in a part of the heavens with which, being perfectly familiar, I was certain that no such brilliant object had before been seen. After a momentary hesitation, the natural consequence of a phenomenon so utterly unexpected, and referring to a map for its configuration with other conspicuous stars in the neighborhood, I became satisfied of its identity with my old acquaintance, η Argus. Its light, was, however, nearly tripled. While yet low, it equalled Rigel, and, when it attained some altitude, was decidedly greater."* Sir John states that it continued to increase until January 2d, 1838, when it was nearly matched with α Centauri. It then faded a little till the close of his observations in April following, but was still as bright as Aldebaran. But in 1842 and 1843 it blazed up brighter than ever, and in March of the latter year was second only to Sirius. During the twenty-five years following, it slowly but

---

* "Astronomical Observations at the Cape of Good Hope," p. 33.

steadily diminished: in 1867 it was barely visible to the naked eye, and the year following it vanished entirely from the unassisted view, and has not yet begun to recover its brightness.

When we speak of this star as the most remarkable of the well-known variables, we refer, not to the mere range of its variations, but to its brilliancy when at its maximum. Several cases of equally great variation are known; but the stars are not so bright, and therefore would not excite so much notice. Thus, the star R Andromedæ varies from the sixth to the thirteenth magnitude in a pretty regular period of 405 days. When at its brightest, it is just visible to the naked eye, while only a large telescope will show it when at its minimum. A number of others range through five or six orders of magnitude, but o Ceti is the only one of these which ever becomes as bright as the second magnitude.

The foregoing stars are the only ones the variations of which would strike the ordinary observer. Among the hundred remaining ones which astronomers have noticed, β Lyræ is remarkable for having two maxima and two minima of unequal brilliancy. If we take it when at its greatest minimum, we find its magnitude to be $4\frac{1}{2}$. In the course of three days, it will rise to magnitude $3\frac{1}{2}$. In the course of the week following, it will first fall to the fourth magnitude, and increase again to magnitude $3\frac{1}{2}$. In three days more it will drop again to its minimum of magnitude $4\frac{1}{2}$; the period in which it goes through all its changes being thirteen days. This period is constantly increasing. The changes of this star can best be seen by comparing it with its neighbor, γ Lyræ. Sometimes it will appear equally bright with the latter, and at other times a magnitude smaller.*

---

* In 1875, Professor Schönfeld, now director of the observatory at Bonn, published a complete catalogue of known variable stars, the total number being 143. The following are the more remarkable ones of his list. The positions are referred to the ecliptic and equinox of 1875:

T Cassiopeiæ: right ascension, 0 hours 16 minutes 29 seconds; declination, 55° 6'.0 N.—This is a case in which a star, having once been observed, was afterwards found to be missing. Examination showed that it had so far diminished as to be no longer visible without a larger telescope, and continued observations

New Stars.—It was once supposed to be no uncommon occurrence for new stars to come into existence and old ones to disappear, the former being looked upon as new creations, and the disappearances as due to the destruction or annihilation of those stars which had fulfilled their end in the economy of nature. The supposed disappearances of stars are, however, found to have no certain foundation in fact, probably owing their origin to errors in recording the position of stars actually existing. It was explained, in treating of Practical Astronomy, that the astronomer determines the position of a body in the celestial vault by observing the clock-time at which it passes the meridian, and the position of the circle of his in-

---

showed it to range from the seventh to the eleventh magnitude with a regular period of 436 days.

B Cassiopeiæ: right ascension, 0 hours 17 minutes 52 seconds; declination, 63° 27'.0 N.—This is supposed to be the celebrated star which blazed out in November, 1572, and was so fully described by Tycho Brahe. But the proof of identity can hardly be considered conclusive, especially as no variation has, of recent years, been noticed in the star.

o Ceti: right ascension, 2 hours 13 minutes 1 second; declination, 3° 32'.7 S.—We have already described the variations of this star.

β Persei, or Algol: right ascension, 3 hours 0 minutes 2 seconds; declination, 40° 28'.4 N.—The variations of this star, which is the most regular one known, have just been described.

R Aurigæ: right ascension, 5 hours 7 minutes 12 seconds; declination, 53° 26'.6 N.—This star is one of very wide and complex variation, changing from the sixth to the thirteenth magnitude in a period of about 465 days.

R Geminorum: right ascension, 6 hours 59 minutes 49 seconds; declination, 22° 53'.8 N.—This star was discovered by Mr. Hind, of England, and ranges between the seventh and the twelfth magnitude in a period of 371 days.

U Geminorum: right ascension, 7 hours 47 minutes 41 seconds; declination, 22° 19'.7 N.—An irregular variable, never visible to the naked eye, remarkable for the rapidity with which it sometimes changes. Schönfeld says that in February, 1869, it increased three entire magnitudes in 24 hours. The periods of its greatest brightness have ranged from 75 to 617 days.

η Argus: right ascension, 10 hours 40 minutes 13 seconds; declination, 59° 1'.6 S.—This remarkable object has already been described.

R Hydræ: right ascension, 13 hours 22 minutes 53 seconds; declination, 22° 38'.0 S.—The variability of this star was recognized by Maraldi, in 1704. It is generally invisible to the naked eye, but rises to about the fifth magnitude at intervals of about 437 days. Its period seems to be diminishing, having been about 500 days when first discovered.

strument when his telescope is pointed at the object. If he happens to make a mistake in writing down any of these numbers—if, for example, he gets his clock-time one minute or five minutes wrong, or puts down a wrong number of degrees for the position of his circle—he will write down the position of the star where none really exists. Then, some subsequent astronomer, looking in this place and seeing no star, may think the star has disappeared, when, in reality, there was never any star there. Where thousands of numbers have to be written down, such mistakes will sometimes occur; and it is to them that some cases of supposed disappearance of stars are to be attributed. There have, however, been several cases of apparently new stars coming suddenly into view, of which we shall describe some of the most remarkable.

T Coronæ: right ascension, 15 hours 54 minutes 16 seconds; declination, 26° 16′.5 N.—This is the "new star" which blazed out in the Northern Crown in 1866, as hereafter described. Of late years it has remained between the ninth and tenth magnitudes without exhibiting any remarkable variations.

T Scorpii: right ascension, 16 hours 9 minutes 36 seconds; declination, 22° 40′.0 S.—This star was discovered by Auwers, in 1860, in the midst of a well-known cluster. It gradually diminished during the following months, and finally disappeared entirely among the stars by which it is surrounded.

— Serpentarii: right ascension, 17 hours 23 minutes 9 seconds; declination, 21° 22′.4 S.—This is supposed to be the celebrated "new star" seen and described by Kepler in 1604, soon to be described.

χ Cygni: right ascension, 19 hours 45 minutes 46 seconds; declination, 32° 36′.0 N.—This star becomes visible to the naked eye at intervals of about 406 days, and then sinks to the twelfth or thirteenth magnitude, so that only large telescopes will show it. Its greatest brightness ranges from the fourth to the sixth magnitude.

η Aquilæ: right ascension, 19 hours 46 minutes 6 seconds; declination, 0° 41′.2 N.—This star varies from magnitude $3\frac{1}{2}$ to $4\frac{3}{4}$, and is therefore one of those which can readily be observed with the naked eye. Its period is 7 days 4 hours 14 minutes 4 seconds.

P Cygni: right ascension, 20 hours 13 minutes 11 seconds; declination, 37° 38′.7 N.—This was supposed to be a new star in 1600, when it was first seen by Janson. During the remainder of the century it varied from the third to the sixth magnitude; but during two centuries which have since elapsed no further variations have been noticed, the star being constantly of the fifth magnitude.

μ Cephei: right ascension, 21 hours 39 minutes 41 seconds; declination, 58° 12′.4 N.—One of the reddest stars visible to the naked eye in the northern hemisphere. Its magnitude is found to vary from the fourth to the fifth in a very irregular manner.

In 1572 an apparently new star showed itself in Cassiopeia. It was first seen by Tycho Brahe on November 11th, when it had attained the first magnitude. It increased rapidly in brilliancy, soon becoming equal to Venus, so that good eyes could discern it in full daylight. In December it began to grow smaller, and continued gradually to fade away until the following May, when it disappeared entirely. This was forty years before the invention of the telescope. Tycho has left us an extended treatise on this most remarkable star.

In 1604 a similar phenomenon was seen in the constellation Ophiuchus. The star was first noticed in October of that year, when it had attained the first magnitude. In the following winter it began to wane, but remained visible during the whole year 1605. Early in 1606 it faded away entirely, having been visible for more than a year. A very full history of this star has been left to us by Kepler.

The most striking recent case of this kind was in May, 1866, when a star of the second magnitude suddenly appeared in Corona Borealis. On the 11th and 12th of that month it was remarked independently by at least five observers in Europe and America, one of the first being Mr. Farquhar, of the United States Patent-office. Whether it really blazed out as suddenly as this would indicate has not been definitively settled. If, as would seem most probable, it was several days attaining its greatest brilliancy, then the only person known to have seen it was Mr. Benjamin Hallowell, a well-known teacher near Washington, whose testimony is of such a nature that it is hard to doubt that the star was visible several days before it was generally known. On the other hand, Schmidt, of Athens, asserts in the most positive manner that the star was not there on May 10th, because he was then scanning that part of the heavens, and would certainly have noticed it. However the fact may have been in this particular case, it is noteworthy that none of the new stars we have described were noticed until they had nearly or quite attained their greatest brilliancy, a fact which gives color to the view that they have all blazed up with great rapidity.

In November, 1876, a new star of the third magnitude was noticed by Schmidt, of Athens, in the constellation Cygnus. It soon began to fade away, and disappeared from the unaided vision in a few weeks. The position of the constellation Cygnus becomes so unfavorable for observation in November that very few people got a sight of this object.

The view that these bodies may be new creations, designed to rank permanently among their fellow-stars, is completely refuted by their transient character, if by nothing else. Their apparently ephemeral existence is in striking contrast to the permanency of the stars in general, which endure from age to age without any change whatever. They are now classified by astronomers among the variable stars, their changes being of a very irregular and fitful character. There is no serious doubt that they were all in the heavens as very small stars before they blazed forth in this extraordinary manner, and that they are in the same place yet. The position of the star of 1572 was carefully determined by Tycho Brahe; and a small telescopic star now exists within 1' of the place computed from his observations, and is probably the same. The star of 1866 was found to have been recorded as one of the ninth magnitude in Argelander's great catalogue of the stars of the northern hemisphere, completed several years before. After blazing up in the way we have described, it gradually faded away to its former insignificance, and has shown no further signs of breaking forth again. There is a wide difference between these irregular variations, or breaking-forth of light, on a single occasion in the course of centuries, and the regular changes of Algol and β Lyræ. But the careful observations of the industrious astronomers who have devoted themselves to this subject have resulted in the discovery of stars of nearly every degree of irregularity between these extremes. Some of them change gradually from one magnitude to another, in the course of years, without seeming to follow any law whatever, while in others some tendency to regularity can be faintly traced. The best connecting link between new and variable stars is, perhaps, afforded by η Argus, which we have just described.

It is probable that the variations of light of which we have spoken are the result of operations going on in the star itself, which, it must be remembered, is a body of the same order of magnitude and brilliancy with our sun, and that these operations are analogous to those which produce the solar spots. It was shown in the chapter on the sun that the frequency of solar spots shows a period of eleven years, during one portion of which there are frequently no spots at all to be seen, while during another portion they are very numerous. Hence, if an observer so far away in the stellar places as to see our sun like a star, could, from time to time, make exact measures of the amount of light it emitted, he would find it to be a variable star, with a period of eleven years, the amount of light being least when we see most spots, and greatest when there are few spots. The variation would, indeed, be so slight that we could not perceive it with any photometric means which we possess, but it would exist nevertheless. Now, the general analogies of the universe, as well as the testimony of the spectroscope, lead us to believe that the physical constitution of the sun and the stars is of the same general nature. We may therefore expect that, as we see spots on the sun which vary in form, size, and number from day to day, so, if we could take a sufficiently close view of the faces of the stars, we should, at least in some of them, see similar spots. It is also likely that, owing to the varying physical constitution of these bodies, the number and extent of the spots might be found to be very different in different stars. In the cases in which the spots covered the larger portion of the surface, their variations in number and extent would alone cause the star to vary in light, from time to time. Finally, we have only to suppose the same kind of regularity which we see in the eleven-year cycle of the solar spots, to have a variation in the brightness of a star going through a regular cycle, as in the case of Algol and Mira Ceti.

The occasional outbursts of stars which we have described, in which their light is rapidly increased a hundred-fold, would seem not to be accounted for on the spot theory, without carrying this theory to an extreme. It would, in fact, if not modified, imply that ninety-nine parts of the surface out of a hundred were ordinarily covered with spots, and that on rare occasions these spots all disappeared. But the spectroscopic observations of the star of 1866 showed an analogy of a little different character with operations going on in our sun. Mr. Huggins found the spectrum of this star to be a continuous one, crossed by bright lines, the position of which indicated that they proceeded partly or wholly from glowing hydrogen. The continuous spectrum was also crossed by dark absorption lines, indicating that the light had passed through an atmosphere of comparatively cool gas. Mr. Huggins's interpretation of this is that there was a sudden and extraordinary outburst of hydrogen gas from the star which, by its own light, as well as by heating up the whole surface of the star, caused the immense accession of brilliancy. Now, we have shown that the red flames seen around the sun during a total eclipse are caused by eruptions of hydrogen from his interior; moreover, these eruptions are generally connected with faculæ, or portions of the sun's disk several times more brilliant than the rest of the photosphere. Hence, it is not unlikely that the blazing-forth of this star arose from an action similar to that which produces the solar flames, only on an immensely larger scale.

We have thus in the spots, faculæ, and protuberances of the sun a few suggestions as to what is probably going on in those stars which exhibit the extraordinary changes of light which we have described. Is there any possibility that our sun may be subject to such outbursts of light and heat as those we have described in the cases of apparently new and temporary stars? We may almost say that the continued existence of the human race is involved in this question; for if the heat of the sun should, even for a few days only, be increased a hundred-fold, the higher orders of animal and vegetable life would be destroyed. We can only reply to it that the general analogies of nature lead us to believe that we need not feel any apprehension of such a catastrophe. Not the slightest certain variation of the solar heat has been detected since the invention of the thermometer, and the general constancy of the light emitted by ninety-nine stars out of every hundred may inspire us with entire confidence that no sudden and destructive variation need be feared in the case of our sun.

§ 4. *Double Stars.*

Telescopic examination shows that many stars which seem single to the naked eye are really double, or composed of a pair of stars lying side by side. There are in the heavens several pairs of stars the components of which are so close together that, to the naked eye, they seem almost to touch each other. One of the easiest and most beautiful of these is in Taurus, quite near Aldebaran. Here the two stars θ^1 Tauri and θ^2 Tauri are each of the fourth magnitude. Another such pair is α Capricorni, in which the two pairs are unequal. Here an ordinary eye has to look pretty carefully to see the smaller star. Yet another pair is ϵ Lyræ, the components of which are so close that only a good eye can distinguish them. These pairs, however, are not considered as double stars in astronomy, because, although to the naked eye they seem so close, yet, when viewed in a telescope of high power, they are so wide apart that they cannot be seen at the same time. The telescopic double stars are formed of components only a few seconds apart; indeed, in many cases, only a fraction of a second. The large majority of those which are catalogued as doubles range from half a second to fifteen seconds in distance. When they exceed the latter limit, they are no longer objects of special interest, because they may be really without any connection, and appear together only because they lie in nearly the same straight line from our system.

The most obvious question which suggests itself here is whether in any case there is any real connection between the two stars of the pair, or whether they do not appear close together, simply because they chance to lie on nearly the same straight line from the earth. That some stars do appear double in this way there is no doubt, and such pairs are called "optically double." But notwithstanding the immense number of visible stars, the chance of many pairs falling within a few seconds of each other is quite small; and the number of close double stars is so great as to preclude all possibility that they appear together only by chance. If any further proof was wanted that the stars of these pairs are really physically connected, and therefore close together in reality as well as in appearance, it is found in the fact that many of them constitute systems in which one revolves round the other, or, to speak more exactly, in which each revolves round the centre of gravity of the pair. Such pairs are called *binary systems*, to distinguish them from those in which no such revolution has been observed. The revolution of these binary systems is generally very slow, requiring many centuries for its accomplishment; and the slower the motion, the longer it will take to perceive and determine it. Generally it has been detected by astronomers of one generation comparing their observations with those of their predecessors; for instance, when the elder Struve compared his observations with those of Herschel, and when Dawes or the younger Struve compared with the elder Struve, a great number of pairs were found to be binary. As every observer is constantly detecting new cases of motion, the number of binary systems known to astronomers is constantly increasing.

A brief account of the manner in which these objects are measured may not be out of place. For the purpose in question, the eye-piece of the telescope must be provided with a "filar micrometer," the important part of which consists of a pair of parallel spider-lines, one of which can be moved sideways by a very fine screw, and can thus be made to pass back and forth over the other. The exact distance apart of the lines can be determined from the position of the screw. The whole micrometer turns round on an axis parallel to the telescope, the centre of which is in the centre of the field of view. To get the direction of one star from the other, the observer turns the micrometer round until the spider-lines are parallel to the line joining the two stars, as shown in Fig. 98, and he then reads the position circle. Knowing what the position circle reads when he turns the wires so that the star shall run along them by its diurnal motion, the difference of the two angles shows the angle which the line joining the two stars makes with the celestial parallel. To obtain the distance apart of the stars, the observer turns the micrometer 90° from the position in Fig. 98, and then turns the screw and moves the telescope, until each star is bisected by one of the wires, as shown in Fig. 99. The position of the wires is then interchanged, and the measure is repeated. The mode in

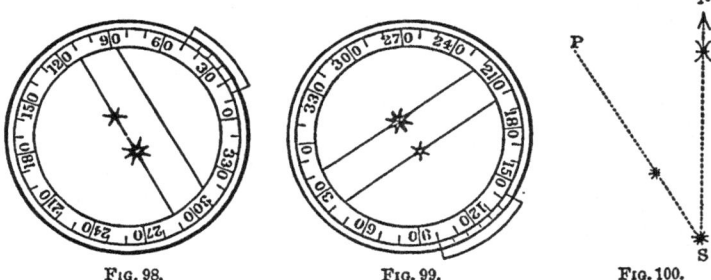

FIG. 98.   FIG. 99.   FIG. 100.

which the direction of one star from another is reckoned is this: Imagine a line, *SN*, in Fig. 100, drawn due north from the brighter star, and another, *SP*, drawn through the smaller star. Then the angle *NSP* which these two lines make with each other, counted from north towards east, is the position angle of the stars, the changes in which show the revolution of one star around the other.

In a few of the binary systems the period is so short that a complete revolution, or more, of the two stars round each other has been observed. As a general rule, the pairs which have the most rapid motion are very close, and therefore of comparatively recent discovery, and difficult to observe. One or two are suspected to have a period of less than thirty years, but they are very hard to measure.

Binary Systems of Short Period.—The following table shows

the periods of revolution in the case of those stars which have been observed through a complete revolution, or of which the periods have been well determined:

42 Comæ	26 years.	ξ Ursæ Majoris	63 years.
ξ Herculis	35 "	η Coronæ Borealis	67 "
Struve, 3121	40 "	α Centauri	77 "
η Coronæ	40 "	μ Ophiuchi	92 "
Sirius	50 "	λ Ophiuchi	96 "
ξ Cancri	58 "	ξ Scorpii	98 "

Two or three others are suspected to move very rapidly, but they are so very close and difficult that it is only on favorable occasions that they can be seen to be double. One of the most remarkable stars in this list is Sirius, the period of which is calculated, not from the observations of the satellite, but from the motion of Sirius itself. It has long been known that the proper motion of this star is subject to certain periodic variations; and, on investigating these variations, it was found by Peters and Auwers that they could be completely represented by supposing that a satellite was revolving around the planet in a certain orbit. The elements of this orbit were all determined except the distance of the satellite, which did not admit of determination. Its direction could, however, be computed from time to time almost as accurately as if it were actually seen with the telescope. But, before the time of which we speak, no one had ever seen it. Indeed, although many observers must have examined Sirius from time to time with good telescopes, it is not likely that they made a careful search in the predicted direction.

Such was the state of the question until February, 1862, when Messrs. Alvan Clark & Sons, of Cambridgeport, were completing their eighteen-inch glass for the Chicago Observatory. Turning the glass one evening on Sirius, for the purpose of trying it, the practised eye of the younger Clark soon detected something unusual. "Why, father," he exclaimed, "the star has a companion!" The father looked, and there was a faint companion due east from the bright star, and distant about 10". This was exactly the predicted direction for that time, though the discoverers knew nothing of it. As the news went round the world, all the great telescopes were pointed on Sirius, and it was now found that when observers knew where the companion was, many telescopes would show it. It lay in the exact direction which theory had predicted for that time, and it was now observed with the greatest interest, in order to see whether it was moving in the direction of the theoretical satellite. Four years' observation showed that this was really the case, so that hardly any doubt could remain that this almost invisible object was really the body which, by its attraction and revolution around Sirius, had caused the inequality in its motion. At the same time, the correspondence has not since proved exact, the observed companion having moved about half a degree per annum more rapidly than the theoretical one. This difference, though larger than was expected, is probably due to the inevitable errors of the very delicate and difficult observations from which the movements of the theoretical companion were computed.

The visibility of this very interesting and difficult object depends almost as much on the altitude of Sirius and the state of the atmosphere as on the power of the telescope. When the images of the stars are very bad, it cannot be seen even in the great Washington telescope, while there are cases of its being seen under extraordinarily favorable conditions with telescopes of six inches aperture or less. These favorable conditions are indicated to the naked eye by the absence of twinkling.

A case of the same kind, except that the disturbing satellite has not been seen, is found in Procyon. Bessel long ago suspected that the position of this star was changed by some attracting body in its neighborhood, but he did not reach a definite conclusion on the subject. Auwers, having made a careful investigation of all the observations since the time of Bradley, found that the star moved around an invisible centre $1''$ distant, which was probably the centre of gravity of the star and an invisible satellite. This satellite has been carefully searched for with great telescopes during the last few years, but without success.

Triple and Multiple Stars. — Besides double stars, groups of three or more stars are frequently found. Such objects are known as triple, quadruple, etc. They commonly occur through one of the stars of a wide pair being itself a close double star, and very often the duplicity of the component has not been discovered till long after it was known to form one star of a pair. For instance, μ Herculis was recognized as a double star by Sir W. Herschel, the companion star being about 30″ distant, and much smaller than μ itself. In 1856, Mr. Alvan Clark, trying one of his glasses upon it, found that the small companion was itself double, being composed of two nearly equal stars, about 1″ apart. This close pair proves to be a binary system of short period, more than half a revolution of the two stars around each other having been made since 1856. Another case of the same kind is γ Andromedæ, which was found by Herschel to have a companion about 10″ distant, while Struve found this companion to be itself double.

Many double and multiple stars are interesting objects for telescopic examination. We give in the Appendix a list of the more interesting or remarkable of them.

§ 5. *Clusters of Stars.*

A very little observation with the telescope will show that while the brighter stars are scattered nearly equally over the whole celestial vault, this is not the case with the smaller ones. A number of stars which it is not possible to estimate are found to be aggregated into clusters, in which the separate stars are so small and so numerous that, with insufficient telescopic power, they present the appearance of a mass of cloudy light. We find clusters of every degree of aggregation. At one extreme we may place the Pleiades, or "seven stars" which form so well-known an object in our winter sky, in which, however, only six of the stars are plainly visible to the naked eye. There is an old myth that this group originally consisted of seven stars, one of which disappeared from the heavens, leaving but six. But a very good eye can even now see eleven when the air is clear, and the telescope shows from fifty to a hundred more, according to its power. We present a view of this group as it appears through a small telescope.

No absolute dividing-line can be drawn between such widely extended groups as the Pleiades and the densest clusters.

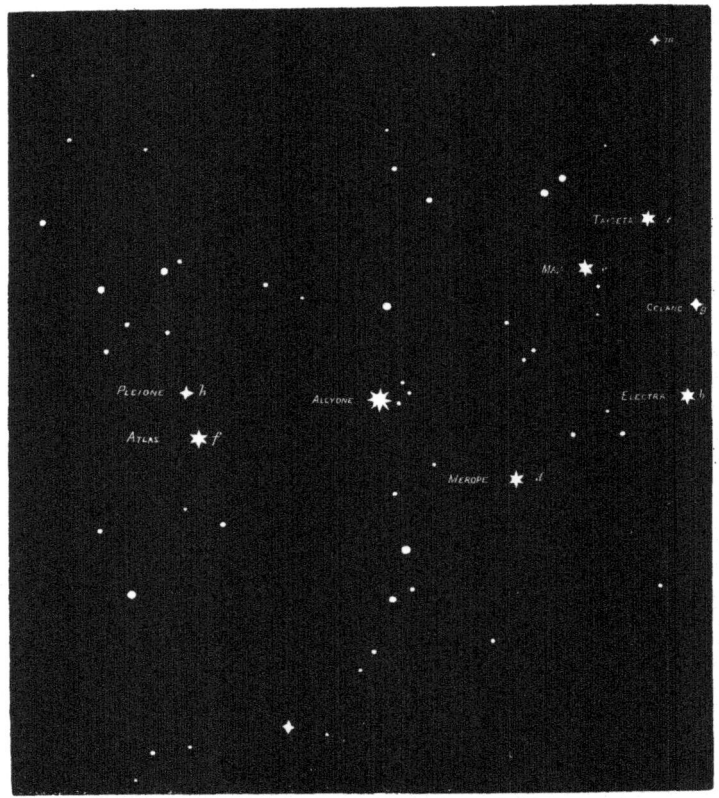

Fig. 101.—Telescopic view of the Pleiades, after Engelmann. The six larger stars are those easily seen by ordinary eyes without a telescope, while the four next in size, having four rays each, can be seen by very good eyes. About an inch from the upper right-hand corner is a pair of small stars which a very keen eye can see as a single star.

The cluster Præsepe, in the constellation Cancer (Map III., right ascension, 8 hours 20 minutes; declination, 20° 10' N.), is plainly visible to the naked eye on a clear, moonless night, as a nebulous mass of light. Examined with a small tele-

CLUSTERS OF STARS. 443

scope, it is found to consist of a group of stars, ranging from the seventh or eighth magnitude upwards. For examination with a small telescope, one of the most beautiful groups is in the constellation Perseus (Map I., right ascension, 2 hours 10 minutes; declination, 57° N.). It is seen to the best advantage with a low magnifying power, between twenty-five and fifty times, and may easily be recognized by the naked eye as a little patch of light.

The heavens afford no objects of more interest to the contemplative mind than some of these clusters. Many of them are so distant that the most powerful telescopes ever made show them only as a patch of star-dust, or a mass of light so faint that the separate stars cannot be distinguished. Their distance from us is such that they are beyond, not only all our means of measurement, but all our powers of estimation. Minute as they appear, there is nothing that we know of to prevent our supposing each of them to be the centre of a group of planets as extensive as our own, and each planet to be as full of inhabitants as this one. We may thus think of them as little colonies on the outskirts of creation itself, and as we see all the suns which give them light condensed into one little speck, we might be led to think of the inhabitants of the various systems as holding intercourse with each other. Yet, were we transported to one of these distant clusters, and stationed on a planet circling one of the suns which compose it, instead of finding the neighboring suns in close proximity, we should only see a firmament of stars around us, such as we see from the earth. Probably it would be a brighter firmament, in which so many stars would glow with more than the splendor of Sirius, as to make the night far brighter than ours; but the inhabitants of the neighboring worlds would as completely elude telescopic vision as the inhabitants of Mars do here. Consequently, to the inhabitants of every planet in the cluster, the question of the plurality of worlds might be as insolvable as it is to us.

To give the reader an idea what the more distant of these star clusters looks like, we present two views from Sir John

Herschel's observations at the Cape of Good Hope. Fig. 102 shows the cluster numbered 2322 in Herschel's catalogue, and known as 47 Toucani. That astronomer describes it as "a most glorious globular cluster, the stars of the fourteenth magnitude immensely numerous. It is compressed to a blaze of light at the centre, the diameter of the more compressed part being 30" in right ascension." Fig. 103 is No. 3504 of Herschel: "The noble globular cluster ω Centauri, beyond all comparison the richest and largest object of the kind in the heavens. The stars are literally innumerable, and as their

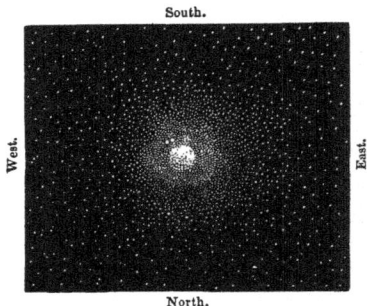 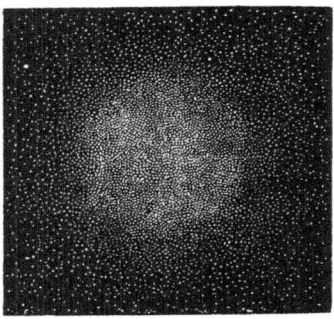

FIG. 102.—Cluster 47 Toucani. Right ascension, 0 hours 18 minutes; declination, 72° 45′ S.

FIG. 103.—Cluster ω Centauri. Right ascension, 13 hours 20 minutes; declination, 46° 52′ S.

total light when received by the naked eye affects it hardly more than a star of the fifth or fourth to fifth magnitude, the minuteness of each star may be imagined."

§ 6. *Nebulæ.*

Nebulæ appear to us as masses of soft diffused light, of greater or less extent. Generally these masses are very irregular in outline, but a few of them are round and well-defined. These are termed *planetary nebulæ*. It may sometimes be impossible to distinguish between star clusters and nebulæ, because when the power of the telescope is so low that the separate stars of a cluster cannot be distinguished, they will present the appearance of a nebula. To the naked eye the cluster Præsepe, described in the last chapter, looks exactly like a nebula, though a very small telescope will resolve it into stars. The early observers with telescopes described many objects as nebulæ which the more powerful instruments of Herschel showed to be clusters of stars. Thus arose the two classes of resolvable and irresolvable nebulæ, the first comprising such as could be resolved into stars, and the second such as could not. It is evident, from what we have just said, that this distinction would depend partly on the telescope, since a nebula which was irresolvable in one telescope might be resolvable in another telescope of greater power. This suggests the question whether all nebulæ may not really be clusters of stars, those which are irresolvable appearing so merely because their distance is so great that the separate stars which compose them cannot be distinguished with our most powerful telescopes. If this were so, there would be no such thing as a real nebula, and everything which appears as such should be classified as a star cluster. The spectroscope, as we shall presently show, has settled this question, by showing that many of these objects are immense masses of glowing gas, and therefore cannot be stars.

Classification and Forms of Nebulæ.—The one object of this class which, more than all others, has occupied the attention of astronomers and excited the wonder of observers, is the great nebula of Orion. It surrounds the middle of the three stars which form the sword of Orion. Its position may be found on Maps II. and III., in right ascension 5 hours 28 minutes, declination 6° S. A good eye will perceive that this star, instead of looking like a bright point, as the other stars do, has an ill-defined, hazy appearance, due to the surrounding nebulæ. This object was first described by Huyghens in 1659, as follows:

"There is one phenomenon among the fixed stars worthy of mention which, so far as I know, has hitherto been noticed by no one, and indeed cannot be well observed except with large telescopes. In the sword of Orion are three stars quite close together. In 1656, as I chanced to be viewing the middle one of these with the telescope, instead of a single star, twelve showed themselves (a not uncommon circumstance). Three of these almost touched each other, and, with four others, shone through a nebula, so that the space around them seemed far brighter than the rest of the heavens, which was entirely clear, and appeared quite black, the effect being that of an opening in the sky, through which a brighter region was visible."*

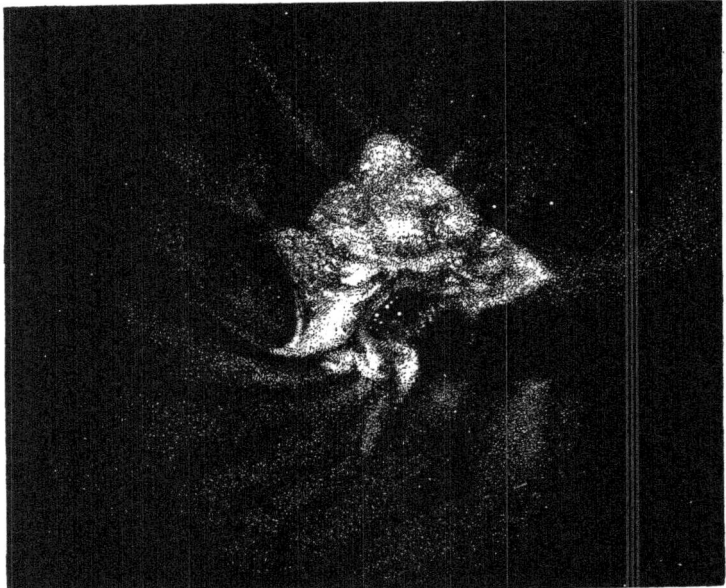

FIG. 104.—The great nebula of Orion, as drawn by Trouvelot with the twenty-six-inch Washington telescope.

Since that time it has been studied with large telescopes by a great number of observers, including Messier, the two

---

* *Systema Saturnium*, p. 8. The last remark of Huyghens seems to have produced the impression that he or some of the early observers considered the nebulæ to be real openings in the firmament, through which they got glimpses of the glory of the empyrean. But it may be doubted whether the old ideas of the firmament and the empyrean were entertained by any astronomer after the invention of the telescope, and there is nothing in the remark of Huyghens to indicate that he thought the opening really existed. His words are rather obscure.

Herschels, Rosse, Struve, and the Bonds. The representation which we give in Fig. 104 is from a drawing made by Mr. Trouvelot with the great Washington telescope. In brilliancy and variety of detail it exceeds any other nebula visible in the northern hemisphere. The central point of interest is occupied by four comparatively bright stars, easily distinguished by a small telescope with a magnifying power of 40 or 50, combined with two small ones, requiring a nine-inch telescope to be well seen. The whole of these form a sextuple group, included in a space a few seconds square, which alone would be an interesting and remarkable object. Besides these, the nebula is dotted with so many stars that they would almost constitute a cluster by themselves.

In the winter of 1864–'65, the spectrum of this object was examined independently by Secchi and Huggins, who found that it consisted of three bright lines, and hence concluded that the nebula was composed, not of stars, but of glowing gas. The position of one of the lines was near that of a line of nitrogen, while another seemed to coincide with a hydrogen line. There is, therefore, a certain probability that this object is a mixture of hydrogen and nitrogen gas, though this is a point on which it is impossible to speak with certainty.

Another brilliant nebula visible to the naked eye is the great one of Andromeda (Maps II. and V., right ascension, 0 hours 35 minutes; declination, 40° N.). The observer can see at a glance with the naked eye that this is not a star, but a mass of diffused light. Indeed, untrained observers have sometimes very naturally mistaken it for a comet.* It was first described by Marius, in 1614, who compared its light to that of a candle shining through horn. This gives a very good idea of the singular impression it produces, which is that of an object not self-luminous, but translucent, and illuminated by a very brilliant light behind it. With a small telescope, it

---

* A ship-captain who had crossed the Atlantic once visited the Cambridge Observatory, to tell Professor Bond that he had seen a small comet, which remained in sight during his entire voyage. The object proved to be the nebula of Andromeda.

448 *THE STELLAR UNIVERSE.*

is easy to imagine it to be a solid like horn; but with a large one, the effect is much more that of a great mass of matter, like fog or mist, which scatters and reflects the light of a brilliant body in its midst. That this impression can be correct, it would be hazardous to assert; but the result of a spectrum

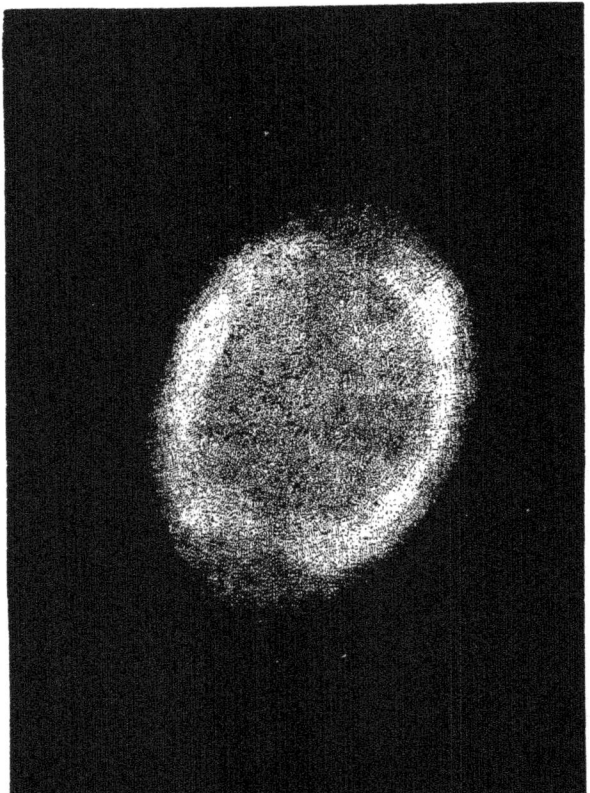

FIG. 105.—The annular nebula in Lyra.   Drawn by Professor E. S. Holden.

analysis of the light of the nebula certainly seems to favor it. Unlike most of the nebulæ, its spectrum is a continuous one, similar to the ordinary spectra from heated bodies, thus indicating that the light emanates, not from a glowing gas, but from matter in the solid or liquid state. This would suggest the idea that the object is really an immense star-cluster, so distant that the most powerful telescopes cannot resolve it. Though we cannot positively deny the possibility of this, yet in the most powerful telescopes the light fades away so softly and gradually that no such thing as a resolution into stars seems possible. Indeed, it looks less resolvable and more like a gas in the largest telescopes than in those of moderate size. If it is really a gas, and if the spectrum is continuous throughout the whole extent of the nebula, it would indicate either that it shone by reflected light, or that the gas was subjected to a great pressure almost to its outer limit, which hardly seems possible. But, granting that the light is reflected, we cannot say whether it originates in a single bright star or in a number of small ones scattered about through the nebula.

Another extraordinary object of this class is the annular, or ring-nebula of Lyra, situated in that constellation, about halfway between the stars β and γ. In the older telescopes it looked like a perfect ring; but the larger ones of modern times show that the opening of the ring is really filled with nebulous light; in fact, that we have here an object of very regular outline, in which the outer portion is brighter than the interior. Its form is neither circular nor exactly elliptic, but egg-shaped, one end being more pointed than the other. A moderate-sized telescope will show it, but a large one is required to see it to good advantage.

It would appear, from a comparison of drawings made at different dates, that some nebulæ are subject to great changes of form. Especially does this hold true of the nebula surrounding the remarkable variable star η Argus. In many other nebulæ changes have been suspected; but the softness and indistinctness of outline which characterize most of these objects, and the great difference of their aspect when seen in telescopes of very different powers, make it difficult to prove a change from mere differences of drawing. One of the strongest cases in favor of change has been made out by Professor Holden from a study of drawings and descriptions of what is called the "Omega nebula," from a resemblance of one of

Fig. 106.—The Omega nebula; Herschel 2008. Right ascension, 18 hours 13 minutes; declination, 16° 14' S. After Holden and Trouvelot.

its branches to the Greek letter Ω. We present a figure of this object as it now appears, from a drawing by Professor Holden and Mr. Trouvelot, with the great Washington telescope. It is the branch on the left-hand end of the nebula which was formerly supposed to have the form of Ω.

As illustrative of the fantastic forms which nebulæ sometimes assume, we present Herschel's views of two more nebulæ. That shown in Fig. 108 he calls the "looped nebula," and describes as one of the most extraordinary objects in the heavens. It cannot be seen to advantage except in the southern hemisphere.

Distribution of the Nebulæ.—A remarkable feature of the distribution of the nebulæ is that they are most numerous where the stars are least so. While the stars grow thicker as we approach the region of the Milky Way, the nebulæ diminish in number. Sir John Herschel remarks that one-third of

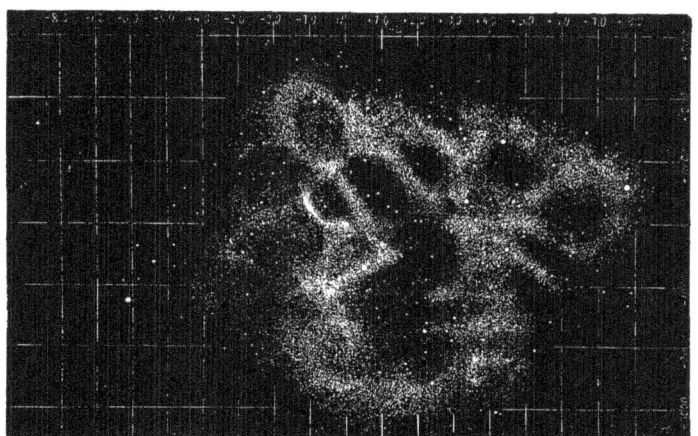

FIG. 107.—Nebula Herschel 3722. Right ascension, 17 hours 56 minutes; declination, 24° 21′ S. After Sir John Herschel.

the nebulous contents of the heavens are congregated in a broad, irregular patch occupying about one-eighth the surface of the celestial sphere, extending from Ursa Major in the north to Virgo in the south. If, however, we consider, not the true nebulæ, but star clusters, we find the same tendency to condensation in the Milky Way that we do in the stars. We thus have a clearly marked distinction between nebulæ and stars as regards the law of their distribution. The law in question can be most easily understood by the non-mathematical reader by supposing the starry sphere in such a position that the Milky Way coincides with the horizon. Then the stars and star clusters will be fewest at the zenith, and will increase in number as we approach the horizon. Also, in the invisible hemisphere the same law will hold, the stars and clusters being fewest under our feet, and will increase as we approach the horizon. But the true nebulæ will then

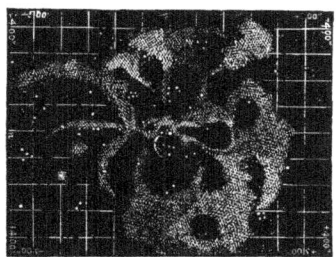

FIG. 108.—The looped nebula; Herschel 2941. Right ascension, 5 hours 40 minutes; declination, 69° 6′ S.

be fewest in the horizon, and will increase in number as we approach the zenith, or as, going below the horizon, we approach the nadir. The positions of the nebulæ and clusters in Sir John Herschel's great catalogue have been studied by Mr. Cleveland Abbe with especial reference to their distance from the galactic circle, and the following numbers show part of his results. Imagine a belt thirty degrees wide extending around the heavens, including the Milky Way, and reaching fifteen degrees on each side of the central circle of the Milky Way. This belt will include nearly one-fourth the surface of the celestial sphere, and if the stars or nebulæ were equally distributed, nearly one-fourth of them would be found in the belt. Instead, however, of one-fourth, we find nine-tenths of the star clusters, but only one-tenth of the nebulæ.

The discovery that the nebulæ are probably masses of glowing gas is of capital importance as tending to substantiate the view of Sir William Herschel, that these masses are the crude material out of which suns and systems are forming. This view was necessarily an almost purely speculative one on the part of that distinguished astronomer; but unless we suppose that the nebulæ are objects of almost miraculous power, there must be some truth in it. A nebulous body, in order to shine by its own light, as it does, must be hot, and must be losing heat through the very radiation by which we see it. As it cools, it must contract, and this contraction cannot cease until it becomes either a solid body or a system of such bodies revolving round each other. We shall explain this more fully in treating of cosmical physics and the nebular hypothesis.

§ 7. *Proper Motions of the Stars.*

To the unassisted eye, the stars seem to preserve the same relative positions in the celestial sphere generation after generation. If Job, Hipparchus, or Ptolemy should again look upon the heavens, he would, to all appearance, see Aldebaran, Orion, and the Pleiades exactly as he saw them thousands of years ago, without a single star being moved from its place. But the refined methods of modern astronomy, in which the telescope is brought in to measure spaces absolutely invisible to the eye, have shown that this seeming unchangeability is not real, and that the stars are actually in motion, only the rate of change is so slow that the eye would not, in most cases, notice it for thousands of years. In ten thousand years quite a number of stars, especially the brighter ones, would be seen to have moved, while it would take a hundred thousand years to introduce a very noticeable change in the aspect of the constellations.

As a general rule, the brighter stars have the greatest proper motions. But this is a rule to which there are many exceptions. The star which, so far as known, has the greatest proper motion of all—namely, Groombridge 1830—is of the seventh magnitude only. Next in the order of proper motion comes the pair of stars 61 Cygni, each of which is of the sixth magnitude. Next are four or five others of the fourth and fifth magnitudes. The annual motions of these stars are as follows:

Groombridge 1830	7″.0	Lalande 21258	4″.4
61 Cygni	5″.2	o^2 Eridani	4″.1
Lalande 21185	4″.7	μ Cassiopeiæ	3″.8
ε Indi	4″.5	α Centauri	3″.7

The first of these stars, though it has the greatest proper motion of all, would require 185,000 years to perform the circuit of the heavens, while μ Cassiopeiæ would require nearly 340,000 years to perform the same circuit. Slow as these motions are, they are very large compared with those of most of the stars of corresponding magnitude. As a general rule, the stars of the fourth, fifth, and sixth magnitudes move only a few seconds in a hundred years, and would therefore require many millions of years to perform the circuit of the heavens.

So far as they have yet been observed, and, indeed, so far as they can be observed for many centuries to come, these motions take place in perfectly straight lines. If each star is moving in some orbit, the orbit is so immense that no curvature can be perceived in the short arc which has been described since accurate determinations of the positions of the stars began to be made. So far as mere observation can inform us, there is no reason to suppose that the stars are severally moving in definite orbits of any kind. It is true that Mädler attempted to show, from an examination of the proper motions of the stars, that the whole stellar universe was revolving around the star Alcyone, of the Pleiades, as a centre—a theory the grandeur of which led to its wide diffusion in popular writings. But not the slightest weight has ever been given it by astronomers, who have always seen it to be an entirely baseless speculation. If the stars were moving in any regular circular orbits whatever having a common centre, we could trace some regularity among their proper motions. But no such regularity can be seen. The stars in all parts of the heavens move in all directions, with all sorts of velocities. It is true that, by averaging the proper motions, as it were, we can trace a certain law in them; but this law indicates, not a particular kind of orbit, but only an apparent proper motion, common to all the stars, which is probably due to a real motion of our sun and solar system.

The Solar Motion.—As our sun is merely one of the stars, and rather a small star too, it may have a proper motion as well as the other stars. Moreover, when we speak of the proper motion of a star, we mean, not its absolute motion, but only its motion relative to our system. As the sun moves, he carries the earth and all the planets along with him; and if we observe a star at perfect rest while we ourselves are thus moving, the star will appear to move in the opposite direction, as we have already shown in explaining the Copernican system. Hence, from an observation of the motion of a single star, it is impossible to decide how much of this apparent motion is due to the motion of our system, and how much to the real motion of the star. If, however, we should observe a great number of stars on all sides of us, and find them all apparently moving in the same direction, it would be natural to conclude that it was really our system which was moving, and not the stars. Now, when Herschel averaged the proper motions of the stars in different regions of the heavens, he found that this was actually the case. In general, the stars moved from the direction of the constellation Hercules, and towards the opposite point of the celestial sphere, near the constellation Argus. This would show that, relatively to the general mass of the stars, our sun was moving in the direction of the constellation Hercules. Herschel's data for this conclusion were, necessarily, rather slender. The subject was afterwards very carefully investigated by Argelander, and then by a number of other astronomers, whose results for the point of the heavens towards which the sun is moving are as follows:

	Right Ascension.	Declination.
Argelander............................	257° 49'	28° 50' N.
O. Struve.............................	261° 22'	37° 36' N.
Lundahl...............................	252° 24'	14° 26' N.
Galloway..............................	260° 1'	34° 23' N.
Mädler.................................	261° 38'	39° 54' N.
Airy and Dunkin.....................	262° 29'	28° 58' N.

It will be seen that while there is a pretty wide range among the authorities as to the exact point, and, therefore, some uncertainty as to where we should locate it, yet, if we lay the different points down on a star-map, we shall find that they all fall in the constellation Hercules, which was originally assigned by Herschel as that towards which we were moving.

As to the amount of the motion, Struve found that if the sun were viewed from the distance of an average star of the first magnitude placed in a direction from us at right angles to that of the solar motion, it would appear to move at the rate of $33''.9$ per century. Dunkin found the same motion to be $33''.5$ or $41''.0$, according to the use he made of stars having large proper motions.

Motion of Groups of Stars.—There are in the heavens several cases of widely extended groups of stars, having a common proper motion entirely different from that of the stars around and among them. Such groups must form connected systems, in the motion of which all the stars are carried along together without any great change in their positions relative to each other. The most remarkable case of this kind occurs in the constellation Taurus. A large majority of the brighter stars in the region between Aldebaran and the Pleiades have a common proper motion of about ten seconds per century towards the east. How many stars are included in this group no one knows, as the motions of the brighter ones only have been accurately investigated. Mr. R. A. Proctor has shown that five out of the seven stars which form the Dipper, or Great Bear, are similarly connected. He proposes for this community of proper motions in certain regions the name of *Star-drift*. Besides those we have mentioned, there are cases of close groups of stars, like the Pleiades, and of pairs of widely separated stars, in which star-drift has been noticed.

Motion in the Line of Sight.—Until quite recently, the only way in which the proper motion of a star could be detected was by observing its change of direction, or the change of the point in which it is seen on the celestial sphere. It is, however, impossible in this way to decide whether the star is or is not changing its distance from our system. If it be moving directly towards us, or directly away from us, we could not see any motion at all. The complete motion of the stars cannot, therefore, be determined by mere telescopic observations. But there is an ingenious method, founded on the undulatory theory of light, by which this motion may be detected with more or less probability by means of the spectroscope, and which was first successfully applied by Mr. Huggins, of England. According to the usual theory of light, the luminosity of a heated body is a result of the vibrations communicated by it to the ethereal medium which fills all space; and if the body be gaseous, it is supposed that a molecule of the gas vibrates at a certain definite rate, and thus communicates only certain definite vibrations to the ether. The rate of vibration is determined by the position of the bright line in the spectrum of the gas. Now, if the vibrating body be moving through the ether, the light-waves which it throws behind it will be longer, and those which it throws in front of it will be shorter, than if the body were at rest. The result will be, that in the former case the spectral lines will be less refrangible, or nearer the red end of the spectrum, and in the latter case nearer the blue end. If the line is not a bright one which the gas emits, but the corresponding dark one which it has absorbed from the light of a star passing through it, the result will be the same. If such a known line is found slightly nearer the blue end of the spectrum than it should be, it is concluded that the star from which it emanates is approaching us, while in the contrary case it is receding from us.

The question may be asked, How can we identify a line as proceeding from a gas, unless it is exactly in the position of the line due to that gas? How do we know but that it may be due to some other gas which emits light of slightly different refrangibility? The reply to this must be, that absolute certainty on this point is not attainable; but that, from the examination of a number of stars, the probabilities seem largely in favor of the opinion that the displaced lines are really due to the gases near whose lines they fall. If the lines were always displaced in one direction, whatever star was examined, the conclusion in question could not be drawn, because it might be that this line was due to some other unknown substance. But as a matter of fact, when different stars are examined, it is found that the lines in question are sometimes on one side of their normal position and sometimes on the other. This makes it probable that they really all belong to one substance, but are displaced by some cause, and the motion of the star is a cause the existence of which is certain, and the sufficiency of which is probable.

Mr. Huggins's system of measurement has been introduced by Professor Airy into the Royal Observatory, Greenwich, where very careful measures have been made during the past two years by Mr. Christie and Mr. Maunder. To show how well the fact of the motion is made out, we give in the tables on the following page the results obtained by Mr. Huggins and by the Greenwich observers for those stars in which the motion is the largest:

STARS RECEDING FROM US.

	By Mr. Huggins.	By Greenwich.
Sirius	20 miles per sec.	25 miles per sec.
α Orionis	22 " "	76 " " receding.
β Orionis	15 " "	
α Geminorum	25 " "	25 miles per sec.
α Leonis	15 " "	30 " "

STARS APPROACHING US.

	By Mr. Huggins.	By Greenwich.
Arcturus	55 miles per sec.	41 miles per sec.
α Lyræ	50 " "	36 " "
α Cygni	39 " "	41 " "
β Geminorum	49 " "	approaching.
α Ursæ Majoris	46 " "	approaching.

There are several collateral circumstances which tend to confirm these results. One is that the general amount of motion indicated is, in a rough way, about what we should expect the stars to have, from their observed proper motions, combined with their probable parallaxes. Another is that those stars in the neighborhood of Hercules are mostly found to be approaching the earth, and those which lie in the opposite direction to be receding from it, which is exactly the effect which would result from the solar motion just described. Again, the five stars in the Dipper which we have described as having a common proper motion are also found to have a common motion in the line of sight. The results of this wonderful and refined method of determining stellar motion, therefore, seem worthy of being received with some confidence so far as the general direction of the motion is concerned. But the displacement of the spectral lines is so slight, and its measurement a matter of such difficulty and delicacy, that we are far from being sure of the exact numbers of miles per second given by the observers. The discordances between the results of Greenwich and those of Mr. Huggins show that numerical certainty is not yet attained.

A necessary result of these motions will be that those stars which are receding from us will, in the course of ages, appear less brilliant, owing to their greater distance, while those which are approaching us will, as they come nearer, appear brighter, always supposing that their intrinsic brightness does not vary. But so immense is the distance of the stars, that many thousands of years will be required to produce any appreciable change in their brightness from this cause. For instance, from the best determinations which have been made, the distance of Sirius from our system is more than a million radii of the earth's orbit. With a velocity of twenty miles per second, it would require more than one hundred and fifty thousand years to pass over this distance.

It will, of course, be understood that the velocities found by the spectroscopic method are not the total velocities with which the stars are moving, but only the rate at which they are approaching to or receding from the earth, or, to speak mathematically, the component of the velocity in the direction of the line of sight. To find the total velocity, this component must be combined with the telescopic velocity found from the observed proper motion of the star, which is the velocity at right angles to the line of sight. None of the stars are moving exactly towards our system, and it is not likely that any will ever pass very near it. In the preceding list, the star α Cygni is the one which is coming most directly towards us. Its telescopic proper motion is so slight that, though we suppose its distance to be two million radii of the earth's orbit, yet its velocity at right angles to the line of sight will hardly amount to one-third of a mile per second. If the spectroscopic determination is correct, then, after an interval which will probably fall between one hundred thousand and three hundred thousand years, α Cygni will pass by our system at something like a hundredth of its present distance, and will, for several thousand years, be many times nearer and brighter than any star is now.

CHAPTER II.

THE STRUCTURE OF THE UNIVERSE.

HAVING in the preceding chapter described those features of the universe which the telescope exhibits to us, we have now, in pursuance of our plan, to inquire what light telescopic discoveries can throw upon the structure of the universe as a whole. Here we necessarily tread upon ground less sure than that which has hitherto supported us, because we are on the very boundaries of human knowledge. Many of our conclusions must be more or less hypothetical, and liable to be modified or disproved by subsequent discoveries. We shall endeavor to avoid all mere guesses, and to state no conclusion which has not some apparent foundation in observation or analogy. The human mind cannot be kept from speculating upon and wondering about the order of creation in its widest extent, and science will be doing it a service in throwing every possible light on its path, and preventing it from reaching any conclusion inconsistent with observed facts.

The first question which we reach in regular order is, How are the forty or fifty millions of stars visible in the most powerful telescopes arranged in space? We know, from direct observation, how they are arranged with respect to direction from our system; and we have seen that the vast majority of small stars visible in great telescopes are found in a belt spanning the heavens, and known as the Milky Way. But this gives us no complete information respecting their absolute position: to determine this, we must know the distance as well as the direction of each star. But beyond the score or so of stars which have a measurable parallax, there is no known way of measuring the stellar distances; so that all we can do is to make more or less probable conjectures, founded on the apparent magnitude of the individual stars and the probable laws of their arrangement. If the stars were all of the same intrinsic brightness, we could make a very good estimate of their distance from their apparent magnitude; but we know that such is not the case. Still, in all reasonable probability, the diversity of absolute magnitude is far less than that of the apparent magnitude; so that a judgment founded on the latter is much better than none at all. It was on such considerations as these that the conjectures of the first observers with the telescope were founded.

§ 1. *Views of Astronomers before Herschel.*

Before the invention of the telescope, any well-founded opinion respecting the structure of the starry system was out of the question. We have seen how strong a hold the idea of a spherical universe had on the minds of men, so that even Copernicus was fully possessed with it, and probably believed the sun to be, in some way, the centre of this sphere. Before any step could be taken towards forming a true conception of the universe, this idea had to be banished from the mind, and the sun had to be recognized as simply one of innumerable stars which made up the universe. The possibility that such might have been the case seems to have first suggested itself to Kepler, though he was deterred from completely accepting the idea by an incorrect estimate of the relative brilliancy of the stars. He reasoned that if the sun were one of a vast number of fixed stars of equal brilliancy scattered uniformly throughout space, there could not be more than twelve which were at the shortest distance from us. We should then have another set at double the distance, another at triple the distance, and so on; and since the more distant they are, the fainter they would appear, we should speedily reach a limit beyond which no stars could be seen. In fact, however, we often see numerous stars of the same magnitude crowded closely together, as in the belt of Orion, while the total number of visible stars is reckoned by thousands. He therefore concludes that the distances of the individual stars from each other are much less than their distances from our sun, the latter being situated near the centre of a comparatively vacant region.

Had Kepler known that it would require the light of a hundred stars of the sixth magnitude to make that of one of the first magnitude, he would not have reached this conclusion. A simple calculation would have shown him that, with twelve stars at distance unity, there would have been four times that number at the double distance, nine times at the treble distance, and so on, until, within the tenth sphere, there would have been more than four thousand stars. The twelve hundred stars on the surface of the tenth sphere would have been, by calculation, of the sixth magnitude, a number near enough to that given by actual count to show him that the hypothesis of a uniform distribution was quite accordant with observations. It is true that, where many bright stars were found crowded together, as in Orion, their distance from each other is probably less than that from our sun. But this agglomeration, being quite exceptional, would not indicate a general crowding together of all the stars, as Kepler seemed to suppose. In justice to Kepler it must be said that he put forth this view, not as a well-founded theory, but only as a surmise, concerning a question in which certainty was not attainable.

Ideas of Kant.—Those who know of Kant only as a speculative philosopher may be surprised to learn that, although he was not a working astronomer, he was the author of a theory of the stellar system which, with some modifications, has been very generally held until the present time. Seeing the Galaxy encircle the heavens, and knowing it to be produced by the light of innumerable stars too distant to be individually visible, he concluded that the stellar system extended much farther in the direction of the Galaxy than it did elsewhere. In other words, he conceived the stars to be arranged in a comparatively thin, flat layer, or stratum, our sun being somewhere near the centre. When we look edgewise along this stratum, we see an immense number of stars, but in the perpendicular direction comparatively few are visible.*

This thin stratum suggested to Kant the idea of a certain resemblance to the solar system. Owing to the small inclinations of the planetary orbits, the bodies which compose this system are spread out in a thin layer, as it were; and we have only to add a great multitude of planets moving around the sun in orbits of varied inclinations to have a representation in miniature of the stellar system as Kant imagined it to exist. Had the zone of small planets between Mars and Jupiter then been known, it would have afforded a striking confirmation of Kant's view by showing a yet greater resemblance of the planetary system to his supposed stellar system. Were the number of these small planets sufficiently increased, we should see them as a sort of Galaxy around the zodiac, a second Milky Way, belonging to our system, and resolvable with the telescope into small planets, just as the Galaxy is resolved into small stars. The conclusion that two systems which were so similar in appearance were really alike in structure would have seemed very well founded in analogy.

As the planets are kept at their proper distances, and prevented from falling into each other or into the sun by the centrifugal force generated by their revolutions in their orbits, so Kant supposed the stars to be kept apart by a revolution around some common centre. The proper motions of the stars were then almost unknown, and the objection was anticipated that the stars were found to occupy the same position in the heavens from generation to generation, and therefore could not be in motion around a centre. To this Kant's reply was that the time of revolution was so long, and the motion so slow, that it was not perceptible with the imperfect means of observation then available. Future generations would, he doubted not, by comparing their observations

---

* The original idea of this theory is attributed by Kant to Wright, of Durham, England, a writer whose works are entirely unknown in this country, and whose authorship of the theory has been very generally forgotten.

with those of their predecessors, find that there actually was a motion among the stars.

This conjecture of Kant, that the stars would be found to have a proper motion, has, as we have seen, been amply confirmed; but the motion is not of the kind which his theory would require. On this theory, all the stars ought to move in directions nearly parallel to that of the Milky Way, just as in the planetary system we find them all moving in directions nearly parallel to the ecliptic. But the proper motions actually observed have no common direction, and follow no law whatever, except that, on the average, there is a preponderance of motions from the constellation Hercules, which is attributed to an actual motion of our sun in that direction. Making allowance for this preponderance, we find the stars to be apparently moving at random in every direction; and therefore they cannot be moving in any regularly arranged orbits, as Kant supposed. A defender of Kant's system might indeed maintain that, as it is only in a few of the stars nearest us that any proper motion has been detected, the great cloud of stars which make up the Milky Way might really be moving along in regular order, a view the possibility of which we shall be better prepared to consider hereafter.

The Kantian theory supposes the system which we have just been describing to be formed of the immense stratum of stars which make up the Galaxy and stud our heavens, and to include all the stars separately visible with our telescopes. But he did not suppose this system, immense though it is, to constitute the whole material universe. In the nebulæ he saw other similar systems at distances so immense that the combined light of their millions of suns only appeared as a faint cloud in the most powerful telescopes. This idea that the nebulæ were other galaxies was more or less in vogue among popular writers until a quite recent period, when it was refuted by the spectroscope, which shows that these objects are for the most part masses of glowing gas. It has, however, not received support among astronomers since the time of Sir William Herschel.

System of Lambert.—A few years after the appearance of Kant's work, a similar but more elaborate system was sketched out by Lambert. He supposed the universe to be arranged in systems of different orders. The smallest systems which we know are those made up of a planet, with its satellites circulating around it as a centre. The next system in order of magnitude is a solar system, in which a number of smaller systems are each carried round the sun. Each individual star which we see is a sun, and has its retinue of planets revolving around it, so that there are as many solar systems as stars. These systems are not, however, scattered at random, but are divided up into greater systems which appear in our telescopes as clusters of stars. An immense number of these clusters make up our Galaxy, and form the visible universe as seen in our telescopes. There may be yet greater systems, each made up of galaxies, and so on indefinitely, only their distance is so immense as to elude our observation.

Each of the smaller systems visible to us has its central body, the mass of which is much greater than that of those which revolve around it. This feature Lambert supposed to extend to other systems. As the planets are larger than their satellites, and the sun larger than its planets, so he supposed each stellar cluster to have a great central body around which each solar system revolved. As these central bodies are invisible to us, he supposed them to be opaque and dark. All the systems, from the smallest to the greatest, were supposed to be bound together by the one universal law of gravitation.

As not the slightest evidence favoring the existence of these opaque centres has ever been found, we are bound to say that this sublime idea of Lambert's has no scientific foundation. Astronomers have handed it over without reservation to the lecturers and essayists.

§ 2. *Researches of Herschel and his Successors.*

Herschel was the first who investigated the structure of the stellar system by a long-continued series of observations, executed with a definite end in view. His plan was that of "star-gauging," which meant, in the first place, the simple enumeration of all the stars visible with a powerful telescope in a given portion of the heavens. He employed a telescope of twenty inches aperture, magnifying one hundred and sixty times, the field of view being a quarter of a degree in diameter. This diameter was about half that of the full moon, so that each count or gauge included all the stars visible in a space having one-fourth the apparent surface of the lunar disk. From the number of stars in any one field of view, he concluded to what relative distance his sight extended, supposing a uniform distribution of the stars throughout all the space included in the cone of sight of the telescope. When an observer looks into a telescope pointed at the heavens, his field of vision includes a space which constantly widens out on all sides as the distance becomes greater; and the reader acquainted with geometry will see that this space forms a cone having its point in the focus of the telescope, and its circular base at the extreme distance to which the telescope reaches. The solid contents of this cone will be proportional to the cube of the distance to which it extends; for instance, if the telescope penetrates twice as far, the cone of sight will be not only twice as long, but the base will be twice as wide in each direction, so that the cone will have altogether eight times the contents, and will, on Herschel's hypothesis, contain eight times as many stars. So, when Herschel found the stars eight times as numerous in one region as in another, he concluded that the stellar system extended twice as far in the direction of the first region.

To count all the stars visible with his telescope, Herschel found to be out of the question. He would have had to point his instrument several hundred thousand times, and count all the visible stars at each pointing. He therefore extended his survey only over a wide belt extending more than half-way round the celestial sphere, and cutting the Galaxy at right angles. In this belt he counted the stars in 3400 telescopic fields. Comparing the average number of stars in different regions with the position of the region relative to the Galaxy, he found that the stars were thinnest at the point most distant from the Galaxy, and that they constantly increased in number as the Galaxy was approached. The following table will give an idea of the rate of increase. It shows the average number of stars in the field of view of the telescope for each of six zones of distance from the Galaxy.

First zone	90° to 75°	from Galaxy	....	4	stars per field.
Second zone	75° " 60°	"	"	5	" "
Third zone	60° " 45°	"	"	8	" "
Fourth zone	45° " 30°	"	"	14	" "
Fifth zone	30° " 15°	"	"	24	" "
Sixth zone	15° " 0°	"	"	53	" "

A similar enumeration was made by Sir John Herschel for the corresponding region on the other, or southern, side of the Galaxy. He used the same telescope, and the same magnifying power. His results were:

First zone	6 stars per field.	Fourth zone	13	stars per field.
Second zone	7 " "	Fifth zone	26	" "
Third zone	9 " "	Sixth zone	59	" "

The reader will, perhaps, more readily grasp the signification of these numbers by the mode of representation which was suggested in describing the distribution of the nebulæ. Let him imagine himself standing under a clear sky at the time when the Milky Way encircles the horizon. Then, the first zone, as we have defined it, will be around the zenith, extending one-sixth of the way to the horizon on every side; the second zone will be next below and around this circular space, extending one-third of the way to the horizon; and so each one will follow in regular order until we reach the sixth, or galactic, zone, which will encircle the horizon to a height of 15° on every side. The numbers we have given show that in the position of the observer which we have supposed the stars would be thinnest around the zenith, and would constantly increase in number as we approached the horizon. The observer being supposed still to occupy the same position, the second table shows the distribution of the stars in the opposite or invisible hemisphere, which he would see if the earth were removed. In this hemisphere the first, or thinnest, zone would be directly opposite the thinnest zone in the observer's zenith; that is, it would be directly under his feet. The successive zones would then be nearer the horizon, the sixth or last encircling it, and extending 15° below it on every side.

The numbers we have given are only averages, and do not give an adequate idea of the actual inequalities of distribution in special regions of the heavens. Sometimes there was not a solitary star in the field of the telescope, while at others there were many hundreds. In the circle of the Galaxy itself, the stars are more than twice as thick as in the average of the first zone, which includes not only this circle, but a space of 15° on each side of it.

Adopting the hypothesis of a uniform distribution of the stars, Herschel concluded from his first researches that the stellar system was of the general form supposed by Kant, extending out on all sides five times as far in the direction of the Galaxy as in the direction perpendicular to it. The most important modification he made was to suppose an immense cleft extending edgewise into the system from its circumference about half-way to the centre. This cleft corresponded to the division in the Milky Way which commences in the summer constellation Cygnus in the north, and passes through Aquila, the Serpent, and Scorpius far into the southern hemisphere. Estimating the distance by the arrangement and apparent magnitude of the stars, he was led to estimate the mean thickness of the stellar stratum from top to bottom as 155 units, and the diameter as 850 units, the unit being the average distance of a star of the first magnitude. Supposing this distance to be that which light would travel over in 16 years—a supposition which is founded on the received estimate of the mean parallax corresponding to stars of that magnitude—then it would take light nearly 14,000 years to travel across the system from one border to the other, and 7000 years to reach us from the extreme boundary.

The foregoing deduction of Herschel was founded on the hypothesis that the stars were equally dense in every part of the stellar system, so that the number of stars in any direction furnished an index to the extent of the stars in that direction. Further study showed Herschel that this assumption might be so far from correct that his conclusions would have to be essentially modified. Binary and other double stars and star clusters evidently offered cases in which several stars were in much closer association than were the stars in general. To show exactly on what considerations this change of view is founded, we remark that if the increase of density in the direction of the Milky Way were quite regular, so that there were no cases of great difference in the thickness of the stars in two adjoining regions, then the original view would have been sound so far as it went. But such irregularities are very frequent, and it would lead to an obvious absurdity to explain them on Herschel's first hypothesis; for instance, when the telescope was directed towards the Pleiades there would be

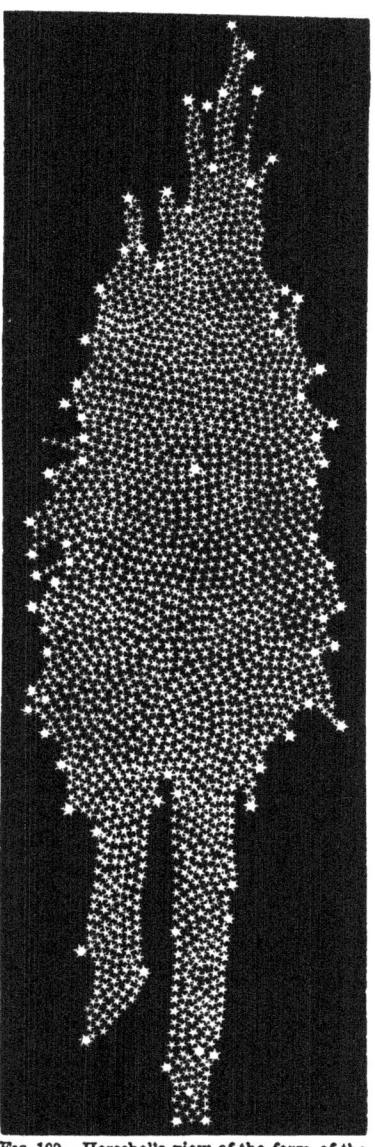

FIG. 109.—Herschel's view of the form of the universe.

found, probably, six or eight times as many stars as in the adjoining fields. But supposing the real thickness of the stars the same, the result would be that in this particular direction the stars extended out twice as far as they did in the neighboring parts of the sky; that is, we should have a long, narrow spike of stars pointing directly from us. As there are many such clusters in various parts of the sky, we should have to suppose a great number of such spikes. In other regions, especially around the Milky Way, there are spaces nearly void of stars. To account for these we should have to suppose long narrow chasms reaching through towards our sun. Thus the stellar system would present the form of an exaggerated star-fish with numerous deep openings, a form the existence of which is beyond all probability, especially if we reflect that all the openings and all the arms have to proceed from the direction of our sun.

The only rational explanation of a group of stars showing itself in a telescope, with a comparatively void space surrounding it, is that we have here a real star cluster, or a region in which the stars are thicker than elsewhere. Now, one can see with the naked eye that the Milky Way is not a continuous uniform belt, but is, through much of its course, partly made up of a great number of irregular cloud-like masses with comparatively dark spaces between them. The conclusion is unavoidable that we have here real aggregations of stars, and not merely a region in which the bounds of the stellar-system are more widely extended. Whether Herschel clearly saw this may be seriously questioned; but however it may have been, he adopted another method of estimating the relative distances of the stars visible in his gauges.

This method consisted in judging of the distances to which his telescope penetrated, not by the number of stars it brought into view, but by their brightness. If all the stars were of the same intrinsic brightness, so that the differences of their apparent magnitude arose only from their various distances from us, then this method would enable us to fix the distance of each separate star. But as we know that the stars are by no means equal in intrinsic brightness, the method cannot be safely applied to any individual star, a fact which Herschel himself clearly saw. It does not follow, however, that we cannot thus form an idea of the relative distances of whole classes or groups of stars. Although it is quite possible that an individual star of the fifth magnitude may be nearer to us than another of the fourth, yet we cannot doubt that the average distance of all the fifth-magnitude stars is greater than the average of those of the fourth magnitude, and greater, too, in a proportion admitting of a tolerably accurate numerical estimate. Such an estimate Herschel attempted to make, proceeding on the following plan:

Suppose a sphere to be drawn around our sun as a centre of such size that it shall be equal to the average space occupied by a single one of the stars visible to the naked eye; that is, if we suppose that portion of the space of the stellar system occupied by the six thousand brighter stars to be divided into six thousand parts, then the sphere will be equal to one of these parts. The radius of this sphere will probably not differ much from the distance of the nearest fixed star, a distance we shall take for unity. Then, suppose a series of larger spheres, all drawn around our sun as a centre, and having the radii 3, 5, 7, 9, etc. The contents of the spheres being as the cubes of their diameters, the first sphere will have $3 \times 3 \times 3 = 27$

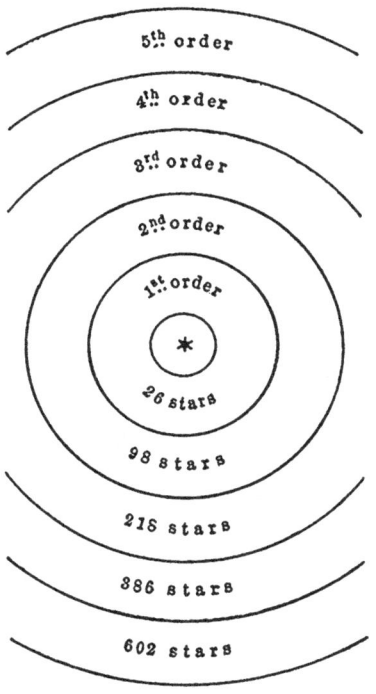

Fig. 110.—Illustrating Herschel's orders of distance of the stars.

times the bulk of the unit sphere, and will therefore be large enough to contain 27 stars; the second will have 125 times the bulk, and will therefore contain 125 stars, and so with the successive spheres. Fig. 110 shows a section of portions of these spheres up to that with radius 11. Above the centre are given the various orders of stars which are situated between the several spheres, while in the corresponding spaces below the centre are given the number of stars which the region is large enough to contain; for instance, the sphere of radius 7 has room for 343 stars, but of this space 125 parts belong to the spheres inside of it: there is, therefore, room for 218 stars between the spheres of radii 5 and 7.

Herschel designates the several distances of these layers of stars as orders; the stars between spheres 1 and 3 are of the first order of distance, those between 3 and 5 of the second order, and so on. Comparing the room for stars between the several spheres with the number of stars of the several magnitudes, he found the result to be as follows:

Order of Distance.	Number of Stars there is room for.	Magnitude.	Number of Stars of that magnitude.
1	26	1	17
2	98	2	57
3	218	3	206
4	386	4	454
5	602	5	1161
6	866	6	6103
7	1178	7	6146
8	1538		

There is evidently no correspondence between the calculated orders of distance and the magnitudes as estimated on the usual scale. But Herschel found that this was because the magnitudes as usually estimated corresponded to an entirely different scale of distance from that which he adopted. In his scale the several distances increased in arithmetical progression; while in the order of magnitudes the increase is in geometrical progression. In consequence, the stars of the sixth magnitude correspond to the eighth, ninth, or tenth order of distances; that is, we should have to remove a star of the first magnitude to eight, nine, or ten times its actual distance to make it shine as a star of the sixth magnitude.

Attempting on this system to measure the extent of the Milky Way, Herschel concluded that it was unfathomable with his twenty-foot telescope, which, he calculated, would penetrate to the 900th order of distances, that is, to stars which were 900 times as far as the average of those of the first magnitude. He does not seem to have made any very extended examination with his forty-foot telescope, but concluded that it would leave him in the same uncertainty in respect to the extent of the Milky Way as the twenty-foot one did. This unrivalled man, to whom it was given to penetrate farther into creation than man had ever done before him, seems to have rested from his labors without leaving any more definite theory of the boundaries of the stellar system than that they extended, at least in the direction of the Milky Way, beyond the utmost limit to which his telescope could penetrate. If we estimate the time it would require light to come from the utmost limit to which he believed his vision to extend, we shall find it to be about fourteen thousand years, or more than double that deduced from his former gauges. We can say with confidence that the time required for light to reach us from the most distant visible stars is measured by thousands of years. But it must be admitted that Herschel's estimate of the extent of the Milky Way may be far too great, because it rests on the assumption that all stars are of the same absolute brightness. If the smallest stars visible in his telescope were, on the average, of the same intrinsic brilliancy as the brighter ones, the conclusion would be well founded. But if we suppose a boundary, it is impossible to decide from Herschel's data whether the minuteness of those stars arises from their great distance or from their small magnitude. Notwithstanding this uncertainty, it has been maintained by some, notably by Mr. Proctor, that the views of Herschel respecting the constitution of the Milky Way, or stellar system, were radically changed by this second method of star-gauging. I see no evidence of any radical change. Although Herschel does not express himself very definitely on the subject, yet, in his last paper on the distribution of the stars (*Philosophical Transactions* for 1817), there are several remarks which seem to imply that he still supposed the stellar system to have the general form shown in Fig. 109, and that, in accordance with that view, he supposed the clustering of stars to indicate protuberant parts of the Milky Way. He did, indeed, apply a different method of research, but the results to which the new methods led were, in their main features, the same as those of the old method.

Since the time of Herschel, one of the most eminent of the astronomers who have investigated this subject is Struve the elder, formerly director of the Pulkowa Observatory. His researches were founded mainly on the numbers of stars of the several magnitudes found by Bessel in a zone thirty degrees wide extending all round the heavens, fifteen degrees on each side of the equator. With these he combined the gauges of Sir William Herschel. The hypothesis on which he based his theory was similar to that employed by Herschel in his later researches, in so far that he supposed the magnitude of the stars to furnish, on the average, a measure of their relative distances. Supposing, after Herschel, a number of concentric spheres to be drawn around the sun as a centre, the successive spaces between which corresponded to stars of the several magnitudes, he found that the farther out he went, the more the stars were condensed in and near the Milky Way. This conclusion may be drawn at once from the fact we have already mentioned, that the smaller the stars, the more they are condensed in the region of the Galaxy. Struve found that if we take only the stars plainly visible to the naked eye—that is, those down to the fifth magnitude—they are no thicker in the Milky Way than in other parts of the heavens. But those of the sixth magnitude are a little thicker in that region, those of the seventh yet thicker, and so on, the inequality of distribution becoming constantly greater as the telescopic power is increased.

From all this, Struve concluded that the stellar system might be considered as composed of layers of stars of various densities, all parallel to the plane of the Milky Way. The stars are thickest in and near the central layer, which he conceives to be spread out as a wide, thin sheet of stars. Our sun is situated near the middle of this layer. As we pass out of this layer, on either side we find the stars constantly growing thinner and thinner, but we do not reach any distinct boundary. As, if we could rise in the atmosphere, we should find the air constantly growing thinner, but at so gradual a rate of progress that we could hardly say where it terminated; so, on Struve's view, would it be with the stellar system, if we could mount up in a direction perpendicular to the Milky Way. Struve gives the following table of the thickness of the stars on each side of the principal plane, the unit of distance being that of the extreme distance to which Herschel's telescope could penetrate:

Distance from Principal Plane.	Density.	Mean Distance between Neighboring Stars.
In the principal plane............	1.0000	1.000
0.05 from principal plane......	0.48568	1.272
0.10 " "	0.33288	1.458
0.20 " "	0.23895	1.611
0.30 " "	0.17980	1.772
0.40 " "	0.13021	1.973
0.50 " "	0.08646	2.261
0.60 " "	0.05510	2.628
0.70 " "	0.03079	3.190
0.80 " "	0.01414	4.131
0.866 " "	0.00532	5.729

This condensation of the stars near the central plane, and the gradual thinning-out on each side of it, are only designed to be the expression of the general or average distribution of those bodies. The probability is that even in the central plane the stars are many times as thick in some regions as in others, and that as we leave the plane, the thinning-out would be found to proceed at very different rates in different regions. That there may be a gradual thinning-out cannot be denied; but Struve's attempt to form a table of it is open to the serious objection that, like Herschel, he supposed the differences between the magnitudes of the stars to arise entirely from their different distances from us. Although where the scattering of the stars is nearly uniform this supposition may not lead us into serious error, the case will be entirely different where we have to deal with irregular masses of stars, and especially where our telescopes penetrate to the boundary of the stellar system. In the latter case we cannot possibly distinguish between small stars lying within the boundary and larger ones scattered outside of it, and Struve's gradual thinning-out of the stars may be entirely accounted for by great diversities in the absolute brightness of the stars.

Among recent researches on this subject, those of Mr. R. A. Proctor are entitled to consideration, from being founded on facts which were not fully known or understood by the investigators whom we have mentioned. The strongest point which he makes is that all views of the arrangement of the stellar system founded upon the theory that the stars are either of similar intrinsic brightness, or approach an equality of distribution in different regions, are entirely illusory. He cites the phenomena of star-drift, described in the last chapter, as proving that stars which had been supposed widely separated are really agglomerated into systems; and claims that the Milky Way may be a collection of such systems, having nothing like the extent assigned it by Herschel.

How far the considerations brought forward by Mr. Proctor should make us modify the views of the subject hitherto held, cannot be determined without further observations on the clustering of stars of different magnitudes. We may, however, safely concede that there is a greater tendency among the stars to be collected into groups than was formerly supposed. A curious result of Mr. J. M. Wilson, of Rugby, England, respecting the orbits of some binary stars, throws light on this tendency. It was found by Struve that although the great common proper motion of the pair of stars 61 Cygni, celebrated for the determinations of their parallax, was such as to leave no reasonable doubt that they were physically connected, yet not the slightest deviation in their courses, arising from their mutual attraction, could be detected. Mr. Wilson has recently confirmed this result by an examination of the whole series of measures on this pair from 1753 to 1874, which do not show the slightest deviation, but seem to indicate that each star of the pair is going on its course independently of the other. But, as just stated, they move too nearly together to permit of the belief that they are really independent. The only conclusion open to us is that each of them describes an immense orbit around their common centre of gravity, an orbit which may be several degrees in apparent diameter, and in which the time of revolution is counted by thousands of years. Two thousand years hence they will be so far apart that no connection between them would be suspected.

It is a question whether we have not another instance of the same kind in the double star Castor, or α Geminorum. Mr. Wilson finds the orbit of this binary to be apparently hyperbolic, a state of things which would indicate that the two stars had no physical connection whatever, but that, in pursuing their courses through space, they chanced to come so close together that they were brought for a while within each other's sphere of attraction. If this be the case, they will gradually separate forever, like two ships meeting on the ocean and parting again. We remark that the course of each star will then be very different from what it would have been if they had not met. We cannot, however, accept the hyperbolic orbit of Mr. Wilson as an established fact, because the case is one in which it is very difficult to distinguish between a large and elongated elliptic orbit and a hyperbolic orbit. The common proper motion of the two objects is such as to lead to the belief that they constitute a pair, the components of which separate to a great distance.

Now, these discoveries of pairs of stars moving around a common centre of gravity, in orbits of immense extent, suggest the probability that there exist in the heavens great numbers of pairs, clusters, and systems of this sort, the members of which are so widely separated that they have never been suspected to belong together, and the widely scattered groups having a common proper motion may very well be systems of this kind.

§ 3. *Probable Arrangement of the Visible Universe.*

The preceding description of the views held by several generations of profound thinkers and observers respecting the arrangement of the visible universe furnishes an example of what we may call the evolution of scientific knowledge. Of no one of the great men whom we have mentioned can it be said that his views were absolutely and unqualifiedly erroneous, and of none can it be said that he reached the entire truth. Their attempts to solve the mystery which they saw before them were like those of a spectator to make out the exact structure of a great building which he sees at a distance in the dim twilight. He first sees that the building is really there, and sketches out what he believes to be its outlines. As the light increases, he finds that his first outline bears but a rude resemblance to what now seems to be the real form, and he corrects it accordingly. In his first attempts to fill in the columns, pilasters, windows, and doors, he mistakes the darker shades between the columns for windows, other lighter shadows for doors, and the pilasters for columns. Notwithstanding such mistakes, his representation is to a certain extent correct, and he will seldom fall into egregious error. The successive improvements in his sketch, from the first rough outline to the finished picture, do not consist in effacing at each step everything he has done, but in correcting it, and filling in the details.

The progress of our knowledge of nature is generally of this character. But in the case now before us, so great is the distance, so dim the light, and so slender our ideas of the principles on which the vast fabric is constructed, that we cannot pass beyond a few rough outlines. Still there are a few features which we can describe with a near approach to certainty, and others respecting which, though our knowledge is somewhat vague, we can reach a greater or less degree of probability. We may include these under the following seven heads:

1st. Leaving the nebulæ out of consideration, and confining ourselves to the stellar system, we may say, with moral certainty, that the great mass of the stars which compose this system are spread out on all sides, in or near a widely extended plane passing through the Milky Way. In other words, the large majority of the stars which we can see with the telescope are contained in a space having the form of a round, flat disk, the diameter of which is eight or ten times its thickness. This was clearly seen by Kant, and has been confirmed by Herschel and Struve. In fact, it forms the fundamental base of the structures reared by these several investigators. When Kant saw, in this arrangement, a resemblance to the solar system, in which the planets all move round near one central plane, he was correct, so far as he went. The space, then, in which we find most of the stars to be contained is bounded by two parallel planes forming the upper and lower surfaces of the disk we have described, the distance apart of these planes being a small fraction of their extent — probably less than an eighth.

2d. Within the space we have described the stars are not scattered uniformly, but are for the most part collected into irregular clusters or masses, with comparatively vacant spaces between them. These collections have generally no definite boundaries, but run into each other by insensible gradations. The number of stars in each collection may range from two to many thousands; and larger masses are made up of smaller ones in every proportion, much as the heavy clouds on a summer's day are piled upon each other.

3d. Our sun, with its attendant planets, is situated near the centre of the space we have described, so that we see nearly the same number of stars in any two opposite quarters of the heavens.

4th. The six or seven thousand stars around us, which are easily seen by the naked eye, are scattered in space with a near approach to uniformity, the only exception being local clusters, the component stars of which are few in number and pretty widely separated. Such are the Pleiades, Coma Berenices, and perhaps the principal stars of many other constellations, which are so widely separated that we do not see any connection among them.

5th. The disk which we have described does not represent the form of the stellar system, but only the limits within which it is mostly contained. The absence of any definite boundary, either to star clusters or the stellar system, and the number of comparatively vacant regions here and there among the clusters, prevent our assigning any more definite form to the system than we could assign to a cloud of dust. The thin and widely extended space in which the stars are most thickly clustered may, however, be called the galactic region.

6th. On each side of the galactic region the stars are more evenly and thinly scattered, but probably do not extend out to a distance at all approaching the extent of the galactic region. If they do extend out to an equal distance, they are very few in number. It is, however, impossible to set any definite boundaries, not only from our ignorance of the exact distance of the smallest stars we can see in the telescope, but because the density of the stars probably diminishes very gradually as we go out towards the boundary.

7th. On each side of the galactic and stellar region we have a nebular region, in which we find few or no stars, but vast numbers of nebulæ. The nebulæ diminish greatly in number as we approach the galactic region, only a very few being found in that region.

The general arrangement of the stars and nebulæ which we have described is seen in Fig. 111, which shows what is probably the general aspect of a section of the visible universe perpendicular to the Milky Way. In the central part of the figure we have the galactic region, in which the stars are mostly aggregated in large masses. Of the arrangement of these masses nothing certain is known; they are, therefore, put in nearly at random. Indeed, it is still an undecided question whether the aggregations of stars which make up the Milky

PROBABLE ARRANGEMENT OF THE VISIBLE UNIVERSE. 481

Way extend all the way across the diameter of the galactic region, or whether they are arranged in the form of a ring, with our sun and his surrounding stars in the centre of it. In the latter case, the masses of stars near the centre should be less strongly marked. This central region being that in which our earth is situated, this uncertainty respecting the density of stars in that region implies an uncertainty whether

FIG. 111.—Probable arrangement of the stars and nebulæ visible with the telescope. In the Galaxy the stars are not evenly scattered, but are agglomerated into clusters.

the stars visible with the naked eye are part of one of the masses which make up the Galaxy, or whether we are in a comparatively thin region. Although this question is still unsolved, it is one which admits of an answer by telescopic research. When we described Sir William Herschel's arrangement of the stars in concentric spheres, we saw that in the more distant spheres the stars were vastly more dense

32

around the galactic belt of each sphere than they were in other parts of it. To answer the question which has been presented, we must compare the densities of the stars at the circumferences of these spheres with the density immediately around us. In other words, the question is, Suppose a human being could dart out in the direction of the Milky Way, and pass through some of the masses of stars composing it, would he find them thicker or thinner than they are in the visible heavens around us?

A question still left open is, whether all the celestial objects visible with the telescope are included within the limits of the three regions we have just indicated, or whether the whole Galaxy, with everything which is included within its limits, is simply one of a great number of widely scattered stellar systems. Since any consideration of invisible galaxies and systems would be entirely idle, the question may be reduced to this: Are the most distant star clusters which the telescope shows us situated within the limits of the stellar system or far without them, a great vacant space intervening? The latter alternative is the popular one, first suggested by Kant, it being supposed that the most distant nebulæ constituted other Milky Ways or stellar systems as extensive as our own.

Although the possibility that this view is correct cannot be denied, yet the arrangement of the star clusters or resolvable nebulæ militates against it. We have shown that the majority of the latter lie near the direction of the plane of the Milky Way, comparatively few being seen near the perpendicular direction. But if these objects were other galaxies, far outside of the one which surrounds us, they would be as likely to lie in one direction as in another, and the probability against the great mass of them lying in one plane would be very great. The most probable conclusion, therefore, is that they constitute part of our stellar system. They may, indeed, be scattered around or outside of the extreme limits within which single stars can be seen, but not at distances so great that they should be considered as separate systems. The most probable conclusion, in the present state of our knowledge, seems to be that the scheme shown in Fig. 111 includes the whole visible universe.

The differences of opinion which now exist respecting the probable arrangement and distance of the stars arise mainly from our uncertainty as to what is the probable range of absolute magnitude of the stars, a subject to which we have already several times alluded. The discovery of the parallax of several stars has enabled us not only to form some idea of this question by comparing the brilliancy of these stars with their known distances, but it has enabled us to answer the interesting question, How does our sun compare with these stars in brightness? The curious result of this inquiry is, that our sun is really a star less than the average, which would modestly twinkle among the smaller of its fellows if removed to the distance from us at which they are placed. Zöllner found, by comparing the light of the sun with that of Capella, or α Aurigæ, that it would have to be removed to 236,000 times its present distance to appear equally bright with that star, which we may take as an average star of the first magnitude. But the greater number of the stars of this magnitude are situated at four or five times this distance; so that if our sun were placed at their average distance, it would probably not exceed the third or fourth magnitude. Still, it would by no means belong among the smallest stars of all, because we do find stars with a measurable parallax which are only of the fifth, sixth, or even the seventh magnitude. Altogether, it appears that the range of absolute brilliancy among the stars extends through eight or ten magnitudes, and that the largest ones emit several thousand times as much light as the smallest. It is this range of magnitude which really forms the greatest obstacle in the way of determining the arrangement of the stars in space.

§ 4. *Do the Stars really form a System?*

We have described the sublime ideas of Kant and Lambert, who, seeing the bodies of our solar system fitted to go through their revolutions without permanent change during an indefinite period of time, reasoned by analogy that the stellar universe was constructed on the same general plan, and that each star had its appointed orbit, round which it would run its course during endless ages. This speculation was not followed up by Herschel and Struve, who, proceeding on a more strictly scientific plan, found it necessary to learn how the stars are now situated before attempting to decide in what kinds of orbits they are moving. In the absence of exact knowledge respecting the structure and extent of the stellar system, it is impossible to say with certainty what will be the state of that system after the lapse of the millions of years which would be necessary for the stars to perform a revolution around one centre. But, as in describing the constitution of the stellar system, we found certain features on which we could pronounce with a high degree of probability, so, in respect to the motions and orbits of the stars, there are some propositions which we may sustain with a near approach to certainty.

Stability of the System.—We may first assert, with a high degree of probability, that the stars do not form a stable system in the sense in which we say that the solar system is stable. By a stable system we mean one in which each star moves round and round in an unchanging orbit, every revolution bringing it back to its starting-point, so that the system as a whole shall retain the same general form, dimensions, and arrangement during innumerable revolutions of the bodies which compose it. It is almost necessary to the existence of such a system that it have a great central body, the mass of which should be at least vastly greater than that of the individual bodies which revolve around it. At least, such a central body could be dispensed with only by the separate stars having a regularity of motion and arrangement which certainly does not exist in the stellar system as we actually see it. The question, then, reduces itself to this: Are there any immense attracting centres around which the separate collections of stars revolve; or is there any centre around which all the stars which compose the visible universe revolve? In all

DO THE STARS REALLY FORM A SYSTEM? 485

human probability, these questions must be answered in the negative. All analogy leads us to believe that if there were any such central masses, they would be not only larger than the other stars, but brighter in a yet greater proportion. It is, of course, possible to conceive of immense dark bodies, such as Lambert supposed to exist, but we cannot but believe the existence of such bodies to be very improbable. Although there is, as we have seen, great diversity among the stars in respect to their magnitudes, there are none of them which seem to have that commanding preëminence above their fellows which the sun presents above the planets which surround him.

But the most conclusive proof that the stars do not revolve round definite attracting centres is found in the variety and irregularity of their proper motions, which we have already described. We have shown (1) that when the motions of great numbers of stars are averaged, there is found a general preponderance of motions from the constellation Hercules, which is supposed to be due to a motion of our sun with his attendant planets in that direction; and (2) that when the motions of stars in the same region are compared, there is often found to be a certain resemblance among them. But this tendency towards a regular law affects only large masses of stars, and does not imply any such regularity in the motions of individual stars as would be apparent if they moved in regular circular orbits, as the planets move round the sun. The motion of each individual star is generally so entirely different from that of its fellows as seemingly to preclude all reasonable probability that these bodies are revolving in definite orbits around great centres of attraction.

The most extraordinary instances of the irregularities of which we speak are found in the stars of unusually rapid proper motion, which are moving forward at such a rate that the gravitation of all the known stars cannot stop them until they shall have passed through and beyond the visible universe. The most remarkable of these, so far as we know, is Groombridge 1830, it having the largest apparent proper motion of any known star. The most careful determinations of its parallax seem to show that its distance is so immense that the parallax is only about a tenth of a second; that is, a line drawn from the sun to the earth would subtend an angle of only a tenth of a second when viewed from this star. But the apparent motion of the star, as we actually see it, is more than seven seconds per annum, or seventy times its parallax. It follows that the star moves over a space of more than seventy times the distance of the sun from us in the space of a year. If, as is likely, the motion of the star is oblique to the line in which we see it, its actual velocity must be yet greater. Leaving this out of account, we see that the star would pass from the earth to the sun in about five days, so that its velocity probably exceeds two hundred miles per second.

To understand what this enormous velocity may imply, we must advert to the theorem of gravitational astronomy that the velocity which a body can acquire by falling towards an attracting centre is, at each point of its path, limited. For example, a body falling from an infinite distance to the earth's surface, and acted on by the attraction of the earth alone, would acquire a velocity of only about seven miles per second. *Vice versa*, a body projected from the earth with this velocity would never be stopped by the earth's attraction alone, but would describe an elliptic orbit round the sun. If the velocity exceeded twenty-seven miles per second, the attraction of the sun himself could never stop it, and it would wander forever through the stellar spaces. The greater the distance from the sun at which the body is started, the less the velocity which will thus carry it forever away from the sun. At the orbit of Uranus the required velocity would be only six miles per second; at Neptune, it would be less than five miles per second; half-way between the sun and a Centauri, it would be a mile in twelve seconds, or a fourth the speed of a cannon-ball. If we knew the masses of each of the stars, and their arrangement in space, it would be easy to compute this limiting velocity for a body falling from an infinite distance to any point of the stellar system. If the motion of a star were found to exceed this limit, it would show that the star did not belong to the visible universe at all, but was only a visitor flying on a course through infinite space at such a rate that the combined attraction of all the stars could never stop it.

Let us now see how the case may stand with our flying star, and what relation its velocity may bear to the probable attraction of all the stars which exist within the range of the telescope. The number of stars actually visible with the most powerful telescopes probably falls short of fifty millions; but, to take a probable outside limit, we shall suppose that within the regions occupied by the farthest stars which the telescope will show, there are fifty millions more, so small that we cannot see them, making one hundred millions in all. We shall also suppose that these stars have, on the average, five times the mass of the sun, and that they are spread out in a layer across the diameter of which light would require thirty thousand years to pass. Then, a mathematical computation of the attractive power exerted by such a system of masses shows that a body falling from an infinite distance to the centre of the system would acquire a velocity of twenty-five miles per second. *Vice versa*, a body projected from the centre of such a system with a velocity of more than twenty-five miles per second in any direction whatever would not only pass entirely through it, but would fly off into infinite space, never to return. If the body were anywhere else than in the centre of the system, the velocity necessary to carry it away would be less than the limit just given. But this calculated limit is only one-eighth the probable velocity of 1830 Groombridge. The force required to impress a given velocity on a body falling through any distance is proportional to the square of the velocity, four times the force being required to give double the velocity, nine times to increase it threefold, and so on. To give eight times the velocity would require sixty-four times the attracting mass. If, then, the star in question belongs to our stellar system, the masses or extent of that system must be many times greater than telescopic observation and astronomical research indicate. We may place the dilemma in a concise form, as follows:

Either the bodies which compose our universe are vastly more massive and numerous than telescopic examination seems to indicate, or 1830 Groombridge is a runaway star, flying on a boundless course through infinite space with such momentum that the attraction of all the bodies of the universe can never stop it.

Which of these is the more probable alternative we cannot pretend to say. That the star can neither be stopped, nor bent far from its course until it has passed the extreme limit to which the telescope has ever penetrated, we may consider reasonably certain. To do this will require two or three millions of years. Whether it will then be acted on by attractive forces of which science has no knowledge, and thus carried back to where it started, or whether it will continue straight forward forever, it is impossible to say.

Much the same dilemma may be applied to the past history of this body. If the velocity of two hundred miles or more per second with which it is moving exceeds any that could be produced by the attraction of all the other bodies in the universe, then it must have been flying forward through space from the beginning, and, having come from an infinite distance, must be now passing through our system for the first and only time.

It may be asked whether, in Lambert's hypothesis of immense attracting bodies, invisible on account of their being dark, we have not at once the centres required to give general stability to the stellar system, and to keep the star of which we have spoken in some regular orbit. We answer, no. To secure such stability, stars equally distant from the attracting centres must move with nearly the same velocity. An attracting centre sufficiently powerful to bring a body moving two hundred miles per second into a regular orbit would draw most of the other stars moving with small velocities into its immediate neighborhood, and thus subvert the system. We thus meet the double difficulty that we have good reason to doubt the existence of these opaque, dark bodies, and that if they did exist, they would not fulfil our requirements.

The general result of our inquiry is that the stellar universe does not seem to possess that form of unvarying stability which we see in the solar system, and that the stars move in irregular courses depending on their situation in respect to the surrounding stars, and probably changing as this situation changes. If there were no motion at all among the stars, they would all fall to a common centre, and universal ruin would be the result. But the motions which we actually see are sufficient to prevent this catastrophe, by supplying each star with a reserve of force which will generally keep it from actual collision with its neighbors. If, then, any one star does fall towards any attracting centre, the velocity which it acquires by this fall will carry it away again in some other direction, and thus it may keep up a continuous dance, under the influence of ever-varying forces, as long as the universe shall exist under its present form.

To those who have been enraptured with the sublime speculations of Kant and Lambert, this may seem an unsatisfactory conclusion; while to those who look upon the material universe as something made to last forever, it may seem improbable. But when we consider the immense periods which would be required for the mutual gravitation of the stars to effect any great change in the stellar system, we may be led to alter such views as these. We have shown that tens of thousands of years would be required to make any great change in the arrangement of the stars which we see with the naked eye. The time required for all the stars visible with the telescope to fall together by their own attraction is to be counted by millions of years. If the universe had existed in its present state from eternity, and were to exist forever, the immensity of these periods would not be at all to the point, because a million of years is no more a part of eternity than a single day. But all modern science seems to point to the finite duration of our system in its present form, and to carry us back to the time when neither sun nor planet existed, save as a mass of glowing gas. How far back that was, it cannot tell us with certainty; it can only say that the period is counted by millions of years, but probably not by hundreds of millions. It also points forward to the time when the sun and stars shall fade away, and nature shall be enshrouded in darkness and death, unless some power now unseen shall uphold or restore her. The time required for this catastrophe cannot be calculated; but it is probably not so great that the stellar system can, in the mean time, be subverted by the mutual gravitation of its members.

It would thus appear as if those nicely arranged adjustments which secure stability and uniformity of motion are not found where they are not necessary to secure the system from subversion during the time it is to last, much as the wheel of an engine which is to make but two or three revolutions while the engine endures need not be adjusted to make thousands of revolutions. The bodies which form our solar system are, on the other hand, like wheels which have to make millions of revolutions before they stop. Unless there is a constant balance between the opposing forces under the influence of which they move, there must be a disarrangement of the movement long before the engine wears out. Thus, although the present arrangement of the stars may be studied without any reference to their origin, yet, when we seek to penetrate the laws of their motion, and foresee the changes of state to which their motions may give rise, we are brought to face the question of their duration, and hence of their beginning and end.

CHAPTER III.

THE COSMOGONY.

THE idea that the world has not endured forever in the form in which we now see it, but that there was a time when it either did not exist at all, or existed only as a mass "without form, and void," is one which we find to have been always held by mankind. The "chaos" of the Greeks—the rude and formless materials, subject to no law, out of which all things were formed by the creative power—corresponds in a striking manner to the nebulous masses of modern astronomy. These old ideas of chaos were expressed by Milton in the second book of "Paradise Lost," before such a thing as a nebula could be said to be known, and he would be a bold astronomer who, in giving a description of the primeval nebulous mass, would attempt to improve on the great poet:

> ——"a dark,
> Illimitable ocean, without bound,
> Without dimension, where length, breadth, and height,
> And time and place, are lost; where eldest Night
> And Chaos, ancestors of Nature, hold
> Eternal anarchy amidst the noise
> Of endless wars, and by confusion stand:
> For hot, cold, moist, and dry, four champions fierce,
> Strive here for mastery, and to battle bring
> Their embryon atoms.
> *   *   *   *   *   *
> Chaos umpire sits,
> And by decision more embroils the fray
> By which he reigns: next him, high arbiter,
> Chance governs all. Into this wild abyss
> The womb of Nature, and perhaps her grave,
> Of neither sea, nor shore, nor air, nor fire,
> But all these in their pregnant causes mixed
> Confusedly, and which thus must ever fight,

> Unless the almighty Maker them ordain
> His dark materials to create more worlds—
> * * * * * *
> Some tumultuous cloud
> Instinct with fire and nitre."

If we classify men's ideas of the cosmogony according to the data on which they are founded, we shall find them divisible into three classes. The first class comprises those formed before the discovery of the theory of gravitation, and which, for this reason, however correct they might have been, had no really scientific foundation. The second are those founded on the doctrine of gravitation, but without a knowledge of the modern theory of the conservation of force; while the third are founded on this theory. It must not be supposed, however, that the ideas of the last-mentioned class are antagonistic to those of the other classes. Kant and Laplace founded the nebular hypothesis on the theory of gravitation alone, the conservation of force being then entirely unknown. It was, therefore, incomplete as it came from their hands, but not necessarily erroneous in its fundamental conceptions.

The consideration of the ancient ideas of the origin of the world belongs rather to the history of philosophy than to astronomy, for the reason that they were of necessity purely speculative, and reflected rather the mode of thought of the minds in which they originated than any definite system of investigating the operations of nature. The Hindoo conception of Brahma sitting in meditation on a lotus-leaf through long ages, and then producing a golden egg as large as the universe, out of which the latter was slowly evolved, is not founded on even the crudest observation, but is purely a result of the speculative tendency of the Hindoo mind. The Jewish cosmogony is the expression of the monotheistic views of that people, and of the identity of their tutelary divinity with the maker of heaven and earth. Hipparchus and Ptolemy showed the scientific turn of their minds by confining themselves to the examination of the universe as it is, without making any vain effort to trace its origin.

Though the systems to which we refer are essentially unscientific, it must not be supposed that they were all erroneous in their results, or that they belong exclusively to ancient times. Thus, the views of Swedenborg, though they belong to the class in question, are remarkably in accordance with recent views of the subject as regards the actual changes which took place during the formation of the planets. A great deal of what is written on the subject at present is to be included in this same ancient class, as being the production of men who are not mathematicians or working astronomers, and who, therefore, cannot judge whether their views are in accordance with mechanical laws and with the facts of observation. Passing over all speculation of this sort, no matter when or by whom produced, we shall consider in historical order the works of those who have actually contributed to placing the laws of cosmogony on a scientific foundation.

§ 1. *The Modern Nebular Hypothesis.*

From a purely scientific point of view, Kant has probably the best right to be regarded as the founder of the nebular hypothesis, because he based it on an examination of the actual features of the solar system, and on the Newtonian doctrine of the mutual gravitation of all matter. His reasoning is briefly this: Examining the solar system, we find two remarkable features presented to our consideration. One is that six planets and nine satellites (the entire number then known) move around the sun in circles, not only in the same direction in which the sun himself revolves on his axis, but very nearly in the same plane. This common feature of the motion of so many bodies could not, by any reasonable possibility, have been a result of chance; we are, therefore, forced to believe that it must be the result of some common cause originally acting on all the planets.

On the other hand, when we consider the spaces in which the planets move, we find them entirely void, or as good as void; for if there is any matter in them, it is so rare as to be without effect on the planetary motions. There is, therefore, no material connection now existing between the planets through which they might have been forced to take up a common direction of motion. How, then, are we to reconcile this common motion with the absence of all material connection? The most natural way is to suppose that there was once some such connection which brought about the uniformity of motion which we observe; that the materials of which the planets are formed once filled the whole space between them. "I assume," says Kant, "that all the materials out of which the bodies of our solar system were formed were, in the beginning of things, resolved in their original elements, and filled all the space of the universe in which these bodies now move." There was no formation in this chaos, the formation of separate bodies by the mutual gravitation of parts of the mass being a later occurrence. But, naturally, some parts of the mass would be more dense than others, and would thus gather around them the rare matter which filled the intervening spaces. The larger collections thus formed would draw the smaller ones into them, and this process would continue until a few round bodies had taken the place of the original chaotic mass.

If we examine the result of this hypothesis by the light of modern science, we shall readily see that all the bodies thus formed would be drawn to a common centre, and thus we should have, not a collection of bodies like the solar system, but a single sun formed by the combination of them all. In attempting to show how the smaller masses would be led to circulate around the larger ones in circular orbits, Kant's reasoning ceases to be satisfactory. He seems to think that the motion of rotation could be produced indirectly by the repulsive forces acting among the rarer masses of the condensing matter, which would give rise to a whirling motion. But the laws of mechanics show that the sum total of rotary motion in a system can never be increased or diminished by the mutual action of its separate parts, so that the present rotary motions of the sun and planets must be the equivalent of that which they had from the beginning.

Herschel's Hypothesis. — It is remarkable that the idea of the gradual transmutation of nebulæ into stars seems to have been suggested to Herschel, not by the relations of the solar system, but by his examinations of the nebulæ themselves. Many of these bodies seemed to him to be composed of immense masses of phosphorescent vapor, and he conceived that these masses must be gradually condensing, each around its own centre, or around those parts where it is most dense, until it should be transmuted into a star or a cluster of stars. On classifying the numerous nebulæ which he discovered, it seemed to him that he could see each stage of this operation going on before his eyes. There were the large, faint, diffused nebulæ, in which the process of condensation seemed to have hardly begun; the smaller but brighter ones, which had been so far condensed that the central parts would soon begin to form into stars; yet others, in which stars had actually begun to form; and, finally, star clusters in which the condensation was complete. As Laplace observes, Herschel followed the condensation of the nebulæ in much the same way that we can, in a forest, study the growth of the trees by comparing those of the different ages which the forest contains at the same time. The spectroscopic revelations of the gaseous nature of the true nebulæ tend to strengthen these views of Herschel, and to confirm us in the opinion that these masses will all at some time condense into stars or clusters of stars.

Laplace's View of the Nebular Hypothesis. — Laplace was led to the nebular hypothesis by considerations very similar to those presented by Kant a few years before. The remarkable uniformity among the directions of rotation of the planets being something which could not have been the result of chance, he sought to investigate its probable cause. This cause, he thought, could be nothing else than the atmosphere of the sun, which once extended so far out as to fill all the space now occupied by the planets. He does not, like Kant, begin with a chaos, out of which order was slowly evolved by the play of attractive and repulsive forces, but with the sun, surrounded by this immense fiery atmosphere. Knowing, from mechanical laws, that the sum total of rotary motion now seen in the planetary system must have been there from the beginning, he conceives the immense vaporous mass forming the sun and his atmosphere to have had a slow rotation on its axis. The mass being intensely hot would slowly cool off, and as it did so would contract towards the centre. As it contracted, its velocity of rotation would, in obedience to one of the fundamental laws of mechanics, constantly increase, so that a time would arrive when, at the outer boundary of the mass, the centrifugal force due to the rotation would counterbalance the attractive force of the central mass. Then, those outer portions would be left behind as a revolving ring, while the next inner portions would continue to contract until, at their boundary, the centrifugal and attractive forces would be again balanced, when a second ring would be left behind, and so on. Thus, instead of a continuous atmosphere, the sun would be surrounded by a series of concentric revolving rings of vapor.

Now, how would these rings of vapor behave? As they cooled off, their denser materials would condense first, and thus the ring would be composed of a mixed mass, partly solid and partly vaporous, the quantity of solid matter constantly increasing, and that of vapor diminishing. If the ring were perfectly uniform, this condensing process would take place equally all around it, and the ring would thus be broken up into a group of small planets, like that which we see between Mars and Jupiter. But we should expect that in general some portions of the ring would be much denser than others, and the denser portions would gradually attract the rarer portions around it until, instead of a ring, we should have a single mass, composed of a nearly solid centre surrounded by an immense atmosphere of fiery vapor. This condensation of the ring of vapor around a single point would have produced no change in the amount of rotary motion originally existing in the ring; the planet, surrounded by its fiery atmosphere, would therefore be in rotation, and would be, in miniature, a reproduction of the case of the sun surrounded by his atmosphere with which we set out. In the same way that the solar atmosphere formed itself first into rings, and then these rings condensed into planets, so, if the planetary atmospheres were sufficiently extensive, they would form themselves into rings, and these rings would condense into satellites. In the case of Saturn, however, one of the rings was so perfectly uniform that there could be no denser portion to draw the rest of the ring around it, and thus we have the well-known rings of Saturn.

If, among the materials of the solar atmosphere, there were any so rare and volatile that they would not unite themselves either into a ring or around a planet, they would continue to revolve around the sun, presenting an appearance like that of the zodiacal light. They would offer no appreciable resistance to the motion of the planets, not only on account of their extreme rarity, but because their motion would be the same as that of the planets which move among them.

Such is the celebrated nebular hypothesis of Laplace which has given rise to so much discussion. It commences, not with a purely nebulous mass, but with the sun surrounded by a fiery atmosphere, out of which the planets were formed. On this theory the sun is older than the planets; otherwise it would have been impossible to account for the slow rotation of the sun upon his axis. If his body had been formed of homogeneous matter extending out uniformly to near the orbit of Mercury, it would not have condensed into a globe revolving on its axis in twenty-five days, but into a flat, almost lens-shaped, body, which would have been kept from forming a sphere by the centrifugal force. But the denser materials being condensed first, perhaps into such a body as we described, the friction of the uncondensed atmosphere would have diminished the rotation of the sun, the rotating energy which he lost being communicated to the embryo planets and throwing them farther away.

In accordance with the hypothesis of Laplace, it has always been supposed that the outer planets were formed first. There is, however, a weak point in Laplace's theory of the formation of rings. He supposed that when the centrifugal and centripetal forces balanced each other at the outer limit of the revolving mass, the outer portions were separated from the rest, which continued to drop towards the centre. If the planetary rings were formed in this way, then, after each ring was thrown off, the atmosphere must have condensed to nearly half its diameter before another would have been thrown off, because we see that each planet is, on the whole, nearly twice as far as the one next within it. But there being no cohesion between particles of vapor, such throwing-off of immense masses of the outside portions of the revolving mass was impossible. The moment the forces balanced, the outer portions of the mass would, indeed, cease to drop towards the sun, and would partially separate from the portions next to it; then these would separate next, and so on; that is, there would be a constant dropping-off of matter from the outer portions, so that, instead of a series of rings, there would have been a flat disk formed of an infinite number of concentrating rings all joined together.

If we examine the subject more closely, we shall see that the whole reasoning by which it is supposed that the inner portions of the mass would drop away from the outer ones needs important modifications. In its primeval state, when it extended far beyond the present confines of the solar system, the rare nebulous atmosphere must have been nearly spherical. As it gradually contracted, and the effect of centrifugal force thus became more marked, it would have assumed the form of an oblate spheroid. When the contraction had gone so far that the centrifugal and attracting forces nearly balanced each other at the outer equatorial limit of the mass, the result would have been that contraction in the direction of the equator would cease entirely, and be confined to the polar regions, each particle dropping, not towards the sun, but towards the plane of the solar equator. Thus, we should have a constant flattening of the spheroidal atmosphere until it was reduced to a thin flat disk. This disk might then separate itself into rings, which would form planets in much the same way that Laplace supposed. But there would probably be no marked difference in the age of the planets; quite likely the smaller inner rings would condense into planets more rapidly than the wide-spread outer ones.

Kant and Laplace may be said to have arrived at the nebular hypothesis by reasoning forward, and showing how, by supposing that the space now occupied by the solar system was once filled by a chaotic or vaporous mass, from which the planets were formed, the features presented by this system could be accounted for. We are now to show how our modern science reaches a similar result by reasoning backward from actions which we see going on before our eyes.

§ 2. *Progressive Changes in our System.*

During the short period within which accurate observations have been made, no actual permanent change has been observed in our system. The earth, sun, and planets remain of the same magnitude, and present the same appearance as always. The stars retain their brilliancy, and, for the most part, the nebulæ their form. Not the slightest variation has been detected in the amount of heat received from the sun, or in the average number and extent of the spots on his surface. And yet we have reason to believe that these things are all changing, and that the time will come when the state of the universe will be very different from that in which we now see it. How a change may be inferred when none is actually visible may be shown by a simple example.

Suppose an inquiring person, walking in what he supposed to be a deserted building, to find a clock running. If he is ignorant of mechanics, he will see no reason why it may not have been running just as he now sees it for an indefinite period, and why the pendulum may not continue to vibrate, and the hands to go through their revolutions, so long as the fabric shall stand. He sees a continuous cycle of motions, and can give no reason why they should not have been going on since the clock was erected, and continue to go on till it shall decay. But let him be instructed in the laws of mechanics, and let him inquire into the force which keeps the hands and pendulum in motion. He will then find that this force is transmitted to the pendulum through a train of wheels, each of which moves many times slower than that in front of it, and that the first wheel is acted upon by a weight, with which it is connected by a cord. He can see a slow motion in the wheel which acts on the pendulum, and perhaps in the one next behind it, while during the short time he has for examination he can see no motion in the others. But if he sees how the wheels act on each other, he will know that they must all be in motion; and when he traces the motion back to the first wheel, he sees that its motion must be kept up by a gradual falling of the weight, though it seems to remain in the same position. He can then say with entire certainty: "I do not see this weight move, but I know it must be gradually approaching the bottom, because I see a system of moving machinery, the progress of which necessarily involves such a slow falling of the weight. Knowing the number of teeth in each wheel and pinion, I can compute how many inches it falls each day; and seeing how much room it has to fall in, I can tell how many days it will take to reach the bottom. When this is done, I see that the clock must stop, because it is only the falling of the weight that keeps its pendulum in motion. Moreover, I see that the weight must have been higher yesterday than it is to-day, and yet higher the day before, so that I can calculate its position backward as well as forward. By this calculation I see backward to a time when the weight was at the top of its course, higher than which it could not be. Thus, although I see no motion, I see with the eye of reason that the weight is running through a certain course from the top of the clock to the bottom; that some power must have wound it up and started it; and that unless the same power intervenes again, the weight must reach the bottom in a certain number of days, and the clock must then stop."

The corresponding progressive change exhibited by the operations of nature consists in a constant transformation of motion into heat, and the constant loss of that heat by radiation into space. As Sir William Thomson has expressed it, a constant "dissipation of energy" is going on in nature. We all know that the sun has been radiating heat into space during the whole course of his existence. A small portion of this heat strikes the earth, and supports life and motion on its surface. All this portion of the sun's heat, after performing its function, is radiated off into space by the earth itself. The portion of the sun's radiant heat received by the earth is, however, comparatively insignificant, since our luminary radiates in every direction equally, while the earth can receive only a part represented by the ratio which its apparent angular magnitude as seen from the sun bears to the whole celestial sphere, which a simple calculation shows to be the ratio of 1 to 2,170,000,000. The stars radiate heat as well as the sun. The heat received from them, when condensed in the focus of a telescope, has been rendered sensible by the thermo-multiplier, and there is every reason to believe that stellar heat and light bear the same proportion to each other that solar heat and light do. Wherever there is white stellar light, there must be stellar heat; and as we have found that the stars in general give more light than the sun, we have reason to believe that they give more heat also. Thus we have a continuous radiation from all the visible bodies of the universe, which must have been going on from the beginning.

Until quite recently, it was not known that this radiation involved the expenditure of a something necessarily limited in supply, and, consequently, it was not known but that it might continue forever without any loss of power on the part of the sun and stars. But it is now known that heat cannot be produced except by the expenditure of force, actual or potential, in some of its forms, and it is also known that the available supply of force is necessarily limited. One of the best-established doctrines of modern science is that force can no more be produced from nothing than matter can: to find it so produced would be as complete a miracle as to see a globe created from nothing before our eyes. Hence, this radiation cannot go on forever unless the force expended in producing the heat be returned to the sun in some form. That it is not now so returned we may regard as morally certain. There is no known law of radiation, except that it proceeds out in straight lines from the radiating centre. If the heat were returned back to the sun from space, it would have to return to the centre from all directions; the earth would then intercept as much of the incoming as of the outgoing heat; that is, we should receive as much heat from the sky at night as from the sun by day. We know very well that this is not the case; indeed, there is no evidence of any heat at all reaching us from space except what is radiated from the stars.

Since, then, the solar heat does not now return to the sun, we have to inquire what becomes of it, and whether a compensation may not at some time be effected whereby all the lost heat will be received back again. Now, if we trace the radiated heat into the wilds of space, we may make three possible hypotheses respecting its ultimate destiny:

1. We may suppose it to be absolutely annihilated, just as it was formerly supposed to be annihilated when it was lost by friction.

2. It may continue its onward course through space forever.

3. It may, through some agency of which we have no conception, be ultimately gathered and returned to the sources from which it emanated.

The first of these hypotheses is one which the scientific thinkers of the present day would not regard as at all philosophical. In our scientific philosophy, the doctrine that force cannot be annihilated is coequal with that that it cannot be created; and the inductive processes on which the latter doctrine is founded are almost as unimpeachable as those from which we conclude that matter cannot be created. At the same time, it might be maintained that all these doctrines respecting the uncreatableness and indestructibility of matter and force can have no proper foundation except induction from experiment, and that the absolute truth of a doctrine like this cannot be proved by induction. Especially may this be claimed in respect of force. The most careful measures of force which we can make under all circumstances show that it is subject to no sensible loss by either transmission or transformation. But this alone does not prove that it can be subject to no loss in a passage through space requiring hundreds of thousands or millions of years. There is also this essential difference between force and matter, that we conceive the latter as made up of individual parts which preserve their identity through all the changes of form which they undergo; while force is something in which we do not conceive of any such identity. Thus, when I allow a drop of water to evaporate from my hand, I can in imagination trace each molecule of water through the air, into the clouds, and down to the earth again in some particular drop of rain, so that, if I only had the means of actually tracing it, I could say, "This cup contains one, or two, or twenty of the identical molecules which evaporated from my hand a week or a month ago." It is on this idea of the separate identity of each molecule of matter that our opinion of the indestructibility of matter is founded, because matter cannot be destroyed without destroying individual molecules, and any cause which could destroy a single molecule might equally destroy all the molecules in the universe.

But neither parts nor identity is possible in force. A certain amount of heat may be expended in simply raising a weight. Here heat has disappeared, and is replaced by a mere change of position—something which cannot be conceived as identical with it. If we let the weight drop, the same amount of heat will be reproduced that was expended in raising the weight; but, though equal in quantity, it cannot be regarded as identical in the way that the water condensed from steam is identical with that which was evaporated to form the steam. If measures showed it to be less in quantity, we could not say there was a destruction of an identical something which previously existed, as we could if the condensed steam were not equal to the water evaporated. Therefore, while the doctrine of the indestructibility of force is universally received as a scientific principle, it can hardly be claimed that induction has established its absolute correctness; and, in a case like the present, where we see something which transcends scientific explanation, the failure of the widest induction may be considered among the possible alternatives.

The second alternative — that the heat radiated from the sun and stars continues its onward course through space forever — is the one most in accord with our scientific conceptions. We actually receive heat from the most distant star visible in our telescopes, and this heat has, according to the best judgment we can form, been travelling thousands of years without any loss whatever. From this point of view, every radiation which has ever emanated from the earth or the sun is still pursuing its course through the stellar spaces, without any other diminution than that which arises from its being spread over a wider area. A very striking presentation of this view is, we believe, due to some modern writer. If an intelligent being had an eye so keen that he could see the smallest object by the faintest light, and a movement so rapid that he could pass from one bound of the stellar system to the other in a few years, then, by viewing the earth from a distance much less than that of the farthest star, he would see it by light which had left it several thousand years before. By simply watching, he would see the whole drama of human history acted over again, except where the actions had been hidden by clouds, or under other obstacles to the radiation of light. The light from every human action performed under a clear sky is still pursuing its course among the stars, and it needs only the powers we have mentioned to place a being in front of the ray, and let him see the action again.

If the hypothesis now under consideration be the correct one, then the heat radiated by the sun and stars is forever lost to them. There is no known way by which the heat thus sent off can be returned to the sun. It is all expended in producing vibrations in the ethereal medium which constantly extend out farther and farther into space.

The third hypothesis, like the first, is a simple conjecture permitted by the necessary imperfection of our knowledge.

All the laws of radiation and all our conceptions of space lead to the conclusion that the radiant heat of the sun can never be returned to it. Such a return can result only from space itself having such a curvature that what seems to us a straight line shall return into itself, as has been imagined by a great German mathematician;* or from the ethereal medium, the vibrations in which constitute heat being limited in extent; or, finally, through some agency as yet totally unknown to science. The first idea is too purely speculative to admit of discussion, while the other two suppositions transcend our science as completely as does that of an actual annihilation of force.

§ 3. *The Sources of the Sun's Heat.*

We may regard it as good as an observed fact that the sun has been radiating heat into void space for thousands or even millions of years, without any apparent diminution of the supply. One of the most difficult questions of cosmical physics—a question the difficulty of which was not seen before the discovery of the conservation of force—has been, How is this sup-

---

* This idea belongs to that transcendental branch of geometry which, rising above those conceptions of space derived from our experience, investigates what may be possible in the relations of parts of space considered in their widest range. It is now conceded that the supposed *a priori* necessity of the axioms of geometry has no really sound logical foundation, and that the question of the limitations within which they are true is one to be settled by experience. Especially is this true of the theorem of parallels, no really valid demonstration either that two parallel straight lines will never meet or never diverge being possible. By rejecting the limitations imposed upon our fundamental geometrical conceptions, yet without admitting anything which positively contradicts them, several geometrical systems have been constructed in recent times, which are included under the general appellation of the *non-Euclidian Geometry*. The most celebrated and remarkable of these systems is that of Riemann, who showed that although we are obliged to conceive of space as *unbounded*, since no position is possible which has not space on all sides of it, yet there is no necessity that we shall consider it as *infinite*. It may return into itself in something the manner of the surface of a sphere, which, though it has no boundary, yet contains only a finite number of square feet, and on which one who travels straight forward indefinitely will finally arrive at his starting-point. Although this idea of the finitude of space transcends our fundamental conceptions, it does not contradict them, and the most that experience can tell us in the matter is that, though space be finite, the whole extent of the visible universe can be but a very small fraction of the sum total of space.

ply of heat kept up? If we calculate at what rate the temperature of the sun would be lowered annually by the radiation from its surface, we shall find it to be $2\frac{1}{3}°$ Fahrenheit per annum, supposing its specific heat to be the same as that of water, and from 5° to 10° per annum, if we suppose it the same as most of the substances which compose our globe. It would, therefore, have entirely cooled off in a few thousand years after its formation if it had no other source of heat than that shown by its temperature.

That the temperature could be kept up by combustion, as terrestrial fires are kept up, is out of the question, as new fuel would have to be constantly added in quantities which cannot possibly exist in the neighborhood of the sun. But an allied source of heat has been suggested, founded on the law of the mechanical equivalency of heat and force. If a body should fall into the sun from a great height, all the force of its fall would be turned into heat, and the heat thus produced would be enormously greater than any that would arise from the combustion of the falling body. An instance of this law is shown by the passage of shooting-stars and aërolites through our atmosphere, where, though the velocity rarely amounts to more than forty miles a second, nearly all such bodies are consumed by the heat generated. Now, the least velocity with which a body could strike the sun (unless it had been merely thrown from the sun and had fallen back) is about 280 miles per second; and if the body fell from a great height, the velocity would be over 350 miles per second. The meteoric theory was founded on this law, and is, in substance, that the heat of the sun is kept up by the impact of meteors upon his surface. The fact that the earth in its course around the sun encounters millions of meteoroids every day is shown by the frequency of shooting-stars, and leads to the result that the solar system is, so to speak, crowded with such bodies revolving in all sorts of erratic orbits. It is therefore to be supposed that great numbers of them fall into the sun; and the question whether the heat thus produced can be equal to that radiated by the sun is one to be settled by calculation. It is thus found that, in order to keep up the solar heat, a mass of matter equal to our planet would have to fall into the sun every century.

This quantity of meteoric matter is so far beyond all reasonable possibility that it requires little consideration to show that the supply of solar heat cannot be thus accounted for. Only a minute fraction of all the meteoroids or other bodies circulating through space or revolving around the sun could strike that luminary. In order to reach the sun, they would have to drop directly to it from space, or be thrown into it through some disturbance of their orbits produced by planetary attraction. If meteors were as thick as this, the earth would be so pelted with them that its whole surface would be made hot by the force of the impact, and all life would be completely destroyed. While, then, the sun may, at some past time, have received a large supply of heat in this way, it is impossible that the supply could always be kept up.

The Contraction Theory. — It is now known that there is really no necessity for supposing the sun to receive heat from any outward source whatever in order to account for the preservation of his temperature through millions of years. As his globe cools off it must contract, and the heat generated by this contraction will suffice to make up almost the entire loss. This theory is not only in accordance with the laws of matter, but it admits of accurate mathematical investigation. Knowing the annual amount of energy which the sun radiates in the form of heat, it is easy, from the mechanical equivalent of the heat thus radiated, to find by what amount he must contract to make it up. It is thus found that, with the present magnitude of the sun, his whole diameter need contract but 220 feet a year to produce all the heat which he radiates. This amounts, in round numbers, to a mile in 25 years, or four miles in a century.

The question whether the temperature of the sun will be raised or lowered by contraction depends on whether we suppose his interior to be gaseous, on the one hand, or solid or liquid, on the other. A known principle of the contraction of gaseous bodies, and one which, at first sight, seems paradoxical, is that the more heat such a body loses, the hotter it will become. By losing heat it contracts, but the heat generated by the contraction exceeds that which it had to lose in order to produce the contraction.* When the mass of gas is so far contracted that it begins to solidify or liquefy, this action ceases to hold, and further contraction is a cooling process. We cannot yet say whether the sun has or has not begun to solidify or liquefy in his interior, and therefore cannot make an exact estimate of the time his heat will last. A rough estimate may, however, be made from the rate of contraction necessary to keep up the present supply of heat. This rate diminishes as the sun grows smaller at such a rate that in five millions of years the sun will be reduced to one-half his present volume. If he has not begun to solidify now, it seems likely that he will then, and his heat must soon after begin to diminish. On the whole, it is quite improbable that the sun can continue the radiation of sufficient heat to support life on the earth ten millions of years more.

The contraction theory enables us to trace the past history of the sun a little more definitely than that of his future. He must have been larger a hundred years ago than he is now by four miles, and yet larger in preceding centuries. Knowing

---

* This curious law of cooling masses of gas was discovered by Mr. J. Homer Lane, of Washington. This gentleman's paper on the theoretical temperature of the sun, in the *American Journal of Science* for July, 1870, contains the most profound discussion of the subject with which I am acquainted. The principle in question may be readily shown in the following way. If a globular gaseous mass is condensed to one-half its primitive diameter, the central attraction upon any part of its mass will be increased fourfold, while the surface upon which this attraction is exercised will be reduced to one-fourth. Hence, the pressure per unit of surface will be increased sixteen times, while the density will be increased only eight times. Hence, if the elastic and gravitating forces were in equilibrium in the primitive condition of the gaseous mass, its temperature must be doubled in order that they may still be in equilibrium when the diameter is reduced one-half. A similar paradox is found in the theorem of celestial mechanics—that the effect of a resisting medium is to accelerate the motion of a planet or comet through it. The effect of the resistance is to make the body approach the sun, and the velocity generated by the approach exceeds that lost by the resistance.

the law of his contraction, we can determine his diameter at any past time, just as in the case of the running clock the height of the weight during preceding days can be calculated. We can thus go back to a time when the globe of the sun extended out to the orbit of Mercury, then to the orbit of the earth, and, finally, when it filled the whole space now occupied by the solar system. We are thus led by a backward process to the doctrine of the nebular hypothesis in a form strikingly similar to that in which it was presented by Kant and Laplace, although our reasoning is founded on natural laws of which those great thinkers had no knowledge.

If we take the doctrine of the sun's contraction as furnishing the complete explanation of the solar heat during the whole period of the sun's existence, we can readily compute the total amount of heat which can be generated by his contraction from any assigned volume. This amount has a limit, however great we may suppose the sun to have been in the beginning: a body falling from an infinite distance would generate only a limited quantity of heat, just as it would acquire only a limited velocity. It is thus found that if the sun had, in the beginning, filled all space, the amount of heat generated by his contraction to his present volume would have been sufficient to last 18,000,000 years at his present rate of radiation. We can say with entire certainty that the sun cannot have been radiating heat at the present rate for more than this period unless he has, in the mean time, received a miraculous accession of energy from some outside source. We use the term "miraculous" to designate any seeming incompatibility with those well-ascertained natural laws which we see in operation around us. These laws teach us that no body can acquire heat except by changes in its own mass akin to contraction of its parts, or by receiving it from some other body hotter than itself. The heat evolved by contraction from an infinite size, or by the falling of all the parts of the sun from an infinite distance, shows the extreme limit of the heat the sun could acquire from internal change, and this quantity, as just stated, would last only 18,000,000 years. In order that the sun should receive heat from another body, it is not merely necessary that that body should be hotter than the sun, but it would have to be so much hotter that the small fraction of its radiant heat which reached the sun would be greater than all that the sun himself radiated. To give an instance of what this condition requires, we remark that the body must radiate more heat than the sun in the proportion that the entire visible celestial sphere bears to the apparent angular magnitude of the body as seen from the sun. For instance, if its apparent diameter were twelve degrees, it would seem to fill about $\frac{1}{3000}$ part of the celestial sphere, and in order to warm the sun at all it would have to radiate more than three thousand times as much heat as the sun did. Moreover, in order to furnish sufficient heat to last the sun any given length of time, it would have to stay in the sun's neighborhood so long that the excess of what the sun received over what he radiated would furnish a supply of heat sufficient for that time. We cannot suppose the sun to have received even a supply of a thousand years of heat in this way without the most extravagant assumptions respecting the volume, the temperature, and the motion of the body from which the heat was received—assumptions which, in addition to their extravagance, would involve the complete destruction of the planets by the heat of the body, and the total disarrangement of their orbits by its attraction, if we suppose them to have been in any way protected from this heat.

The foregoing computation of the limit of time the sun can have been radiating heat is founded on the supposition that the amount of heat radiated has always been the same. If we suppose this amount to have been less formerly than now, the period of the sun's existence may have been longer, and in the contrary case it may have been shorter. The amount in question depends on several causes, the effect of which cannot be accurately computed—namely, the magnitude, temperature, and condition of the solar globe. Supposing a uniform radiation, the diameter of this globe was twice as great nine millions of years ago as it is now. Its surface was then of four times its present extent, so that, if it was of the same nature and at the same temperature as now, there would have been four times the radiation. But its density would have been only one-eighth as great as at present, and its temperature would have been lower. These circumstances would tend to diminish its radiation, so that it is quite possible that the total amount of heat radiated was no greater than at present. The probability would seem to be on the side of a greater total radiation, and this probability is strengthened by geological evidence that the earth was warmer in its earlier ages than now. If we reflect that a diminution of the solar heat by less than one-fourth its amount would probably make our earth so cold that all the water on its surface would freeze, while an increase by much more than one-half would probably boil the water all away, it must be admitted that the balance of causes which would result in the sun radiating heat just fast enough to preserve the earth in its present state has probably not existed more than 10,000,000 years. This is, therefore, near the extreme limit of time that we can suppose water to have existed on the earth in the fluid state.

§ 4. *Secular Cooling of the Earth.*

An instance of a progressive loss of heat, second in importance only to the loss from the sun itself, and, indeed, connected with it, is afforded by the secular cooling of the earth. As we have shown in a preceding chapter, the interior of the earth is hotter than the surface, and wherever there is such a difference of temperature as this, there must be a conduction of heat from the hotter to the colder parts. In order that heat may thus be conducted, there must be a supply of heat inside. The increase of heat downwards into the earth cannot, therefore, terminate suddenly, but must extend to a great depth.

Whatever view we may take of the question of the earth's fluidity, it must be admitted that it was hotter in former ages than now. To borrow an illustration from Sir William Thomson, the case is much the same as if we should find a hot stone in a field. We could say, with entire certainty, that the stone had been in the fire, or some other hot place, within a limited period of time. Respecting the origin of this heat, two hypotheses have prevailed—one, founded on the nebular theory, that the earth was originally condensed as a molten mass, and has not yet cooled off; the other, that it received its heat from some external source. The latter was the view of Poisson, who accounted for the increase of temperature by supposing that the solar system had, at some former period, passed through a hotter region of space than that in which it is now found. This view is, however, now known to be entirely untenable, for several reasons. Space itself cannot be warm, and the earth could have derived heat only from passing near a hot body. A star passing near enough to heat up the earth would have totally disarranged the planetary orbits, by its attraction, and destroyed all life on the surface of the globe by its heat.

Thus, tracing back the earth's heat, we are led back to the time when it was white-hot; and then, again, to when it was enveloped in the fiery atmosphere of the sun; and again, when it was itself a mass of fiery vapor. Respecting the time required for it to cool off, we cannot make any exact calculation, as we have done in the case of the sun, because the circumstances are entirely different. Owing to the solidity of at least the outer crust of the earth, the heat which it loses bears no known relation to its interior temperature. In fact, were we to compute how long the earth might have been able to radiate heat at its present rate, we may find it to be counted by hundreds or thousands of millions of years. The kernel of the difficulty lies in the fact that when a solid crust once formed over the molten earth, there was a sudden change in the rate of cooling. As long as the globe was molten, there would be constant currents between its surface and the interior, the cooling superficial portion constantly sinking down, and being replaced by fresh hot matter from the interior. But when a continuous solid crust was once formed, the heat could reach the surface only by conduction through the crust, and the latter, though only a few feet thick, would operate as a screen to prevent the further loss of heat. There would, as the crust cooled, be enormous eruptions of molten matter from the interior; but these would rapidly cool, and thus help to thicken the crust.

A fact not to be lost sight of, and which in some way assimilates the earth to the sun, is that of the heat lost by the earth by far the greater part is made up, not by a lowering of the temperature of the earth, but by its contraction. It is true that there must be some lowering of temperature, but for each degree that the temperature is lowered there will probably be a hundred degrees of heat evolved by the contraction of our globe. Considering only the earth, it is difficult to set an exact limit to the time it may have been cooling since its crust was formed.

The sudden change produced in the radiation of a molten body by the formation of a solid crust over its surface may afford us some clue to the probable termination of the heat-giving powers of the sun. Whenever the latter so far cools off that a continuous solid crust is formed over its surface, it will rapidly cease to radiate the heat necessary to support life on the globe. At its present rate of radiation, the sun will be as dense as the earth in about 12,000,000 years; and it is quite likely to be long before that time that we are to expect the permanent formation of such a crust.

The general cosmical theory which we have been considering accounts for the supposed physical constitution of Jupiter, which has been described in treating of that planet. On the nebular hypothesis, as we have set it forth, the ages of the several planets do not greatly differ. The smaller planets would, therefore, cool off sooner than the larger ones. It is possible that, owing to the great masses of Jupiter and Saturn, their rate of cooling has been so slow that no solid crust is yet formed over them. In this case they would appear self-luminous, were they not surrounded by immense atmospheres, filled with clouds and vapors, which shut off a great part of the internal heat, and thus delay the cooling process.

§ 5. *General Conclusions respecting the Nebular Hypothesis.*

It would seem from what has been said that the widest inductions of modern science agree with the speculations of thinking minds in past ages, in presenting the creation of the material universe to our view as a process rather than act. This process began when the present material universe was a mass of fiery vapor, filling the stellar spaces; it is still going on in its inevitable course, and it will end when sun and stars are reduced to dark and cold masses of dead matter. The thinking reader will, at this stage of the inquiry, very naturally inquire whether this view of the cosmogony is to be received as an established scientific fact, or only as a result which science makes more or less probable, but of the validity of which opinions may reasonably differ. We consider that the latter is the more correct view. All scientific conclusions necessarily rest on the postulate that the laws of nature are absolutely unchangeable, and that their operations have never been interfered with by the action of any supernatural cause; that is, by any cause not now in operation in nature, or operating in any way different from that in which it has always done The question of the correctness of this postulate is one of philosophy and common-sense rather than of science; and all we can say in its favor is that, as a general rule, the better men understand it, the more difficulty they find in doubting it. And all we can say in favor of the nebular hypothesis amounts to this: that the operations of nature, in their widest range, when we trace them back, seem to lead us to it, as the mode of running of the clock leads to the conclusion that it was once wound up.

Helmholtz, Thomson, and others have, as we have explained, made it evident that by tracing back the cooling processes we now see going forward in nature, we are led to a time when the planets were enveloped in the fiery atmosphere of the sun, and were therefore themselves in a molten or vaporous form. But the reverse problem, to show that a nebulous mass would or might condense into a system possessing the

CONCLUSIONS RESPECTING THE NEBULAR HYPOTHESIS.

wonderful symmetry of our solar system—the planets revolving round the sun, and the satellites round their primaries in nearly circular orbits—has not been solved in a manner at all satisfactory. We have seen that Kant's ideas were in some respects at variance with the laws of mechanics which have since been discovered. Laplace's explanation of how the planets might have been formed from the atmosphere of the sun is not mathematical enough to be conclusive. In the absence of a mathematical investigation of the subject, it seems more likely that the solar atmosphere would, under the conditions supposed by Laplace, condense into a swarm of small bodies like the asteroids, filling the whole space now occupied by the planets. Again, when we examine the actual nebulæ, we find very few of them to present that symmetry of outline which would lead to their condensation into a system so symmetrical as that to which our planet belongs. The double stars, revolving in orbits of every degree of eccentricity, and the rings of Saturn, composed apparently of a swarm of small particles, offer better examples of what we should expect from the nebular hypothesis than do the planets and satellites of our system.

These difficulties may not be insurmountable. The greatest of them, perhaps, is to show how a ring of vapor surrounding the sun could condense into a single planet encircled by satellites. The conditions under which such a result is possible require to be investigated mathematically. At the present time we can only say that the nebular hypothesis is indicated by the general tendencies of the laws of nature; that it has not been proved to be inconsistent with any fact; that it is almost a necessary consequence of the only theory by which we can account for the origin and conservation of the sun's heat; but that it rests on the assumption that this conservation is to be explained by the laws of nature, as we now see them in operation. Should any one be sceptical as to the sufficiency of these laws to account for the present state of things, science can furnish no evidence strong enough to overthrow his doubts until the sun shall be found growing smaller by actual measurement, or the nebulæ be actually seen to condense into stars and systems.

§ 6. *The Plurality of Worlds.*

When we contemplate the planets as worlds like our own, and the stars as suns, each, perhaps, with its retinue of attendant planets, the idea naturally suggests itself that other planets as well as this may be the abode of intelligent beings. The question whether other planets are, as a general rule, thus peopled is one of the highest interest to us, not only as involving our place in creation, but as showing us what is really greatest in the universe. Many thinking people regard the discovery of evidences of life in other worlds as the great ultimate object of telescope research. It is, therefore, extremely disappointing to learn that the attainment of any direct evidence of such life seems entirely hopeless—so hopeless, indeed, that it has almost ceased to occupy the attention of astronomers. The spirit of modern science is wholly adverse to speculation on questions for the solution of which no scientific evidence is attainable, and the common answer of astronomers to all questions respecting life in other worlds would be that they knew no more on the subject than any one else, and, having no data to reason from, had not even an opinion to express. Still, in spite of this, many minds will speculate; and although science cannot answer the great question for us, she may yet guide and limit our speculations. It may, therefore, not be unprofitable to show within what limits speculation may not be discordant with the generalizations of science.

First, we see moving round our sun eight large planets, on one of which we live Our telescopes show us other suns, in such numbers that they defy count, amounting certainly to many millions. Are these suns, like our own, centres of planetary systems? If our telescopes could be made powerful enough to show such planets at distances so immense as those of the fixed stars, the question would at once be settled; but all the planets of our system would disappear entirely from the reach of the most powerful telescopes we can ever hope to make at a distance far less than that which separates us from the nearest fixed star. Observation can, therefore, afford us no information on the subject. We must have recourse to cosmological considerations, and these may lead to the conclusion that if the whole universe condensed from a nebulous mass, the same cause which led our sun to be surrounded by planets would operate in the case of other suns. But we have just shown that the symmetry of form and arrangement seen in our system is something we could rarely expect to result from the condensation of masses so irregular as those which make up the large majority of the nebulæ, while the irregular orbits of the double stars show us what we should rather expect to be the rule. It is, therefore, quite possible that retinues of planets revolving in circular orbits may be rare exceptions, rather than the rule, among the stars.

Next, granting the existence of planets without number, what indications can we have of their habitability? There is one planet besides our own for which the telescope settles this point—namely, the moon. This body has neither air nor water, and, consequently, nothing on which organic life can be supported. The speculations sometimes indulged in respecting the possible habitability of the other side of the moon, which we can never see, are nothing more than plays of the imagination. The primary planets are all too distant to enable us to form any certain judgment of the nature of their surfaces, and the little we can see indicates that their constitution is extremely varied. Mars has every appearance of being like our earth in many particulars, and is, therefore, the planet which we should most expect to find inhabited. Most of the other planets give indications of being surrounded by immense atmospheres, filled with clouds and vapors, through which sight cannot penetrate, and we can reach no certain knowledge of what may be under these clouds. On the whole, we may consider the chances to be decidedly against the idea that any considerable fraction of the heavenly bodies are fitted to be the abode of such animals as we have on the earth, and that the number of them which have the requisites for supporting civilization is a very small fraction indeed of the whole.

This conclusion rests on the assumption that the conditions of life are the same in other worlds as in our own. This assumption may be contested, on the ground that we can set no limits to the power of the Creator in adapting life to the conditions which surround it, and that the immense range of adaptation on our globe—some animals living where others are immediately destroyed—makes all inferences founded on the impossibility of our earthly animals living in the planets entirely inconclusive. The only scientific way of meeting this argument is to see whether, on our earth, there are any limits to the adaptability in question. A cursory examination shows that while there are no well-defined limits to what may be considered as life, the higher forms of animal life are very far from existing equally under all conditions, and the higher the form, the more restricted the conditions. We know that no animal giving evidence of self-consciousness is developed except under the joint influence of air and water, and between certain narrow limits of temperature; that only forms of life which are intellectually very low are developed in the ocean; that there is no adapting power exercised by nature on our globe whereby man can maintain a high degree of intellectual or bodily vigor in the polar regions; that the heats of the torrid zone also impose restrictions upon the development of our race. The conclusion which we may draw from this is that, if great changes should occur on the surface of our globe, if it should be cooled down to the temperature of the poles, or heated up to that of the equator, or gradually be covered with water, or deprived of its atmosphere, the higher present forms of animal life would refuse to adapt themselves to the new state of things, and no new forms of life of equal elevation would take the place of those destroyed by the change. There is not the slightest reason for believing that anything more intelligent than a fish would ever live under water, or anything more intellectual than the Esquimaux ever be supported in regions as cold as the poles. If we apply this consideration to the question which now occupies us, we are led to the conclusion that, in view of the immense diversity of conditions which probably prevails in the universe, it would be only in a few favored spots that we should expect to find any very interesting development of life.

An allied consideration will lead us to nearly the same conclusion. Enthusiastic writers not only sometimes people the planets with inhabitants, but calculate the possible population by the number of square miles of surface, and throw in a liberal supply of astronomers who scan our earth with powerful telescopes. The possibility of this it would be presumption to deny; but that it is extremely improbable, at least in the case of any one planet, may be seen by reflecting on the brevity of civilization on our globe, when compared with the existence of the globe itself as a planet. The latter has probably been revolving in its orbit ten millions of years; man has probably existed on it less than ten thousand years; civilization less than four thousand; telescopes little more than two hundred. Had an angel visited it at intervals of ten thousand years to seek for thinking beings, he would have been disappointed a thousand times or more. Reasoning from analogy, we are led to believe that the same disappointments might await him who should now travel from planet to planet, and from system to system, on a similar search, until he had examined many thousand planets.

It seems, therefore, so far as we can reason from analogy, that the probabilities are in favor of only a very small fraction of the planets being peopled with intelligent beings. But when we reflect that the possible number of the planets is counted by hundreds of millions, this small fraction may be really a very large number, and among this number many may be peopled by beings much higher than ourselves in the intellectual scale. Here we may give free rein to our imagination, with the moral certainty that science will supply nothing tending either to prove or to disprove any of its fancies.

ADDENDUM TO PART III., CHAPTER II.

As this work is passing through the press, Professor Henry Draper, of New York, has made an addition to the theory of the solar spectrum which can hardly fail to add a quite new feature to the spectral analysis of the sun and, perhaps, to that of other heavenly bodies. Hitherto the solar spectrum has been universally considered as a continuous one, like that from a glowing solid, crossed by dark lines produced by the absorption of vapors surrounding the sun. This is the view explained on page 225. Professor Draper's point is that, in addition to this, the spectrum is crossed by the bright lines and bands arising from glowing gases, and that these lines admit of being recognized in certain parts of the spectrum if the proper steps are taken to bring them out. That bright lines might well exist in the spectrum no one would deny, because the gases of the chromosphere must produce them. But it has always been supposed that they must be so excessively faint as to be entirely invisible when projected on the spectrum of the sun itself, and so no one is known to have sought for them with especial care. Dr. Draper's course has been to photograph side by side the solar spectrum between the lines G and H, and the corresponding part of the spectrum of oxygen rendered luminous by the electric spark. The result is that out of thirteen bright lines of oxygen, some of them double or treble, nearly all have corresponding lines in the solar spectrum. The coincidence is so striking that it seems hardly possible to avoid the conclusion that a considerable part of the violet light of the sun's spectrum arises from glowing oxygen in the photosphere.

The reason why these lines are brought out here when they are not found in other parts of the spectrum is to be found in the extreme faintness of the violet part of the continuous spectrum, whereby the bright lines are not obscured by the dazzling brilliancy of the background of continuous spectrum. If it be asked why these bright lines have not been noticed before, the answer is, that the dark lines are here so broad and numerous as to cut up the continuous spectrum into very narrow lines of very irregular brightness, besides which absorption bands or half shades are numerous. Again, the lines of oxygen do not appear to be so narrow and sharply defined as those of the metallic vapors, and this makes it more difficult to distinguish them from spaces between the dark bands.

The full confirmation of this discovery must be sought for in careful measures of the relative brilliancy of the oxygen lines in various parts of the spectrum, in order to determine whether the violet lines are bright enough to be seen on the background of continuous spectrum, and in a more minute study of the solar spectrum in this region. It may then well rank as the most important advance in spectrum analysis since Lockyer and Janssen discovered the spectrum of the solar protuberances.

APPENDIX.

I.

LIST OF THE PRINCIPAL GREAT TELESCOPES OF THE WORLD.

A. *Reflecting Telescopes.*

Owner, and Place.	Construction.*	Aperture.	When built, and by whom.
The Earl of Rosse, Parsonstown, Ireland..................	Newtonian.	6 feet.	Earl of R., 1844.
Mr. Lassell, Maidenhead, England......................	Newtonian.	4 feet.	Mr. Lassell, †1860.
The Observatory of Melbourne, Australia.................	Cassegr.	4 feet.	Mr. Grubb, 1870.
The Observatory of Paris........	Newt., S. G.	47 in.	M. Martin and M. Eichens, 1875.
The Earl of Rosse, Parsonstown, Ireland..................	Newtonian.	3 feet.	The owner.
Professor Henry Draper, Dobbs Ferry, New York............	Cass., S. G.	28 in.	The owner.
The Observatory of Toulouse, France...................	S. G.	31.5 in.	M. Foucault.
The Observatory of Marseilles, France...................	S. G.	31.5 in.	M. Foucault and M. Eichens.
Mr. Lassell, Maidenhead, England......................	Newtonian.	2 feet.	The owner.

B. *Refracting Telescopes.*

Owner, and Place.	Aperture.	Maker, and Date.
U. S. Naval Observatory, Washington.........	26 in.	A. Clark and Sons, 1873.
The Imperial Observatory, Vienna‡............	26 in.	Mr. Grubb, 1877.
Mr. R. S. Newall, Gateshead, England.:........	25 in.	T. Cooke and Sons, 1870.
The Observatory of Strasburg, Germany‡	19 in.	Merz and Mähler.
The Observatory of Chicago.................	18.5 in.	A. Clark and Sons, 1862.
Mr. Van der Zee, Buffalo, New York..........	18 in.	Mr. Fitz, of N. Y.
The Observatory of Harvard College, Cambridge, Mass.........................	15 in.	Merz and Mähler, 1843.

* In this column, "Cassegr." signifies the Cassegranian construction, described on page 124. S. G. signifies that the mirror is of silvered glass.
† Mr. Lassell's four-foot telescope is, the writer believes, dismantled.
‡ These telescopes are still unfinished.

Owner, and Place.	Aperture.	Maker, and Date.
The Royal Observatory, Pulkowa, Russia.......	15 in.	Merz and Mähler, 1840.
Mr. William Huggins, London, England*......	15 in.	Mr. Grubb.
Lord Lindsay, Aberdeen, Scotland............	15 in.	Mr. Grubb.
The Observatory of Lisbon, Portugal..........	14.8 in.	Merz and Mähler.
The Observatory, Markree Castle, England.....	14 in.	......
Hamilton College, Clinton, New York..........	13.5 in.	Mr. Spencer.
The Paris Observatory†.....................	13 in.	M. Eichens.
The Allegheny Observatory, Pennsylvania.....	13 in.	......
Mr. L. M. Rutherfurd, New York.............	13 in.	The owner.
The Dudley Observatory, Albany, New York....	13 in.	Mr. Fitz, of N. Y.
The Royal Observatory, Greenwich, England‡.................................	12.5 in.	Merz and Sons, 1860. Troughton and Simms.
Michigan University, Ann Arbor.............	12.5 in.	Mr. Fitz, of N. Y.
Vassar College, Poughkeepsie, New York......	12.3 in.	Mr. Fitz, of N. Y. A. Clark and Sons.
The Physical Observatory, Oxford, England....	12.2 in.	Mr. Grubb.
The Imperial Observatory, Vienna............	12 in.	A. Clark and Sons, 1876.
The Cambridge Observatory, England.........	12 in.	M. Cauchoix.
The Royal Observatory, Dublin...............	12 in.	M. Cauchoix.
Professor Henry Draper, Dobbs Ferry, New York.................................	12 in.	A. Clark and Sons, 1876.
The Pritchett Institute, Glasgow, Missouri	12 in.	A. Clark and Sons, 1876.
Mr. S. V. White, Brooklyn, New York	12 in.	A. Clark and Sons.
The Radcliffe Observatory, Oxford, England....	12 in.	M. Cauchoix.
The Observatory, Bothkamp, Germany........	11.7 in.	Schroeder.
The Observatory, Cordova, South America......	11.2 in.	Mr. Fitz, of N. Y.
The Observatory, Munich, Germany...........	11 in.	Merz.
The Observatory, Copenhagen, Denmark.......	11 in.	Merz.
The Observatory of Cincinnati, Ohio..........	11 in.	Merz.
Middletown University, Connecticut..........	11 in.	A. Clark and Sons.

Besides these, the following telescopes are projected: A reflector of seven or eight feet aperture, by Grubb of Dublin, for the Lick Observatory of California, and a refractor of 28 or 29 inches aperture, by Alvan Clark and Sons, for Yale College.

---

* This telescope belongs to the Royal Society, but is in possession of Mr. Huggins.
† The object-glass is an old one, but the mounting is new, by Eichens.
‡ The object-glass is by Merz, of Munich, the mounting by Troughton and Simms.

II.

LIST OF THE MORE REMARKABLE DOUBLE STARS.

COMPILED BY S. W. BURNHAM.

Name.	Right Ascen. 1880.			Declination 1880.			Positi'n Angle.	Distance.	Magnitudes.		Notes.
	H.	M.	S.	°	′		°	″			
35 Piscium...		8	47	8	9		149.8	11.53	6.2	7.8	{ White, Σ. Pale-white: violet, Smyth.
38 " ...		11	13	8	12		237.6	4.59	7.0	8.0	
42 " ...		16	13	12	49		338.0	29.73	6.8	10.7	{ Yellow: blue-green, Herschel.
51 " ...		26	12	6	18		82.3	27.42	5.0	9.0	White: ashy.
55 " ...		33	36	20	47		192.7	6.37	5.0	8.2	{ Yellow: deep-red, Dembowski.
η Cassiopeæ...		41	43	57	11		140.0	5.86	4.0	7.6	Yellow: purple.
36 Andromedæ.		48	32	22	59		358.9	1.34	6.2	6.8	Binary, 349.1 years.
φ Piscium....	1	7	14	23	57		227.5	7.98	4.7	10.1	White: blue.
42 Ceti......		13	41	−1	8		351.4	1.25	6.2	7.2	
Polaris......		13	45	88	40		210.1	18.27	2.0	9.0	
ε Sculptoris...		40	1	−25	39		69.6	5.53	6.0	10.0	White: dull red.
α Piscium....		55	50	2	11		322.2	3.12	2.8	3.9	
γ Andromedæ.		56	32	41	45		62.4	10.33	3.0	5.0	{ Yellow: blue. B again double, 0″.5.
ι Trianguli....	2	5	25	29	45		80.5	3.68	5.0	6.4	Yellow: blue.
ι Cassiopeæ...		19	10	66	52		265.1	2.01	4.2	7.1	A and B. }
......		...		...			107.3	7.62		8.1	A and C. }
84 Ceti.......		35	4	−1	12		324.7	4.63	6.0	9.2	Yellow: ashy.
γ Ceti........		37	5	2	44		289.2	2.67	3.0	6.8	Yellow: blue.
ε Arietis......		52	21	20	52		201.9	1.26	5.7	6.0	Binary.
ζ Persei......	3	46	35	31	32		207.6	12.47	2.7	9.3	{ Light-green: ashy. Other small stars in the field.
ε Persei......		49	48	39	40		9.2	8.81	3.1	8.3	Pale-white: lilac.
39 Eridani....	4	8	41	−10	33		153.7	6.26	6.0	9.1	Yellow: blue.
φ Tauri		12	58	27	4		245.5	53.78	5.0	8.0	Red: bluish.
ρ Orionis.....	5	7	1	2	43		63.4	7.05	4.7	8.5	Yellow: blue.
β "		8	46	−8	20		198.8	9.14	1.0	8.0	
23 "		16	32	3	26		28.1	31.71	5.0	7.0	
η "		18	27	−2	30		83.8	1.11	4.0	5.0	Discovered by Dawes.
λ "		28	32	9	51		40.3	4.23	4.0	6.0	Yellow: purple.
θ¹ "		29	23	−5	28		...	...	...	...	{ Sextuple. In the great nebula of Orion.
σ "		32	43	−2	40		236.5	11.00	4.1	10.3	A and B. }
......		...		...			84.5	12.86		7.5	A and C. }
ζ Orionis.....		34	42	−2	0		151.3	2.55	2.0	5.7	Yellow: light-purple.
11 Monocerotis.	6	23	0	−6	57		130.0	7.25	5.0	5.5	A and B. }
......		...		...			101.7	2.46		6.0	B and C. }
12 Lyncis.....		35	38	59	34		153.7	1.53	5.2	6.1	A and B. }
......		...		...			304.2	8.67		7.4	A and C. }

Name.	Right Ascen. 1880.			Declination 1880.			Positi'n Angle.	Distance.	Magnitudes.		Notes.
	H.	M.	S.	°	′		°	″			
56 Aurigæ....		38	5	43	42		17.1	55.38	6.0	9.0	White: blue.
μ Canis Maj...		50	36	−13	53		343.5	3.22	4.7	8.0	
δ Geminorum.	7	12	57	22	12		196.9	7.14	3.2	8.2	
Castor.......		26	57	32	9		239.3	5.49	2.7	3.7	
5 Navis......		42	19	−11	54		17.5	3.32	5.3	7.4	
ζ Cancri.....	8	5	19	18	1		130.1	0.74	5.0	5.7	A and B. }
......		....		....			132.0	5.48	5.5		A and C. }
38 Lyncis....	9	11	23	37	19		240.2	2.69	4.0	6.7	
γ Leonis.....	10	13	20	20	27		111.2	3.18	2.0	3.5	Yellow: greenish.
35 Sextantis..		37	7	5	23		240.5	6.72	6.1	7.2	Yellow: blue.
ξ Ursæ Maj...	11	11	48	32	13		317.6	1.09	4.0	4.9	Binary.
65 Ursæ Maj..		48	51	47	9		36.4	3.71	6.0	8.3	Yellow: blue.
ς Comæ......		58	8	22	8		240.6	3.73	6.0	7.5	" "
24 "	12	29	6	19	2		271.9	20.42	4.7	6.2	" "
γ Virginis....		35	36	−0	47		159.3	4.77	3.0	3.0	Binary.
35 Comæ.....		47	23	21	54		25.3	1.43	5.0	7.8	A and B. }
......		....		....			124.7	28.60		9.0	A and C. }
84 Virginis...	13	37	2	4	9		235.3	3.39	5.8	8.2	Yellow: blue.
ζ Boötis......	14	35	25	14	15		303.2	1.02	3.5	3.9	
ε "		39	45	27	35		320.6	2.63	3.0	6.3	Yellow: blue or green.
ξ "		45	51	19	36		301.6	5.44	4.7	6.6	Yellow: reddish purple.
44 "		59	51	48	7		239.8	4.80	5.2	6.1	Yellowish: bluish.
μ "	15	19	58	37	48		171.9	108.46		4.0	A and B. } Binary.
......		....		....			141.9	0.69	6.7	7.3	B and C. }
δ Serpentis...		29	5	10	56		196.9	2.56	3.0	4.0	Binary.
ξ Libræ......		57	46	−11	3		173.1	1.06	4.9	5.2	A and B. } Binary.
......		....		....			70.3	7.05		7.2	A and C. }
Antares......	16	22	2	−26	10		268.7	3.46	1.0	7.0	Red: green.
36 Ophiuchi..	17	7	59	−26	25		227.3	5.55	6.0	6.0	
α Herculis....		9	10	14	32		118.5	4.65	3.0	6.1	Yellow: emerald.
ρ "		19	33	37	15		307.2	3.60	4.0	5.1	
70 Ophiuchi..		59	23	2	33		83.7	3.48	4.1	6.1	Yellow: purple. Binary.
ε¹ Lyræ......	18	40	22	39	33		26.0	3.03	4.6	6.3	}
ε² "		40	24	39	29		155.2	2.57	4.9	5.2	}
ζ "		40	38	37	29		149.7	43.71	4.2	5.5	
β·Cygni.....	19	25	53	27	42		55.7	34.29	3.0	5.3	Golden yellow: blue.
ζ Sagittæ....		43	39	18	51		312.8	8.49	5.7	8.8	Light-green: blue.
ε Draconis...		48	34	69	58		354.5	2.79	4.0	7.6	Yellow: blue.
θ Sagittæ....	20	4	39	20	33		326.7	11.40	6.0	8.3	
49 Cygni.....		36	11	31	53		49.4	2.74	6.0	8.1	Yellow: blue.
ε Equulei.....		53	5	3	50		283.9	0.06	5.2	6.2	A and B. }
......		....		....			76.2	10.83		7.1	A, B, and C. }
12 Aquarii....		57	44	−6	18		189.6	2.66	5.6	7.7	Yellowish: blue.
61 Cygni.....	21	1	14	38	8		115.6	19.55	5.3	5.9	
β Cephei.....		27	6	70	2		250.0	13.57	3.0	8.0	Light-green: blue.
41 Aquarii....	22	7	40	−21	40		119.4	4.08	6.0	8.5	Yellow: blue.
53 "		20	3	−17	21		304.5	8.20	6.0	6.3	White: yellow.
ζ "		22	39	−0	38		334.5	3.40	4.0	4.1	Binary.
ψ "	23	9	35	−9	44		312.2	49.63	4.5	8.5	Yellow: blue.
σ Cassiopeæ..		52	56	55	5		323.4	3.01	5.4	7.5	White: blue.

NOTE.—The sign *minus* (−) before declinations means *south*; without the sign, it is *north*.

III.

LIST OF THE MORE INTERESTING AND REMARKABLE NEBULÆ AND STAR CLUSTERS.

Object.	R. A. 1880.	Dec. 1880.
	H. M.	° ′
47 Toucani cluster...................................	0 19	72 45 S.
Great nebula of Andromeda.......................	0 36	40 37 N.
Nebula..	0 42	25 57 S.
"	3 29	36 32 S.
Tempel's variable nebula...........................	3 39	23 23 N.
Hind's variable nebula.............................	4 15	19 14 N.
Globular cluster.....................................	5 9	68 55 S.
" "	5 10	40 11 S.
Great nebula of Orion..............................	5 29	5 29 S.
Chacornac's variable nebula......................	5 30	21 8 N.
Nebula around ε Orionis...........................	5 30	1 17 S.
Looped nebula.......................................	5 39	69 10 S.
Cluster and nebula Mess. 46.....................	7 36	14 32 S.
Star cluster...	7 48	38 13 S.
" " Mess. 67................................	8 45	12 15 N.
Planetary nebula....................................	9 11	36 7 S.
" "	9 18	57 47 S.
Nebula..	9 45	69 38 N.
Planetary nebula....................................	10 2	39 51 S.
" "	10 19	18 2 S.
" "	11 8	55 40 N.
Spiral nebula...	12 13	15 5 N.
" "	12 17	16 29 N.
Nebula..	12 34	10 57 S.
"	12 36	33 12 N.
Star cluster...	13 7	18 48 N.
Bifid nebula..	13 18	42 23 S.
Cluster around ω Centauri	13 20	46 41 S.
Spiral or ring nebula...............................	13 25	47 49 N.
Spiral nebula...	13 30	29 16 S.
" "	13 32	17 16 S.
Cluster..	13 37	28 59 N.

Object.	R. A. 1880.	Dec. 1880.
	H. M.	° '
Cluster...	15 12	2 33 N.
"	15 38	37 23 S.
Resolvable nebula.................................	16 10	22 41 S.
Great Cluster of Hercules.........................	16 37	36 42 N.
Cluster...	16 41	1 44 S.
"	16 51	3 54 S.
"	16 52	44 29 S.
"	16 54	29 56 S.
Small annular nebula..............................	17 14	38 21 S.
" " "	17 22	23 39 S.
Cluster...	17 31	3 10 S.
Trifid nebula.......................................	17 55	23 2 S.
Nebulous cluster...................................	17 57	24 21 S.
Hooked nebula.....................................	18 14	16 13 S.
Cluster...	18 29	24 0 S.
Annular nebula of Lyra...........................	18 49	32 53 N.
Variable nebula....................................	19 5	0 50 N.
Dumb-bell nebula..................................	19 54	22 24 N.
Small annular nebula..............................	20 11	30 12 N.
Planetary nebula...................................	20 17	19 44 N.
Nebula around κ Cygni............................	20 40	30 17 N.
Planetary nebula...................................	20 58	11 50 S.
Cluster...	21 27	1 22 S.
"	21 34	23 43 S.
Blue planetary nebula.............................	23 20	41 53 N.

To facilitate the finding of the above nebulæ and clusters, their positions are marked on the star-maps with small circles.

IV.

PERIODIC COMETS SEEN AT MORE THAN ONE RETURN.

	Last Return to Perihelion Observed.	Least Distance from Sun.	Greatest Distance from Sun.	Inclination.	Longitude of Node.	Distance from Node to Perihelion.	Periodic Time.
				° ′	° ′	° ′	Years.
Encke's...	1875, April 13	0.342	4.10	13 5	334 30	183 30	3.3035
Biela's...	1852, Sept. 23	0.860	6.19	12 33	245 52	223 13	6.620
Faye's...	1873, July 18	1.686	5.92	11 22	209 42	200 15	7.413
Bronsen's...	1873, Oct. 10	0.621	5.66	29 49	101 41	13 57	5.561
D'Arrest's...	1857, Nov. 28	1.17	5.72	13 56	148 28	174 32	6.39
Winnecke's...	1875, March 12	0.78	5.50	11 17	111 33	165 10	5.55
Tuttle's...	1871, Dec. 2	1.03	10.51	54 17	269 17	206 47	13.78
Tempel's...	1873, May 9	1.771	4.81	9 46	78 43	159 18	5.96
Halley's...	1835, Nov. 15	0.586	35.3	162 15	57 15	112 43	76.00

V.
ELEMENTS OF THE ORBITS OF THE EIGHT MAJOR PLANETS FOR 1850.

Name.	Mean Motion in 365¼ Days.	Mean Distance from the Sun.		Eccentricity of Orbit.	Longitude of Perihelion.	Inclination to Ecliptic.	Longitude of the Node.	Mean Longitude of Planet, 1849. Declination, 31.0.	Authority.
		Astronomical Units.	Millions of Miles.		° ′ ″	° ′ ″	° ′ ″	° ′ ″	
Mercury....	5381016.2925	0.3870988	35¾	.20560478	75 7 13.8	7 0 7.71	46 33 8.6	323 11 23.53	Leverrier.
Venus..... {	2106641.3980	0.7233322	66¼	.00684331	129 27 14.4	3 23 34.83	75 19 52.2	243 57 44.34	Leverrier.
	2106641.3040	0.723	.....	.00684311	129 27 42.9	3 23 35.01	75 19 53.1	243 57 43.82	G. W. Hill.
Earth..... {	1295977.4260	1.0	92⅜	.01677110	100 21 21.4	.....	.....	99 48 18.66	Leverrier.
	1295977.4212	1.0	.....	.01677120	100 21 41.0	.....	.....	99 48 17.71	Hansen.
Mars......	689050.8013	1.5236914	141	.09326113	333 17 53.5	1 51 2.28	48 23 53.0	83 9 16.92	Leverrier.
Jupiter....	109256.6197	5.202800	480	.0482519	11 54 58.2	1 18 41.37	98 56 16.9	159 56 12.94	Leverrier.
Saturn.... {	43996.0508	9.538852	881	.0559428	90 6 56.5	2 29 39.80	112 20 52.9	14 50 28.49	Leverrier.
	43996.209	9.5388	.....	.0560470	90 3 59.8	2 29 39.20	112 20 0.0	14 49 43.50	G. W. Hill.
Uranus....	15424.797	19.18338	1771	.0463592	170 38 48.7	0 46 20.92	73 14 37.6	29 12 43.73	Newcomb.
Neptune...	7865.862	30.05437	2775	.0089903	46 9 13.1	1 46 58.75	130 7 18.3	334 30 5.75	Newcomb.

	Masses.	Mean Angular Semi-diameters.		Angular Diameters at Distance Unity.		Mean Diameter in Miles.	Density.		Gravity at Surface ⊕=1.	Axial Rotation.	Periodic Time.	Orbital Velocity in Miles per Second.
		″	At Dist.	Polar.	Equatorial.		Water=1.	Earth=1.				
Sun.......	Unity.	960.0	1.00	′ ″ 32 0.00	′ ″ 32 0.00	860,000	1.444	0.2552	27.71	25 to 26 days.	Days.	.....
Mercury...	1/5000000 (?)	3.34	.....	0 6.68	0 6.68	2,992	6.85	1.21	0.46	24ʰ 5ᵐ (?)	87.97	29.55
Venus.....	1/425000	8.55	.....	0 17.10	0 17.10	7,660	4.81	0.850	0.82	23ʰ 21ᵐ (?)	224.70	21.61
Earth.....	1/312500	8.84	.....	0 17.64	0 17.70	7,918	5.66	1.000	1.00	23ʰ 56ᵐ 4.09	365.26	18.38
Mars......	1/2680000	4.69	.....	0 9.36	0 9.42	4,211	4.17	0.737	0.39	24ʰ 37ᵐ 22.7	686.98	14.99
											Years.	
Jupiter....	1/1047.88	18.26	5.20	0 184.2	0 195.8	86,000	1.378	0.2435	2.64	9ʰ 55ᵐ 20.0	11.86	8.06
Saturn....	1/3501.6	8.10	9.54	0 146.3	0 162.8	70,500	0.750	0.1325	1.18	10ʰ 14ᵐ	29.46	5.95
Uranus....	1/24000	1.84	19.2	0 70.7	0 70.7	31,700	1.28	0.226	0.90	Unknown.	84.02	4.20
Neptune...	1/19380	1.28	30.0	0 77.0	0 77.0	34,500	1.15	0.204	0.89	Unknown.	164.78	3.36

ELEMENTS OF THE SATELLITES OF JUPITER.

Satellite.	Mass. (Jupiter = 1.)	Mean Daily Tropical Motion.	Synodic Period, or Interval between Eclipses.				Paris Mean Time of First Superior Mean Conjunction in 1880.				Mean Distance from Jupiter.	
			Days.	Hrs.	Min.	Sec.		Hrs.	Min.	Sec.	In Arc at Distance = 5.20273.	In Miles.
		°					Days.				″	
I.......	.000016877	203.48993385	1	18	28	35.945375 = 1.7698605	Jan. 1,	6	16	45.1	111.82	260,000
II......	.000023227	101.374762063	3	13	17	53.735233 = 3.5540942	Jan. 2,	17	3	2.7	177.81	414,000
III.....	.000088437	50.317646432	7	3	59	35.854197 = 7.1663872	Jan. 3,	20	27	5.3	283.63	661,000
IV......	.000042475	21.571109430	16	18	5	6.928330 = 16.7763524	Jan. 0,	15	6	37.3	498.85	1,162,000

ELEMENTS OF THE SATELLITES OF SATURN.

	Mean Daily Motion.	Mean Distance from Saturn.	Longitude of Peri-Sat.	Eccentricity.	Inclination to Ecliptic.		Longitude of Node.	
	°	″	°	′		°	′	°	′
Mimas......	381.953	....	?	?	28	00	168	00
Enceladus..	262.721	....	?	?	28	00	168	00
Tethys.....	190.69773	42.70	?	?	28	10	167	38
Dione......	131.534930	54.60	?	?	28	10	167	38
Rhea.......	79.690216	76.12	?	?	28	11	166	34
Titan......	22.577033	176.75	257	16	.0286	27	34	167	56
Hyperion...	16.914	214.22	40	00	.125	28	00	168	00
Japetus....	4.538036	514.64	351	25	.0282	18	44	142	53

ELEMENTS OF THE SATELLITE OF NEPTUNE.

Mean daily motion...........	61°.25679
Periodic time................	5ᵈ.87690
Distance (log. Δ = 1.47814)...	16″.275
Inclination of orbit.........	145° 7′
Longitude of node (1830).....	184° 30′
Increase in 100 years........	1° 24′

ELEMENTS OF THE SATELLITES OF URANUS.

	Periodic Times.	Mean Dist. log. Δ = 1.28310.	Inclination of Orbit (1850).	Motion in 100 Years.	Longitude of Node.	Motion in 100 Years.
	Days.	″				
Ariel.......	2.520383	13.78	97° 51′	−0″.8	165° 29′	+ 1° 24′
Umbriel.....	4.144181	19.20				
Titania.....	8.705897	31.48				
Oberon......	13.463269	42.10				

VI.

ELEMENTS OF THE SMALL PLANETS.

COMPILED BY D. P. TODD.

Sign and Name.	Year of Discovery.	Discoverer.	Greatest Distance.	Least Distance.	Daily Motion.	Periodic Time.	Eccentricity.	Longitude of Perihelion.	Longitude of Node.	Inclination.	Mean Distance.
					″	Yrs.		°	°	°	
(1) Ceres.........	1801	Piazzi.........	2.98	2.56	770.2	4.61	0.077	150.0	80.8	10.6	2.769
(2) Pallas.........	1802	Olbers.........	3.43	2.11	768.9	4.62	0.238	122.0	172.8	34.7	2.771
(3) Juno..........	1804	Harding........	3.35	1.98	814.1	4.36	0.257	54.9	170.9	13.0	2.668
(4) Vesta.........	1807	Olbers.........	2.57	2.15	977.8	3.63	0.089	250.9	103.5	7.1	2.361
(5) Astræa........	1845	Hencke........	3.06	2.10	856.9	4.14	0.186	134.9	141.5	5.3	2.579
(6) Hebe..........	1847	Hencke........	2.92	1.93	939.9	3.78	0.203	15.2	138.7	14.8	2.424
(7) Iris...........	1847	Hind..........	2.94	1.83	962.6	3.69	0.231	41.4	259.8	5.5	2.386
(8) Flora.........	1847	Hind..........	2.55	1.86	1086.3	3.27	0.156	32.9	110.3	5.9	2.201
(9) Metis.........	1848	Graham........	2.68	2.09	962.3	3.69	0.123	71.1	68.5	5.6	2.387
(10) Hygeia.......	1849	Gasparis.......	3.49	2.80	636.4	5.58	0.109	238.3	285.5	3.8	3.144
(11) Parthenope .	1850	Gasparis.......	2.70	2.21	924.0	3.84	0.100	317.9	125.2	4.6	2.452
(12) Victoria.....	1850	Hind..........	2.84	1.82	994.8	3.57	0.219	301.7	235.6	8.4	2.334
(13) Egeria.......	1850	Gasparis.......	2.80	2.35	857.9	4.14	0.087	120.2	43.2	16.5	2.577
(14) Irene........	1851	Hind..........	3.01	2.17	851.0	4.17	0.163	180.3	86.8	9.1	2.591
(15) Eunomia....	1851	Gasparis.......	3.14	2.15	825.4	4.30	0.187	27.9	293.9	11.7	2.644
(16) Psyche......	1852	Gasparis.......	3.33	2.52	710.8	4.99	0.139	15.1	150.6	3.1	2.921
(17) Thetis.......	1852	Luther........	2.79	2.15	912.4	3.89	0.129	261.3	125.4	5.6	2.473
(18) Melpomene..	1852	Hind..........	2.80	1.80	1020.1	3.48	0.218	15.1	150.1	10.2	2.296
(19) Fortuna.....	1852	Hind..........	2.83	2.05	930.1	3.82	0.159	31.1	211.5	1.5	2.442
(20) Massalia	1852	Gasparis.......	2.75	2.06	948.9	3.74	0.143	99.1	206.6	0.7	2.409
(21) Lutetia......	1852	Goldschmidt...	2.83	2.04	933.6	3.80	0.162	327.1	80.5	3.1	2.435
(22) Calliope.....	1852	Hind..........	3.20	2.62	715.2	4.96	0.101	59.9	66.6	13.7	2.909
(23) Thalia.......	1852	Hind..........	3.24	2.02	832.4	4.27	0.231	123.8	67.7	10.2	2.629
(24) Themis......	1853	Gasparis.......	3.52	2.75	639.0	5.56	0.124	144.1	35.8	0.8	3.136
(25) Phocæa......	1853	Chacornac.....	3.01	1.79	954.2	3.72	0.255	302.8	214.2	21.6	2.400
(26) Proserpine..	1853	Luther.........	2.89	2.42	819.7	4.33	0.087	236.4	45.9	3.6	2.656
(27) Euterpe.....	1853	Hind..........	2.76	1.94	986.7	3.60	0.174	88.0	93.9	1.6	2.347
(28) Bellona	1854	Luther........	3.20	2.35	766.6	4.63	0.153	122.4	144.7	9.4	2.777
(29) Amphitrite..	1854	Marth.........	2.71	2.34	869.0	4.09	0.074	56.4	356.7	6.1	2.525
(30) Urania......	1854	Hind..........	2.66	2.06	975.4	3.64	0.127	32.1	308.1	2.1	2.365

ELEMENTS OF THE SMALL PLANETS.

Sign and Name.	Year of Discovery.	Discoverer.	Greatest Distance.	Least Distance.	Daily Motion.	Periodic Time.	Eccentricity.	Longitude of Perihelion.	Longitude of Node.	Inclination.	Mean Distance.
					"	Yrs.		°	°	°	
(31) Euphrosyne.	1854	Ferguson......	3.85	2.45	635.3	5.59	0.223	93.4	31.5	26.5	3.148
(32) Pomona.....	1854	Goldschmidt...	2.80	2.37	852.6	4.16	0.083	193.4	220.7	5.5	2.587
(33) Polyhymnia.	1854	Chacornac.....	3.83	1.89	733.3	4.84	0.340	342.4	9.2	1.9	2.861
(34) Circe.......	1855	Chacornac.....	2.97	2.40	805.8	4.41	0.107	148.7	184.8	5.4	2.686
(35) Leucothea...	1855	Luther.........	3.66	2.32	685.0	5.18	0.224	202.4	355.8	8.2	2.994
(36) Atalanta....	1855	Goldschmidt ..	3.57	1.92	780.0	4.55	0.302	42.9	359.4	18.7	2.745
(37) Fides.......	1855	Luther	3.11	2.17	826.4	4.30	0.177	66.5	8.3	3.1	2.642
(38) Leda........	1856	Chacornac.....	3.16	2.32	782.1	4.54	0.154	101.2	296.4	7.0	2.740
(39) Lætitia......	1856	Chacornac.....	3.08	2.46	769.8	4.61	0.111	3.2	157.4	10.4	2.770
(40) Harmonia...	1856	Goldschmidt...	2.37	2.16	1039.3	3.42	0.047	0.9	93.6	4.3	2.267
(41) Daphne.....	1856	Goldschmidt...	3.51	2.02	773.3	4.59	0.270	220.0	179.2	16.0	2.761
(42) Isis.........	1856	Pogson........	2.99	1.89	930.9	3.81	0.226	318.0	84.5	8.6	2.440
(43) Ariadne.....	1857	Pogson........	2.57	1.83	1085.0	3.27	0.167	278.0	264.9	3.5	2.203
(44) Nysa........	1857	Goldschmidt...	2.79	2.06	940.5	3.78	0.151	112.2	131.1	3.7	2.423
(45) Eugenia.....	1857	Goldschmidt...	2.94	2.50	791.0	4.49	0.082	229.0	148.2	6.6	2.720
(46) Hestia......	1857	Pogson........	2.94	2.11	884.0	4.02	0.165	354.2	181.5	2.3	2.526
(47) Aglaia.......	1857	Luther.........	3.25	2.50	725.9	4.89	0.130	312.8	4.3	5.0	2.880
(48) Doris.......	1857	Goldschmidt...	3.33	2.89	646.4	5.49	0.071	70.3	185.2	6.5	3.112
(49) Pales.......	1857	Goldschmidt...	3.81	2.36	655.3	5.42	0.235	31.6	290.7	3.1	3.084
(50) Virginia.....	1857	Ferguson......	3.41	1.90	821.6	4.32	0.285	10.1	173.8	2.8	2.652
(51) Nemausa....	1858	Laurent	2.52	2.21	975.4	3.64	0.067	175.2	175.9	10.0	2.365
(52) Europa.....	1858	Goldschmidt...	3.35	2.70	651.2	5.45	0.109	107.1	129.7	7.4	3.026
(53) Calypso.....	1858	Luther.........	3.15	2.08	836.5	4.25	0.204	93.0	144.0	5.1	2.620
(54) Alexandra..	1858	Goldschmidt...	3.25	2.17	795.6	4.46	0.199	294.3	313.8	11.8	2.709
(55) Pandora.....	1858	Searle.........	3.15	2.37	774.0	4.59	0.142	12.1	10.9	7.2	2.760
(56) Melete......	1857	Goldschmidt...	3.21	1.98	848.1	4.19	0.236	294.6	194.1	8.0	2.596
(57) Mnemosyne.	1859	Luther	3.50	2.81	633.0	5.61	0.109	54.1	200.2	15.2	3.155
(58) Concordia...	1860	Luther	2.81	2.59	799.6	4.44	0.042	189.2	161.4	5.0	2.700
(59) Elpis.......	1860	Chacornac.....	3.03	2.40	794.0	4.47	0.117	18.4	170.4	8.6	2.713
(60) Echo........	1860	Ferguson......	2.83	1.95	953.3	3.70	0.184	98.6	192.1	3.6	2.393
(61) Danaë......	1860	Goldschmidt...	3.47	2.50	687.5	5.16	0.162	344.1	334.2	18.2	2.987
(62) Erato.......	1860	Foerster.......	3.67	2.59	640.9	5.54	0.173	38.5	125.7	2.2	3.130
(63) Ausonia.....	1861	Gasparis.......	2.69	2.10	955.6	3.72	0.124	270.4	338.0	5.8	2.398
(64) Angelina....	1861	Tempel........	3.02	2.34	808.3	4.39	0.128	123.7	311.3	1.3	2.681
(65) Cybele	1861	Tempel........	3.80	3.05	558.9	6.35	0.110	260.8	153.8	3.5	3.428
(66) Maia	1861	Tuttle.........	3.09	2.21	824.6	4.32	0.165	46.4	8.3	3.1	2.650
(67) Asia........	1861	Pogson........	2.67	1.97	941.5	3.77	0.186	306.4	202.6	6.0	2.422
(68) Leto........	1861	Luther........	3.30	2.26	765.3	4.64	0.188	345.2	45.0	8.0	2.781
(69) Hesperia....	1861	Schiaparelli....	3.49	2.47	639.9	5.15	0.170	108.5	187.2	8.5	2.980
(70) Panopæa....	1861	Goldschmidt...	3.09	2.14	839.6	4.23	0.183	299.8	48.3	11.6	2.614
(71) Niobe.......	1861	Luther........	3.23	2.28	775.4	4.58	0.173	221.3	316.5	23.3	2.756
(72) Feronia.....	1861	Peters.........	2.54	1.99	1040.1	3.41	0.120	308.0	207.8	5.4	2.266
(73) Clytia......	1862	Tuttle.........	2.78	2.55	815.4	4.35	0.042	57.9	7.9	2.4	2.665
(74) Galatea.....	1862	Tempel........	3.44	2.12	765.6	4.64	0.238	8.6	197.9	4.0	2.780
(75) Eurydice....	1862	Peters.........	3.49	1.85	812.3	4.37	0.306	335.5	359.9	5.0	2.672

APPENDIX.

Sign and Name.	Year of Discovery.	Discoverer.	Greatest Distance.	Least Distance.	Daily Motion.	Periodic Time.	Eccentricity.	Longitude of Perihelion.	Longitude of Node.	Inclination.	Mean Distance.
					″	Yrs.		°	°	°	
(76) Freia	1862	D'Arrest	4.00	2.82	563.7	6.30	0.174	92.8	212.2	2.0	3.409
(77) Frigga	1862	Peters	3.03	2.31	812.2	4.37	0.134	60.4	2.0	2.5	2.672
(78) Diana	1863	Luther	3.16	2.08	835.3	4.25	0.205	121.8	334.1	8.6	2.623
(79) Eurynome	1863	Watson	2.92	1.97	928.9	3.82	0.194	44.4	206.7	4.6	2.444
(80) Sappho	1864	Pogson	2.76	1.84	1019.8	3.48	0.200	355.3	218.7	8.6	2.296
(81) Terpsichore	1864	Tempel	3.45	2.25	736.2	4.82	0.211	48.7	2.7	7.9	2.853
(82) Alcmene	1864	Luther	3.38	2.15	771.4	4.60	0.221	132.4	27.0	2.9	2.766
(83) Beatrix	1865	Gasparis	2.64	2.22	936.7	3.79	0.086	191.8	27.5	5.0	2.430
(84) Clio	1865	Luther	2.92	1.80	976.9	3.63	0.236	339.3	327.5	9.4	2.363
(85) Io	1865	Peters	3.16	2.15	820.7	4.33	0.191	322.6	203.9	11.9	2.654
(86) Semele	1866	Tietjen	3.76	2.46	646.3	5.49	0.210	29.7	88.1	4.8	3.112
(87) Sylvia	1866	Pogson	3.76	3.21	546.0	6.50	0.079	335.4	76.1	10.9	3.482
(88) Thisbe	1866	Peters	3.21	2.32	770.2	4.61	0.160	309.3	277.6	5.2	2.769
(89) Julia	1866	Stephan	3.01	2.09	870.8	4.08	0.180	353.4	311.7	16.2	2.551
(90) Antiope	1866	Luther	3.68	2.61	636.2	5.58	0.169	301.1	71.4	2.3	3.145
(91) Ægina	1866	Stephan	2.87	2.31	851.8	4.17	0.108	80.3	11.1	2.1	2.589
(92) Undina	1867	Peters	3.51	2.86	623.7	5.69	0.102	330.8	102.9	9.9	3.187
(93) Minerva	1867	Watson	3.14	2.37	776.5	4.57	0.140	274.7	5.1	8.6	2.754
(94) Aurora	1867	Watson	3.44	2.89	630.7	5.63	0.086	46.0	4.6	8.1	3.163
(95) Arethusa	1867	Luther	3.52	2.63	657.7	5.40	0.144	31.2	244.3	12.9	3.076
(96) Ægle	1868	Coggia	3.48	2.62	666.2	5.33	0.140	163.2	322.8	16.1	3.050
(97) Clotho	1868	Tempel	3.36	1.98	814.2	4.36	0.258	65.6	160.7	11.8	2.668
(98) Ianthe	1868	Peters	3.20	2.18	804.8	4.41	0.189	147.6	354.4	15.6	2.689
(99) Dike	1868	Borelly	3.46	2.13	758.7	4.68	0.238	240.6	41.7	13.9	2.797
(100) Hekate	1868	Watson	3.60	2.58	652.5	5.44	0.164	307.7	128.2	6.4	3.092
(101) Helena	1868	Watson	2.94	2.23	854.2	4.16	0.138	327.4	343.7	10.2	2.584
(102) Miriam	1868	Peters	3.47	1.86	817.0	4.35	0.303	354.6	212.0	5.1	2.662
(103) Hera	1868	Watson	2.92	2.48	799.1	4.44	0.080	321.0	136.3	5.4	2.701
(104) Clymene	1868	Watson	3.70	2.60	635.0	5.59	0.174	58.2	44.0	2.9	3.149
(105) Artemis	1868	Watson	2.79	1.96	970.1	3.66	0.175	242.8	183.0	21.5	2.374
(106) Dione	1868	Watson	3.73	2.59	631.6	5.62	0.181	27.0	63.4	4.6	3.160
(107) Camilla	1868	Pogson	4.00	3.12	528.2	6.72	0.123	112.8	175.7	9.8	3.560
(108) Hecuba	1869	Luther	3.54	2.88	616.4	5.76	0.103	173.5	352.4	4.4	3.212
(109) Felicitas	1869	Peters	3.50	1.89	802.0	4.43	0.300	56.0	4.9	8.0	2.695
(110) Lydia	1870	Borelly	2.94	2.52	785.4	4.52	0.077	336.8	57.2	6.0	2.733
(111) Ate	1870	Peters	2.86	2.32	849.9	4.18	0.105	108.7	306.2	4.9	2.593
(112) Iphigenia	1870	Peters	2.74	2.12	934.7	3.80	0.128	338.2	324.0	2.6	2.433
(113) Amalthea	1871	Luther	2.58	2.17	968.8	3.66	0.087	198.7	123.2	5.0	2.376
(114) Cassandra	1871	Peters	3.05	2.30	810.4	4.38	0.140	153.1	164.4	4.9	2.676
(115) Thyra	1871	Watson	2.84	1.92	966.9	3.67	0.194	43.0	309.1	11.6	2.379
(116) Sirona	1871	Peters	3.16	2.37	770.9	4.60	0.143	152.8	64.4	3.6	2.767
(117) Lomia	1871	Borelly	3.06	2.92	686.0	5.18	0.023	48.8	349.6	15.0	2.991
(118) Peitho	1872	Luther	2.83	2.05	931.9	3.81	0.161	77.6	47.5	7.8	2.438
(119) Althæa	1872	Watson	2.79	2.36	855.0	4.15	0.083	12.4	204.0	5.8	2.580
(120) Lachesis	1872	Borelly	3.27	2.97	643.5	5.52	0.047	214.0	342.9	7.0	3.121

ELEMENTS OF THE SMALL PLANETS.

Sign and Name.	Year of Discovery.	Discoverer.	Greatest Distance.	Least Distance.	Daily Motion.	Periodic Time.	Eccentricity.	Longitude of Perihelion.	Longitude of Node.	Inclination.	Mean Distance.
					''	Yrs.		°	°	°	
(121) Hermione..	1872	Watson........	3.88	3.03	552.5	6.43	0.122	1.3	77.0	7.6	3.455
(122) Gerda'......	1872	Peters.........	3.34	3.10	614.1	5.78	0.037	208.9	179.0	1.6	3.220
(123) Brunhilda..	1872	Peters.........	3.00	2.38	803.4	4.42	0.115	72.9	308.5	6.5	2.692
(124) Alceste.....	1872	Peters.........	2.84	2.42	832.0	4.27	0.078	245.7	188.4	2.9	2.630
(125) Liberatrix..	1872	Prosper Henry.	4.09	1.98	671.0	5.29	0.347	251.3	171.3	6.1	3.035
(126) Velleda.....	1872	Paul Henry....	2.70	2.18	931.0	3.81	0.106	347.8	23.1	2.9	2.440
(127) Johanna ...	1872	Prosper Henry.	2.92	2.59	775.9	4.58	0.066	122.6	31.8	8.3	2.755
(128) Nemesis....	1872	Watson........	3.10	2.40	777.5	4.57	0.126	16.6	76.5	6.3	2.751
(129) Antigone...	1873	Peters.........	3.47	2.28	727.4	4.88	0.207	241.0	138.0	12.2	2.876
(130) Electra.....	1873	Peters.........	3.77	2.47	643.9	5.51	0.208	20.5	146.0	22.9	3.120
(131) Vala........	1873	Peters.........	2.62	2.22	942.8	3.77	0.082	258.6	65.3	4.6	2.419
(132) Æthra......	1873	Watson........	3.59	1.61	845.1	4.20	0.380	152.4	260.0	25.0	2.603
(133) Cyrene	1873	Watson........	3.48	2.64	662.2	5.36	0.137	248.3	321.2	7.2	3.062
(134) Sophrosyne.	1873	Luther	2.87	2.27	864.6	4.12	0.117	60.9	346.5	11.6	2.567
(135) Hertha.....	1874	Peters.........	2.93	1.93	937.1	3.79	0.205	319.8	344.0	2.3	2.429
(136) Austria.....	1874	Palisa.........	2.48	2.09	1025.9	3.46	0.085	316.5	186.2	9.6	2.287
(137) Melibœa....	1874	Palisa.........	3.78	2.48	639.7	5.55	0.208	310.3	204.3	13.8	3.133
(138) Tolosa	1874	Perrotin	2.83	2.06	928.8	3.82	0.158	311.4	54.9	3.2	2.444
(139) Juewa......	1874	Watson........	2.96	2.67	751.6	4.72	0.051	115.5	358.6	8.3	2.814
(140) Siwa	1874	Palisa.........	3.32	2.14	785.9	4.52	0.216	300.6	107.0	3.2	2.732
(141) Lumen.....	1875	Paul Henry....	3.31	2.10	795.6	4.46	0.223	22.6	319.1	11.5	2.709
(142) Polana.....	1875	Palisa.........	2.64	2.14	962.0	3.69	0.105	227.4	292.6	2.3	2.387
(143) Adria	1875	Palisa.........	2.93	2.57	777.0	4.57	0.066	223.8	333.8	11.5	2.752
(144) Vibilia	1875	Peters.........	3.27	2.03	822.4	4.32	0.233	8.3	76.8	4.9	2.650
(145) Adeona.....	1875	Peters.........	3.27	2.12	802.5	4.42	0.213	118.1	77.7	14.4	2.694
(146) Lucina	1875	Borelly........	2.89	2.53	796.3	4.46	0.067	237.7	84.4	12.7	2.708
(147) Protogeneia	1875	Schulhof.......	3.22	3.03	642.2	5.53	0.030	84.7	252.5	2.0	3.125
(148) Gallia	1875	Prosper Henry.	3.28	2.26	769.8	4.61	0.185	36.0	145.1	25.3	2.770
(149) Medusa.....	1875	Perrotin.......	2.39	1.88	1139.2	3.12	0.119	....	160.1	1.1	2.132
(150) Nuwa	1875	Watson........	3.38	2.59	685.2	5.18	0.132	357.5	207.5	2.2	2.981
(151) Abundantia.	1875	Palisa.........	2.84	2.33	854.2	4.16	0.100	215.9	40.0	7.9	2.584
(152) Atala.......	1875	Paul Henry....	3.39	2.87	640.1	5.55	0.082	80.0	41.5	12.2	3.132
(153) Hilda	1875	Palisa.........	4.60	3.31	451.9	7.86	0.163	285.0	228.3	7.8	3.954
(154) Bertha.....	1875	Prosper Henry.	3.54	2.90	613.8	5.78	0.100	168.7	37.6	20.8	3.221
(155) Scylla	1875	Palisa.........	....	....	....	....	....	....	....	....	....
(156) Xantippe...	1875	Palisa.........	3.84	2.24	670.2	5.30	0.264	155.9	246.2	7.5	3.037
(157) Dejanira ...	1875	Borelly........	3.16	2.02	853.4	4.16	0.220	109.7	62.4	11.8	2.586
(158) Koronis....	1876	Knorre.........	3.86	2.12	686.2	5.17	0.292	355.2	282.8	1.4	2.990
(159) Æmilia.....	1876	Paul Henry....	3.49	2.76	642.2	5.53	0.116	100.7	135.1	6.1	3.125
(160) Una........	1876	Peters.........	2.90	2.57	783.6	4.53	0.061	56.8	9.8	8.8	2.737
(161) Athor	1876	Watson	2.69	2.06	963.8	3.66	0.133	312.9	18.5	9.2	2.376
(162) Laurentia ..	1876	Prosper Henry.	3.54	2.52	673.1	5.28	0.169	143.0	38.2	6.3	3.029
(163) Erigone....	1876	Perrotin.......	2.71	2.00	982.1	3.62	0.149	93.3	158.8	4.7	2.354
(164) Eva	1876	Paul Henry....	3.37	1.73	870.1	4.08	0.321	2.8	77.4	24.8	2.553
(165) Loreley.....	1876	Peters.........	3.36	2.90	641.2	5.54	0.073	....	304.0	11.2	3.129

Sign and Name.	Year of Discovery.	Discoverer.	Greatest Distance.	Least Distance.	Daily Motion.	Periodic Time.	Eccentricity.	Longitude of Perihelion.	Longitude of Node.	Inclination.	Mean Distance.
					″	Yrs.		°	°	°	
(166) Rhodope...	1876	Peters.........	3.37	2.07	791.0	4.49	0.239	....	129.2	11.7	2.720
(167) Urda.......	1876	Peters.........	4.22	2.22	614.5	5.78	0.312	32.7	170.1	1.7	3.218
(168) Sibylla.....	1876	Watson........	3.60	3.15	571.5	6.21	0.067	....	209.6	4.6	3.378
(169) Zelia.......	1876	Prosper Henry.	2.67	2.05	979.9	3.62	0.131	....	354.6	5.5	2.358
(170)	1877	Perrotin.......	2.72	2.38	870.8	4.08	0.065	....	301.3	14.3	2.551
(171)	1877	Borelly........	3.52	2.76	636.8	5.58	0.121	....	101.9	2.6	3.143
(172)	1877	Borelly........	2.65	2.11	965.9	3.68	0.113	....	331.8	10.0	2.381

REMARKS ON THE PRECEDING ELEMENTS OF THE PLANETS.

Masses.—The masses of many of the planets are still very uncertain, because exact observations have not yet been made long enough to permit of their satisfactory determination. The mass of Mercury may be estimated as uncertain by $\frac{1}{8}$ of its entire amount; that of Mars by $\frac{1}{12}$; those of Venus, the earth, Uranus, and Neptune by $\frac{1}{100}$; while those of Jupiter and Saturn are probably correct to $\frac{1}{1000}$.

The value of the earth's mass which we have given does not include that of the moon. The mass of the latter is estimated at $\frac{1}{81.44}$ that of the earth.

The masses of Jupiter, Saturn, Uranus, and Neptune which we have cited are all derived from observations of the satellites of these planets. The masses derived from the perturbations of the planets do not differ from them by amounts exceeding the uncertainty of the determinations. The most noteworthy deviation is in the case of Saturn, of which Leverrier has found the mass to be $\frac{1}{3529.6}$, a result entirely incompatible with the observations of the satellites.

Diameters.—These are also uncertain in many cases, especially in those of the outer planets, Uranus and Neptune. The densities which we have assigned to these last-mentioned planets, depending on their masses and diameters, must be regarded as uncertain by half their entire amounts.

Elliptic Elements.—Of these it may be said that in general they are very accurate for the planets nearest the sun, but diminish in precision as we go outward, those of Neptune being doubtful by one or more minutes.

Elements of the Small Planets.—These are only given approximately, in order that the reader may see the relations of the group at a glance. They are mostly taken from the *Berliner Astronomisches Jahrbuch*, which gives annually the latest elements known. The elements of the twenty or thirty last ones are very uncertain.

VII.

DETERMINATIONS OF STELLAR PARALLAX.

THE following is a list of the stars the parallaxes of which are known to be investigated, with the results obtained by the different investigators. The years are generally those in which the observations are supposed to have been made, but in the case of one or two of the earlier determinations they may be those of the publication of results. In the references the following abbreviations are used:

- A. G. *Publicationen der Astronomischen Gesellschaft.*
- A. N. *Astronomische Nachrichten.*
- B. M. *Monatsbericht* (of the Berlin Academy of Sciences).
- C. R. *Comptes Rendus* (of the French Academy of Sciences).
- D. O. *Astronomical Observation, etc., at Dunsink,* by Francis Brünnow. 2 Parts. Dublin, 1870 and 1874.
- Mel. *Mélanges Mathématiques et Astronomiques, Académie de St. Pétersbourg.*
- M. N. *Monthly Notices of the Royal Astronomical Society.*
- M. R. A. S. *Memoirs of the Royal Astronomical Society.*
- M. P. *Mémoires de l'Académie de Sciences de St. Pétersbourg.*
- P. M. *Recueil des Mémoires des Astronomes de Poulkowa, publié par W. Struve.* St. Pétersbourg, 1853, vol. i.
- R. O. *Radcliffe Observations, Oxford.*

Star's Name.	Astronomer, and Date.	Parallax.	Probable Error.	Reference.
Groombridge No. 34	Auwers, from heliometer measures, 1863-'66	0.292″	±.036″	B. M., 1867.
Pole Star	Lindenau, from R. A.'s, 1750-1816	0.144	....	P. M., p. 65.
	W. Struve, Dorpat, 1818-'21	0.075	....	....
	Struve and Preuss, from R.A.'s, 1822-'38	0.172	....	....
	Lundahl, from Dorpat declinations	0.147	±.030	....
	Peters, from declinations, 1842-'44	0.067	....	P. M., p. 121.
	Lindhagen	0.025	±.018	P. M., p. 264.
Capella	Peters, from declinations, 1842	0.046	±.20	P. M., p. 136.
	Struve, with Pulkowa equat., 1855	0.305	±.043	Mel., II., p. 400.
Sirius	Henderson, 1833	0.34	....	P. M., p. 64.

APPENDIX.

Star's Name.	Astronomer, and Date.	Parallax.	Probable Error.	Reference.
Sirius.........	Maclear, 1837..........................	0.16″	″	....
	Henderson, from his own and Maclear's observations	0.23	....	M.R.A.S., xi., 243.
	Gyldén, from Maclear's obs., 1836-'37.	0.193	±.087	Mel., III., 595.
	Abbe, from Cape obs., 1856-'63	0.273	±.102	M.N., xxviii., p.2.
Castor.........	Johnson, with Oxford heliometer, 1854-'55............................	0.210	±.062	R. O., xvi., p. (xi).
Ursæ Maj.	Peters, from declinations, 1842	0.133	±.106	P. M., p. 136.
Lalande No. 21185........	Winnecke, with heliometer, 1857-'68..	0.501	±.011	A. G., No. xi.
Lalande No. 21258........	Auwers, 1860-'62	0.271	±.011	A. N., No. 1411.
	Krueger, 1862 (?)	0.260	±.020	M. N., xxiii., 173.
Groombridge No. 1830 ...	Peters, from declinations, 1842	0.226	±.141	P. M., p. 136.
	Faye, at the Paris Observatory.......	1.08*	....	C. R., xxiii.
	Wichman, from Schlüter's observations, 1842-'43	0.180	±.018	A.N., vol. 36, p.29.
	Wichmann from his own obs., 1851†	0.085 / 0.089	±.018 / ±.023	} Ib., p. 33.
	Struve, 1847-'49......................	0.034	±.029	P. M., p. 291.
	Johnson, with heliometer, 1854-'55....	0.033	±.028	R. O., xvi., p. (xxii).
	Auwers, from Johnson's obs.	0.023	±.033	B. M., 1874.
	Brünnow, 1870-'71....................	0.09	±.01	D. O., II., p. 23.
Oeltzen Arg. N., No. 17415	Krueger, 1862 (?)	0.247	±.021	M. N., xxiii., 173.
β Centauri.....	Moesta, from declinations, 1860-'64 ...	0.213	±.069	A. N., 1688.
α Bootis	Peters, from declinations, 1842........	0.127	±.073	P. M., p. 136.
	Johnson, Oxford heliometer, 1845-'55.	0.138	±.052	R. O., xvi., p. (xxiii).
α Centauri.....	Henderson, from his meridian obs. at the Cape of Good Hope, 1832-'33.	....	....	....
	α¹ Centauri, from right ascensions ...	0.92	±.35	
	α¹ Centauri, from direct declinations.	1.42	±.19	
	α¹ Centauri, from reflected decs.	1.96	±.47	
	α² Centauri, from right ascensions	0.48	±.34	M. R. A. S., xi., p. 67-68.
	α² Centauri, from direct declinations.	1.05	±.18	
	α² Centauri, from reflected decs.	1.21	±.64	
	Mean of all for both stars	1.16	±.11	
	Peters, from the same obs., finds	1.14	±.11	P. M., p. 62.
	Henderson, from Maclear's observations, 1839-'40	0.913	....	M. R. A. S., xii., p. 370.
	Peters, from the same observations ..	0.976	±.064	P. M., p. 63.
	Maclear, from decs., 1842-'44 and 1848.	0.919	±.034	M.R.A.S., xx., 98.
	Moesta, from declinations, 1860-'64...	0.880	±.068	A. N., 1688.
p Ophiuchi,....	Krueger, 1858-'59.....................	0.169	±.010	A. N., 1212.

* This result is probably erroneous.

† These results of Wichmann are parallaxes relative to the mean of certain stars of comparison. He concluded that one of the latter had a large parallax which made the parallax of 1830 Gr. 0″.72; but this view was afterwards proved wrong.

DETERMINATIONS OF STELLAR PARALLAX.

Star's Name.	Astronomer, and Date.	Parallax.	Probable Error.	Reference.
p Ophiuchi	Krueger, 1858–'62	0.162″	±.007″	A. N., 1403.
α Lyræ	Airy, Troughton's circle, 1836	0.224	}	{ M. R. A. S., x., p. 269–270.
	Airy, Jones's circle, 1836	−0.102		
	Struve, 1837–'40	0.262	....	P. M., p. 58.
	Peters, from declinations, 1842	0.103	±.053	P. M., p. 136.
	Struve, 1851–'53	0.147	±.009	M. P., vii., vol. i.
	Johnson, 1854–'55	0.154	±.046	
	Brünnow, 1868–'69	0.212	±.010	D. O., Part I.
	Brünnow, 1870	0.188	±.033	D. O., Part II.
α Cygni	Peters, from declinations, 1842	−0.082	±.043	P M., p. 136.
61 Cygni	{ Bessel, with Königsberg heliometer, 1838 }	0.314	....	....
	Bessel, from subsequent obs., 1840	0.348	....	....
	Peters, from declinations, 1842	0.349	±.080	P. M., p. 136.
	{ Johnson, with Oxford heliometer, 1852–'53 }	0.392	....	R. O., vol. xiv.
	Auwers, from Johnson's obs.	0.42	....	....
	Struve, 1852–'53	0.506	±.028	M. P., VII., I., 45.
	Auwers, from Königsberg heliometer.	0.564	±.016	A. N., 1411–'16.

VIII.

SYNOPSIS OF PAPERS ON THE SOLAR PARALLAX, 1854–'77.

The following is believed to be a nearly complete list of the determinations of the solar parallax which have appeared since the discovery of the error of the old parallax in 1854. No papers have been included except those which relate immediately to the determination in question.

1. HANSEN, 1854—*M. N. R. A. S.*, xv., p. 9.

Statement that he finds the coefficient of the parallactic equation of the moon to be $125''.705$ — a value greater than that deduced from the solar parallax as given by the transits of Venus.

2. LEVERRIER, 1858—*Annales de l'Observatoire de Paris*, iv., p. 101.

Discussion of solar parallax from lunar equation of the earth, giving $8''.95$. (In this paper Mr. Stone has found two small numerical errors: correcting them, there results $8''.85$. There is also a doubt about the theory, which might allow the result $8''.78$.)

3. FOUCAULT, 1862—*Comptes Rendus*, lv., p. 501.

Experimental determination of the velocity of light, leading to the value of the solar parallax, $8''.86$.

4. HALL, 1863—*Washington Observations for* 1863, p. lx.

Solar parallax, deduced from observations of Mars with equatorial instruments, in 1862: result, $8''.8415$.

5. FERGUSON, 1863—*Washington Observations for* 1863, p. lxv.

Solar parallax, deduced from observations with meridian instruments at Washington, Albany, and Santiago. Results various and discordant, owing to incompleteness of the work.

6. STONE, 1863—*M. N. R. A. S.*, xxiii., p. 183; *Mem. R. A. S.*, xxxiii., p. 97.

Discussion of fifty-eight corresponding observations of Mars (twenty-one pairs) at Greenwich, Cape, and Williamstown, leading to $8''.943$.

7. HANSEN, 1863—*M. N. R. A. S.*, xxiii., p. 243.

Deduction of the value 8″.97 from the parallactic inequality of the moon.

8. HANSEN, 1863—*M. N. R. A. S.*, xxiv., p. 8.

A more accurate computation from the same data gives 8″.9159.

9. WINNECKE, 1863—*Astr. Nachr.*, lix., col. 261.

Comparison of twenty-six corresponding observations (thirteen pairs) at Pulkowa and the Cape of Good Hope. Parallax, 8″.964.

10. POWALKY, 1864—Doctoral Dissertation, translated in *Connaissance des Temps*, 1867.

Discussion of the transit of Venus, 1769. Result, 8″.832, or 8″.86 when the longitude of Chappe's station is left arbitrary.

11. STONE, 1867—*M. N. R. A. S.*, xxvii., p. 239.

Attention directed to a slight lack of precision in Hansen's first paper (No. 7). Deduction also from its data of the result 8″.916—agreeing with that from Hansen's second paper.

12. STONE, 1867—*M. N. R. A. S.*, xxvii., p. 241.

Correction of one of the numerical errors in Leverrier's determination. Result, 8″.91.

13. STONE, 1867—*M. N. R. A. S.*, xxvii., p. 271.

Determination of the parallactic inequality of the moon from 2075 observations at Greenwich. Inequality, 125″.36. Solar parallax, 8″.85.

14. NEWCOMB, 1867—*Washington Observations*, 1865, Appendix II.

Discussion of the principal methods employed in determining the solar parallax, and of all the meridian observations of Mars during the opposition of 1862. Result, 8″.848.

15. STONE, 1867—*M. N. R. A. S.*, xxviii., p. 21.

Comparison of Newcomb's and Leverrier's determinations of the solar parallax, leading to the detection of another small error in the latter.

16. STONE, 1868—*M. N. R. A. S.*, xxviii., p. 255.

Rediscussion of the observations of the transit of Venus, 1769. Only observations of ingress and egress at the same station are used, and certain alterations are made in the usual interpretation of the observations by Chappe in California, and Captain Cook and his companions at Otaheite. The result of these alterations is that the parallax is increased to 8″.91.

17. NEWCOMB, 1868—*M. N. R. A. S.*, xxix., p. 6.

Criticism of Mr. Stone's interpretation of Chappe's observation of egress in 1769.

18. STONE, 1868—*M. N. R. A. S.*, xxix., p. 8.

Reply to the preceding paper.

19. FAYE, 1869—*Comptes Rendus*, lxviii., p. 42.

Examination of the observations and interpretations in Mr. Stone's paper, concluding that all that we can decide from these observations is that the solar parallax is between $8''.7$ and $8''.9$.

20. STONE, 1869—*M. N. R. A. S.*, xxix., p. 236.

Reply to Faye, criticism of Powalky's paper, and further discussions having for their object to show that the results of his paper agree with the scattered observations of ingress and egress in Europe and America.

21. ANONYMOUS, 1869—*Vierteljahrsschrift der Astr. Gesel.*, iv., p. 190.

General review of recent papers on the solar parallax, dealing more especially with the work of Stone and Powalky.

22. POWALKY, 1870—*Astr. Nachr.*, lxxvi., col. 161.

From a second discussion of the transit of Venus, 1769, he deduces $8''.7869$.

23. POWALKY, 1871—*Astr. Nachr.*, lxxix., col. 25.

From the mass of the earth as given by the motion of the node of Venus, $8''.77$. But the adopted mass of Venus enters into the result in such a way as to make it decidedly uncertain.

24. LEVERRIER, 1872—*Comptes Rendus*, lxxv., p. 165.

Determination of the solar parallax from the mass of the earth as derived from the motions of the planets, and the diminution of the obliquity of the ecliptic. Result, $8''.86$. (The distinguished author of this paper does not distinctly state in what way he has allowed for the fact that it is the combined mass of the earth and moon which is derived from the perturbations of the planets, while it is the mass of the earth alone which enters into the formula for the solar parallax. His presentation of the formulæ seems to need a slight correction, which will diminish the parallax to $8''.83$.)

25. CORNU, 1874–'76—*Annales de l'Observatoire de Paris*, xiii.

Redetermination of the velocity of light, leading to the parallax $8''.794$, if Struve's constant of aberration ($20''.445$) is used.

SYNOPSIS OF PAPERS ON SOLAR PARALLAX, 1854–'77. 541

26. GALLE, 1875—*Breslau, Maruschke & Berendt.*

"Ueber eine Bestimmung der Sonnen Parallaxe aus correspondirenden Beobachtungen des Planeten Flora, im October und November 1873." Discussion of observations made at nine northern observatories, and the Cape, Cordoba, and Melbourne, in the southern hemisphere. Result, 8".873.

27. PUISEUX, 1875—*Comptes Rendus,* lxxx., p. 933.

Computation of four contact observations of the transit of Venus in 1874, made at Peking and St. Paul's Island. Result, 8".879.

28. LINDSAY and GILL, 1877—*M. N. R. A. S.,* xxxvii., p. 308.

Reduction of observations of Juno with a heliometer at Mauritius, in 1874. The result is 8".765; or 8".815 when a discordant observation is rejected.

IX.

LIST OF ASTRONOMICAL WORKS, MOST OF WHICH HAVE BEEN CONSULTED AS AUTHORITIES IN THE PREPARATION OF THE PRESENT WORK.

The following comprises: 1. A few of the leading works of the great astronomers of the past, and of the investigators of the present, arranged nearly in the order of time. In the case of works before 1800, the supposed date of composition, or the years within which the author flourished, are given. The list is presented for the benefit of those teachers and students who wish to be acquainted with these authorities, and cannot refer to such works as the *Bibliographie Astronomique* of Lalande, or the Pulkowa *Catalogus Librorum*.

2. Modern telescopic researches upon the physical aspects of the planets which have been employed in the preparation of Part III. of the present work.

3. Recent works on special departments of astronomy, which may be useful to those who wish to pursue special subjects with greater fulness than that with which they are treated in elementary works.

In the first two classes the selection is, for the most part, limited to works which have been consulted as authorities in the preparation of this treatise. In the case of Hevelius, however, some writings are added which I have not used, nor even seen, with the object of making the list of his larger works complete. Writings which have appeared in periodicals and the transactions of learned societies are necessarily omitted from the list, owing to their great number.

The prices given for some of the older books are those for which they are commonly sold by antiquarian dealers in Germany.

B.C. 250. ARISTARCHUS: *De Magnitudinibus et Distantiis Solis et Lunæ*. Pisa, 1572. $1.

A.D. 150. PTOLEMY, CLAUDE: ΜΕΓΑΛΗΣ ΣΥΝΤΑΞΕΩΣ ΒΙΒΛ. ΙΓ, commonly called *The Almagest*.

> The most recent edition is by the Abbé Halma, in Greek, with French translation. Two vols., 4to. Paris, 1813–'16. Commonly sells for $8 to $10.

LIST OF ASTRONOMICAL WORKS

880. ALBATEGNIUS: *De Scientia Stellarum Liber.* Bonn, 1645.

1543. COPERNICUS: *De Revolutionibus Orbium Cœlestium.*

The first edition of the great work of Copernicus is rare. The second (Basel, 1566) sells for $4. Two fine editions have been published in Germany in recent times. Price $7 to $10.

1597. TYCHO BRAHE: *Astronomiæ Instauratæ Mechanica.* Noriberg, 1602. $3.

Contains description of Tycho's instruments and methods of observing.

————— *Astronomiæ Instauratæ Progymnasmata.*

————— *De Mundi Ætherei Recentioribus Phænomenis.* Frankfort, 1610.

These two volumes generally go under the title of the former. A later edition (1648) was issued under the misleading title *Opera Omnia.* The selling price is $6 for the two.

1596–1630. } KEPLER, JOHANNES: *Opera Omnia.* Edidit Dr. Ch. Frisch. 8 vols., 8vo. Frankfort, 1858–'71.

A recent and complete edition of Kepler's voluminous writings. Price from $25 to $30. Generally cheaper at second-hand.

1590–1636. } GALILEO GALILEI: *Opere.* 13 vols., 8vo. Milan, 1811. Price about $10.

A much better edition, published in 4to, about 1845, is more expensive. Galileo wrote almost entirely in Italian.

1603. BAYER, JOHANNES: *Uranometria.*

Bayer's celebrated star-charts, in which the stars were first named with Greek letters. Three or more editions were published, the second being in 1648, the third in 1661. $2 50.

RICCIOLUS: *Almagestum Novum.* 2 vols. in one, folio. Bonn. 1651.

————— *Astronomia Reformata.* Folio. Bonn, 1665.

Two ambitious works, remarkable rather for their voluminousness than for their value. The author being an ecclesiastic, had to profess a disbelief in the Copernican system.

1630. BULLIALDUS: *Astronomia Philolaica.* Folio. Paris, 1645.

The last three works are cited as probably the most voluminous compendiums of astronomy of the seventeenth century. They can all be purchased for $3 or $4 each.

1611. FABRITII, J.: *De Maculis in Sole Observatis.*

1655. BORELLI: *De Vero Telescopii Inventore.* Hague, 1655. $1.

1647–1690. } HEVELIUS, J.: *Selenographia, sive Lunæ Descriptio.* Folio.

The earliest great work on the geography of the moon and the aspects of the planets. Profusely illustrated. $4 to $5.

HEVELIUS, J.: *Mercurius in Sole Visus.* Folio, 1662. $1.

Contains also Horrox's observation of the transit of Venus in 1639.

——————— *Cometographia.* Folio, 1668.

The first great modern treatise on the subject of comets.

——————— *Machina Cœlestis, Pars Prior.* Folio, 1673.

Contains descriptions of his instruments, and a disquisition on the practical astronomy of his time.

——————— *Machina Cœlestis, Pars Posterior.* Folio, 1679.

A very rare book, almost the entire edition having been destroyed by fire. A copy was sold for $50 in 1872.

——————— *Annus Climactericus.* Dautzic, 1685.

——————— *Prodromus Astronomiæ.* Dantzic, 1690.

——————— *Firmamentum Sobiescianum.* Dantzic, 1690.

These works comprise star-catalogues, star-maps, etc. $3 50.

1659. HUYGHENS: *Systema Saturnium.* Hague, 1659.

——————— *Horologium Oscillatorium.* Paris, 1673.

The latter work contains the theory of the pendulum clock. These two and most of the other important works of Huyghens were published in Leiden in 1751, under the title of *Opera Mechanica, Geometrica, Astronomica et Miscellanea,* nominally in four volumes, but the paging is continuous throughout the series, the total number of pages being 776. Leiden, 1751. $5.

1687. NEWTON, ISAAC: *Philosophiæ Naturalis Principia Mathematica.* 4to. London, 1687.

A number of editions of Newton's *Principia* have appeared. One of the most common is that of Le Seur and Jacquier, 3 vols. in 4. Geneva, 1739. It is accompanied by an extended commentary. Sells for about $4. A very fine edition was issued in 1871, by Sir William Thomson, in Glasgow. There is also an English translation by Motte, which has gone through several editions in England and one in America.

BREWSTER, SIR D.: *Memoirs of the Life, Writings, and Discoveries of Sir Isaac Newton.* 2 vols., 8vo. Edinburgh, 1855.

1720. FLAMSTEED, J.: *Historia Cœlestis Britannica.* 3 vols., folio. London, 1725. $10.

Contains Flamsteed's observations and star-catalogue.

1728. BLANCHINI, F.: *Hesperi et Phosphori nova Phænomena sive Observationes circa Planetam Veneris.* Folio. Rome, 1728.

1740. CASSINI: *Élémens d'Astronomie.* 4to. Paris, 1740. $1.

1741. WEIDLER, JO.: *Historia Astronomiæ.* Small 4to. Wittemberg, 1741. $2.

BERNOUÏLLI, JOHN: *Opera Omnia.* 4 vols., 4to. Lausanne, 1742. $5.

LIST OF ASTRONOMICAL WORKS.

  LE MONNIER: *La Théorie des Comètes.* 1 vol., 8vo. Paris, 1743. $1.

1760. KANT, IMMANUEL: *Schriften zur Physischen Geographie.* 8vo. Leipzig, 1839.

1780. PINGRE: *Cométographie; ou Traité Historique et Théorique des Comètes.* 2 vols., 4to. Paris, 1783.

  The most complete historical and general treatise on comets which has appeared.

1780–1790. BAILLY: *Histoire de l'Astronomie Ancienne depuis son Origine jusqu'à l'Établissement de l'École d'Alexandrie.* 1 vol., 4to. Paris, 1781. $10.

  ———— *Histoire de l'Astronomie Moderne depuis la Fondation de l'École d'Alexandrie, jusqu'à l'Époque de MDCCXXX.* 3 vols., 4to. Paris, 1779. $6.

  ———— *Traité de l'Astronomie Indienne et Orientale.* 1 vol., 4to. Paris, 1787.

  These histories by Bailly are considered very unsound, the author having a greatly exaggerated opinion of the knowledge of the ancients.

1800. LALANDE, J. DE: *Bibliographie Astronomique; avec l'Histoire de l'Astronomie depuis 1781 jusqu'à 1802.* 4to. Paris, 1803. $3.

1817. LAPLACE, P. S.: *Traité de Mécanique Céleste.* 4 vols., 4to. Paris, 1799–1805. $60.

  This work is now expensive, all the editions being exhausted. A new edition is soon to be issued.

  ———— *Exposition du Système du Monde.* 1 vol., 4to. $2.

  The latter work gives a very clear popular exposition of the laws of the celestial motions.

  DELAMBRE: *Histoire de l'Astronomie Ancienne.* 2 vols., 4to. Paris, 1817. $4.

  ———— *Histoire de l'Astronomie du Moyen Age.* 1 vol., 4to. Paris, 1819. $3.

  ———— *Histoire de l'Astronomie Moderne.* 2 vols., 4to. Paris, 1821. $5.

  Histoire de l'Astronomie au dix-huitième Siècle. 1 vol., 4to. Paris, 1827. $3.

  These histories by Delambre consist principally of abstracts of the writings of all eminent astronomers, accompanied by a running commentary, but without any attempt at logical arrangement. Each work is taken up and passed through in regular order, but it is only in the introductory essays that general views of the progress of the science are found.

ENCKE, J. F.: *Die Entfernung der Sonne von der Erde aus dem Venusdurchgange von 1761 hergeleitet.* 12mo. Gotha, 1822.

———— *Der Venusdurchgang von 1769.* 12mo. Gotha, 1824.

These two little books contain Encke's researches on the solar parallax leading to the result $8''.5776$, and the distance of the sun 95,300,000 miles.

IDELER, DR. LUDWIG: *Handbuch der Mathematischen und Technischen Chronologie.* 2 vols., 8vo. Berlin, 1825.

An exhaustive and commendable work on the measures of time adopted in various countries, especially in ancient times.

WHEWELL, WM.: *History of the Inductive Sciences.* London.

HERSCHEL, SIR JOHN: *Results of Astronomical Observations made during the Years 1834, '5, '6, '7, '8, at the Cape of Good Hope.* 1 vol., 4to. London, 1847.

STRUVE, F. G. W.: *Études d'Astronomie Stellaire.* St. Petersburg, 1847.

GRANT, ROBERT: *History of Physical Astronomy, from the Earliest Ages to the Middle of the Nineteenth Century.* 8vo. London, 1852.

BIOT, J. B.: *Études sur l'Astronomie Indienne et Chinoise.* 8vo. Paris, 1862.

LOVERING, JOSEPH: *On the Periodicity of the Aurora.* Memoirs of the American Academy of Arts and Sciences. Boston, 1859 and 1865.

OLBERS, W., and GALLE, J. G.: *Die leichtste und bequemste Methode die Bahn eines Cometen zu berechnen.* 8vo. Leipzig, 1864.

This work contains a table of all orbits of comets computed, brought up to the end of 1863.

ZÖLLNER, DR. J. C. F.: *Ueber die Natur der Kometen.* 8vo. Leipzig, 1872.

DÜHRING, DR. E.: *Kritische Geschichte der Principien der Mechanik.* 8vo. Berlin, 1873.

TODHUNTER, I.: *History of the Mathematical Theories of Attraction and the Figure of the Earth, from the Time of Newton to that of La Place.* 2 vols., 8vo. London, 1873.

II.—WORKS ON THE PHYSICAL ASPECTS OF THE PLANETS.

SCHROETER, J. H.: *Beiträge zu den Neuesten Astronomischen Entdeckungen. Herausgegeben von Bode.* 3 vols., 8vo. Berlin, 1788–1800. $5.

LIST OF ASTRONOMICAL WORKS. 547

SCHROETER, J. H.: *Selenotopographische Fragmente zur genauern Kenntniss der Mondfläche.* 4to. Lilienthal, 1791. $3.

——— *Aphroditographische Fragmente zur genauern Kenntniss des Planeten Venus.* 4to. Helmstedt, 1796. $6.

Schroeter's style was intolerably prolix and diffuse, so that a clear idea of the results he really attained involves no small labor.

BEER, W., and MÄDLER, J. H.: *Physische Beobachtungen des Mars bei seiner Opposition im September* 1830. 12mo. Berlin, 1830.

——— *Der Mond nach seinen kosmischen und individuellen Verhältnissen, oder Allgemeine vergleichende Selenographie.* 4to. Berlin, 1837. $7.

This volume is accompanied by a large map of the moon, and is the most complete and celebrated work on selenography which has yet appeared.

BEER, W., and MÄDLER, J. H.: *Beitrage zur physischen Kenntniss der himmlischen Körper im Sonnensysteme.* 4to. Weimar, 1841.

ZÖLLNER: *Photometrische Untersuchungen mit besonderer Rücksicht auf die physische Beschaffenheit der Himmelskörper.* 8vo. Leipzig, 1865.

ENGELMANN: *Ueber die Helligkeitsverhältnisse der Jupiterstrabanten.* 8vo. Leipzig, 1871.

VOGEL, H. C., and LOHSE: *Beobachtungen angestellt auf der Sternwarte des Kammerherrn von Bülow zu Bothkamp.* 3 pts., 4to. Leipzig, 1872–'75.

III.—RECENT TREATISES ON SPECIAL SUBJECTS.

THE SUN.

PROCTOR, R. A.: *The Sun: Ruler, Fire, Light, and Life of the Planetary System.* 8vo. London, 1871.

LOCKYER, J. N.: *Contributions to Solar Physics.* 8vo, London, 1874.

SECCHI, A.: *Le Soleil.* 2 vols., 8vo, with Atlas. Paris, 1875–'77.

The latter is the most complete and beautifully illustrated treatise on the sun which has yet appeared.

THE MOON.

NASMYTH and CARPENTER: *The Moon.* London, 1874.

Contains very beautiful illustrations of lunar scenery.

PROCTOR, R. A.: *The Moon: Her Motions, Aspects, Scenery, and Physical Condition.* 8vo. London, 1873.

This work is illustrated with several of Mr. Rutherfurd's photographs.

NEISON, EDMUND: *The Moon, and the Condition and Configurations of its Surface.* Illustrated. 8vo. London, 1876.
Principally devoted to selenography.

TRANSITS OF VENUS.

FORBES, GEORGE: *Transits of Venus.* London, 1874.

PROCTOR, R. A.: *Transits of Venus. A Popular Account of Past and Coming Transits.* 8vo. London, 1875.

THEORETICAL AND PRACTICAL ASTRONOMY.

LOOMIS, ELIAS: *An Introduction to Practical Astronomy, with a Collection of Astronomical Tables.* 8vo. New York, 1855.
Contains much information for the amateur astronomer.

SAWITCH: *Abriss der Practischen Astronomie.* 2 vols., 8vo. Hamburg, 1850.

BRUNNOW, F.: *Practical and Spherical Astronomy.* 8vo. London and New York, 1865.

CHAUVENET, W.: *Manual of Spherical and Practical Astronomy.* 2 vols., 8vo. Philadelphia, 1863.
The most complete and exhaustive treatise on the subject which has yet appeared.

WATSON, J. C.: *Theoretical Astronomy.* 8vo. Philadelphia, 1868.

X.

GLOSSARY OF TECHNICAL TERMS OF FREQUENT OCCURRENCE IN ASTRONOMICAL WORKS.

The following list is believed to include all the technical terms used in the present work, as well as a number of others which the reader of astronomical literature will frequently meet with. The words in parentheses which sometimes follow a term express its literal signification.

Aberration (*a wandering-away*). Generally applied to a real or apparent deviation of the course of a ray of light. Especially (1) an apparent displacement of a star, owing to the progression motion of light combined with that of the earth in its orbit, p. 211; (2) the defects of action of a lens in not bringing all rays to the same focus. The *spherical aberration* of a lens results in the rays which pass through the glass near its edge coming to a shorter focus than those which pass near its centre, while the *chromatic aberration* is the separation of the light of different colors.

Achromatic (*without color*). Applied to an object-glass in which rays of different colors are brought to the same focus. See p. 114.

Aerolite. A meteoric stone or other body falling from the celestial spaces.

Albedo. Degree of whiteness, or proportion of incident light reflected by a non-luminous body. When the albedo of a body is said to be 0.6, it means that it reflects $\frac{6}{10}$ of the incident light.

Alidade. A movable frame carrying the microscopes or verniers of a graduated circle. Not generally used in instruments of recent construction.

Altitude. The apparent angular elevation of a body above the horizon, usually expressed in degrees and minutes. At the horizon the altitude is zero, at the zenith it is 90°

Annular (*ring-shaped*). Having the appearance or form of a ring.

Anomaly. The angular distance of a planet from that point of its orbit in which it is nearest to the sun, or, in the ancient astronomy, to the earth. Draw two straight lines from the sun, one to the nearest point of the orbit, or the perihelion, and the other to the planet, and the angle between these lines will be the anomaly of the planet.

Anomalistic. Pertaining to the anomaly. The anomalistic year is the period between two consecutive returns of the earth to its perihelion. It is about 4′ 15″ longer than the sidereal year.

APPENDIX.

Ansæ (*handles*). The apparent ends of the rings of Saturn, which look like handles projecting from the planet.

Aperture of a Telescope. The diameter of the glass or mirror which admits the rays of light, clear of all obstacles.

Aphelion. The part of the orbit of a planet in which it is farthest from the sun.

Apogee. The point of an orbit in which the planet is farthest from the earth. In the ancient astronomy the planets were said to be in apogee when beyond the sun, and therefore at their greatest distance from the earth; but the term is now applied only to the most distant point of the moon's orbit.

Apsis (pl. *Apsides*). The two points of an orbit which are nearest to, and farthest from, the centre of motion, called, respectively, the lower and higher apsis. The *line of apsides* is that which joins these two points, and so forms the major axis of an elliptic orbit. The term is now nearly superseded by the more special terms *aphelion, perihelion, perigee,* etc. See *Elements*.

Armillary Sphere. A combination of circles used before the invention of the telescope for determining the relative directions or apparent positions of the heavenly bodies on the celestial sphere. It is now entirely out of use. See p. 105.

Astrolabe. A simple form of armillary sphere used by the ancient astronomers.

Azimuth. The angular distance of a point of the horizon from the north or south. The azimuth of a horizontal line is its deviation from the true north and south direction. The azimuth of the east and west points is 90°.

Binary System. A double star, in which the two components are found to revolve round each other.

Binocular (*two-eyed*). Applied to a telescope or microscope in which both eyes can be used at once, as an opera-glass.

Black Drop. A distortion of Mercury or Venus at the time of internal contact with the limb of the sun. See p. 179.

Centesimal. Reckoning by hundreds. Applied to those denominational systems in which each unit is one hundred times that next below it. The centesimal division of the angle is one in which the quadrant is divided into 100 degrees or grades, the grade into 100 minutes, and the minute into 100 seconds.

Chronograph (*time-mark*). An instrument for measuring time by marking on a moving paper (see p. 155). Time is then represented by space passed over.

Circle, Great. A circle which divides the sphere into two equal hemispheres, as the equator and the ecliptic.

GLOSSARY OF TECHNICAL TERMS. 551

Colures. The four principal meridians of the celestial sphere, all of which pass from the pole, and one of which passes through each equinox, and one through each solstice. They mark the circles of 0^h, 6^h, 12^h, and 18^h of right ascension, respectively.

Conjunction (*a joining*). The nearest apparent approach of two heavenly bodies which seem to pass each other in their course. They are commonly considered as in conjunction when they have the same longitude. The term is applied especially in the case of a planet and the sun. The nearest approach is called superior conjunction when the planet is beyond the sun, inferior when it is this side of it. Mercury and Venus are, of course, the only planets which can be in inferior conjunction.

Cosmical. Relating to creation at large, in contradistinction to terrestrial, which relates to the earth. By a cosmical phenomenon is meant one which has its origin outside the earth and its atmosphere.

Culmination. The passage of a heavenly body over the meridian of a place. This passage may be considered as occurring twice in a day, once above the pole, and again below it, twelve hours later. The former is called the *upper*, the latter the *lower*, culmination. The upper culmination of the sun occurs at noon, the lower at midnight.

Cusps (*points*). The pointed ends of the seeming horns of the moon or of a planet when it presents the appearance of a crescent.

Cycle (*circle*). A period of time at the end of which any aspect or relation of the heavenly bodies recurs, as the Metonic cycle.

Declination. The angular distance of a heavenly body from the equator. When north of the equator, it is said to be in north declination; otherwise, in south declination.

Deferent. In the ancient astronomy the mean orbit of a planet which was supposed to carry the epicycle. It is represented by the dotted circles in Figs. 10 and 11, pp. 38 and 39.

Dichotomy (*a cutting in two*). The aspect of a planet when half illuminated, as the moon at first and last quarter.

Digit. The twelfth part of the diameter of the sun or moon, formerly used to express the magnitude of eclipses. See p. 28.

Dip of the Horizon. At sea, the depression of the apparent horizon below the true level, owing to the height of the observer's eye above the water.

Direct Motion. A motion from west to east among the stars, like that of the planets in general.

Eccentric. In the ancient astronomy, a circle of which the centre was displaced from the centre of motion. See p. 42, Fig. 13.

Eccentricity. See *Elements*.

Ecliptic. The apparent path of the sun among the stars, described in Part I., Chap. I., § 3. See p. 13.

Egress (*a going forth*). The end of the apparent transit of one body over another, when the former seems to leave the latter.

Elements. In general, the data for predicting an astronomical phenomenon. Especially, the quantities which determine the motion of a planetary body. The independent elements of a planet are six in number, namely:

1. The *mean distance*, or half the longer axis, AP, of the ellipse in which the planet moves round the sun, the latter being in the focus at S.

2. The *eccentricity*, the ratio of the distance CS between the centre and focus of the ellipse to the mean distance.

These two elements determine the size and form of the elliptic orbit of the planet.

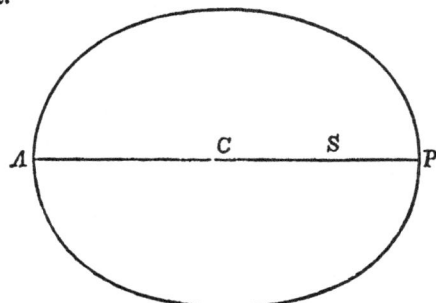

Fig. 112.—Diagram illustrating elliptic elements of a planet.

3. The longitude of the ascending node, which gives the direction of the line in which the plane of the orbit intersects that of the ecliptic, or the angle which this line makes with the vernal equinox.

4. The inclination of the plane of the orbit to that of the ecliptic.

5. The longitude of the perihelion, P, for which is taken the longitude of the node, *plus* the angular distance from the node to the perihelion, as seen from the sun.

These three quantities determine the position of the orbit in space.

6. The mean longitude of the planet at some given epoch, or the time at which it passed the perihelion, P.

To these six the time of revolution, or mean angular motion in a day or year, is usually added; but as this can always be determined from the mean distance, and *vice versa*, by Kepler's third law, the two are not regarded as independent elements.

The quantities we have described are usually represented by algebraic symbols, as follows:

a, the mean distance.
e, the eccentricity.
θ or Ω, the longitude of the node.
i or ϕ, the inclination.

ϖ or π, the longitude of the perihelion.
ϵ, the mean longitude at some epoch.
n, the mean motion.
ω, the distance from node to perihelion.

GLOSSARY OF TECHNICAL TERMS. 553

Ellipticity. Deviation from a truly circular or spherical form, so as to become an ellipse or spheroid. An orbit is said to be more elliptic the more it deviates from a circle.

Elongation. The apparent angular distance of a body from its centre of motion, as of Mercury or Venus from the sun, or of a satellite from its primary.

Emersion (*a coming out*). The reappearance of an object after being eclipsed or otherwise hidden from view.

Ephemeris. A table giving the position of a heavenly body from day to day, in order that observers may know where to look for it. Applied also to an astronomical almanac giving a collection of such tables.

Epicycle. In the ancient astronomy, a small circle the centre of which moves round on the circumference of a larger one, especially the circle in which the three outer planets seemed to perform an annual revolution in consequence of the revolution of the earth around the sun.

Equation of the Centre. The angular distance by which a planet moving in an ellipse is ahead of or behind the mean position which it would occupy if it moved uniformly. It arises from the eccentricity of the ellipse, vanishes at perihelion and aphelion, and attains its greatest value nearly half-way between those points.

Equation of Time. See p. 164.

Equator. The great circle half-way between the two poles in the earth or heavens. The celestial equator is the line EF in Fig. 3, p. 12. See also pp. 62, and 146, 147.

Equatoreal. A telescope mounted so as to follow a star in its apparent diurnal course, as described on p. 119.

Equinox. Either of the two points in which the sun, in its apparent annual course among the stars, crosses the equator. So called because the days and nights are, when the sun is at those points, equal.

Evection. An inequality in virtue of which the moon oscillates about $1\frac{1}{4}°$ on each side of her mean position in a period of 31 days 19 hours.

Eye-piece, of a telescope. The small glasses nearest to the eye, which magnify the image. See pp. 110 and 118.

Faculæ (*small torches*). Groups of small shining spots on the surface of the sun which are brighter than other parts of the photosphere. They are generally seen in the neighborhood of the dark spots, and are supposed to be elevated portions of the photosphere.

Filar (*made of thread*). Applied to micrometers made of spider lines.

Focus (*a fireplace*). A point in which converging rays all meet. The focus of a telescope is the point at which the image is formed. See p. 109.

Geocentric. Referred to the centre of the earth. The geocentric position of a heavenly body is its position as seen or measured from the earth's centre.

Geodesy. The art or science of measuring the earth without reference to the heavenly bodies.

Gnomon. In the old astronomy, the style of a sundial or any object the shadow of which is measured in order to learn the position of the sun.

Golden Number. The number of the year in the Metonic cycle, counted from 1 to 19. See p. 48.

Heliacal (*relating to the sun*). Applied in the ancient astronomy to those risings or settings of bright stars which took place as near to sunrise or sunset as they could be observed.

Heliocentric. Referred to the sun as a centre. Applied to the positions of the heavenly bodies as seen from the sun's centre.

Heliometer. An instrument in which the object-glass is sawed into two equal parts, each of the parts forming an independent image of a heavenly body in the focus. When the two parts are together in their original position, these images coincide, but by sliding one part on the other they may be separated as far as is desired for the purposes of measurement. It is much used in Germany for measuring distances too great for the application of a filar micrometer.

Heliostat. An instrument in which a mirror is moved by clock-work in such a way as to reflect the rays of the sun in a fixed direction, notwithstanding the diurnal motion.

Heliotrope. An instrument invented by Gauss for throwing a ray of sunlight in the direction of a distant station. It is much used in geodetic measurements.

Hour Angle. The distance of a heavenly body from the meridian, measured by the angle at the pole. It is commonly expressed in time by the number of hours, minutes, etc., since the body crossed the meridian.

Immersion (*a plunging in*). The disappearance of a body in the shadow of another, or behind it.

Inclination, of an orbit. See *Elements*.

Ingress (*a going in*). The commencement of the transit of one body over the face of another.

Latitude. The angular distance of a heavenly body from the ecliptic, as declination is distance from the equator.

Libration (*a slow swinging, as of a balance*). The seeming slight oscillations of the moon around her axis, by which we sometimes see a little on one side of her, and sometimes on the other.

Longitude. If a perpendicular be dropped from a body to the ecliptic, its celestial longitude is the distance of the foot of the perpendicular from the vernal equinox counted towards the east.

Lunation. The period from one change of the moon to the next. Its duration is $29\frac{1}{2}$ days, or, more exactly, 29.5305879 days.

Mass, of a body. The quantity of matter contained in it, as measured

GLOSSARY OF TECHNICAL TERMS.

by its weight at a given place. Mass differs from weight in that the latter is different in different places even for the same body, depending on the intensity of gravity, whereas the *mass* of a body is necessarily the same everywhere.

Mean Distance. See *Elements*.

Meridian. The terrestrial meridian of a place is the north and south vertical plane passing through that place, or, the great circle in which this plane intersects the celestial sphere. It passes through the pole, the zenith, and the north and south points of the horizon. Celestial meridians are great circles passing from one pole of the heavens to the other in all directions, as shown in Fig. 45, p. 147. Every celestial meridian coincides with the terrestrial meridian of some point on the earth.

Metonic Cycle. See p. 48.

Micrometer (*small measurer*). Any instrument for the accurate measurement of very small distances or angles.

Nadir. The point of the celestial sphere directly beneath our feet, or the direction exactly downwards.

Node. The point in which an orbit intersects the ecliptic, or other plane of reference. See *Elements*, and p. 23.

Nutation. A very small oscillation of the direction of the earth's axis. It arises from the fact that the forces which produce the precession of the equinoxes do not act uniformly, and may therefore be considered as the inequality of precession arising from the inequality of the force which produces it.

Oblate. Applied to a round body which differs from a sphere in being flattened at the poles, as in the case of the earth.

Obliquity of the Ecliptic. The inclination of the plane of the equator to that of the ecliptic, which is equal to half the difference between the greatest meridian altitude of the sun, which occurs about June 21st, and the least, which occurs about December 21st. At the beginning of 1850 its value was about $23° \ 27\frac{1}{2}'$, and it is diminishing at the rate of about $47''$ per century.

Occultation (*a hiding*). The disappearance of a distant body through the interposition of a nearer one of greater angular magnitude. Applied especially to the case of the moon passing over a star or planet, and to that of Jupiter hiding one of his satellites.

Opposition. The relation of two bodies in opposite directions. The planets are said to be in opposition when their longitude differs 180° from that of the sun, so that they rise at sunset, and set at sunrise.

Orbit. The path described by a planet around the sun, or by a satellite around its primary planet.

Parallax. The difference of direction of a heavenly body as seen from two points, as the centre of the earth and some point on its surface. See Part II., Chap. III., § 1.

Parallels. Imaginary circles on the earth or in the heavens parallel to the equator, and having the pole as their centre. The parallel of 40° N. is one which is everywhere 40° from the equator and 50° from the north pole. See Fig. 45, p. 147.

Penumbra. A partial shadowing. Applied generally in cases where light is partially, but not entirely, cut off.

Peri- (*near*). A general prefix to denote the point at which a body revolving in orbit comes nearest its centre of motion; as, *perihelion*, the point nearest the sun; *perigee*, that nearest the earth; *peri-Saturnium*, that nearest the planet Saturn, etc.

Perturbation. A disturbance in the regular elliptic or other motion of a heavenly body, produced by some force additional to that which causes its regular motion. The perturbations of the planets are caused by their attraction on each other.

Photometer (*light-measurer*). An instrument for estimating the intensity of light. The number of kinds of photometers is very great.

Precession of the Equinoxes. A motion of the pole of the equator around that of the ecliptic in about 26,000 years. See pp. 19, 62, 88.

Prime Vertical. The vertical circle passing due east and west through the zenith, and therefore intersecting the horizon in its east and west points.

Quadrature. The positions of the moon when she is 90° from the sun, and therefore in her first or last quarter.

Radiant Point. That point of the heavens from which the meteors all seem to diverge during a meteoric shower. See p. 390.

Refraction (*a breaking*). The bending of a ray of light by passing through a medium. *Astronomical refraction* means the refraction of the light of a heavenly body caused by the atmosphere, as described on p. 300.

Retrograde (*backward*). Applied to the motion of a planet from east to west among the stars.

Saros. A period or cycle of 18 years 11 days, in which eclipses recur. See p. 30.

Scintillation (*a twinkling*). The twinkling of the stars.

Secular (*relating to the ages*). Applied to those changes in the planetary orbits which require immense periods for their completion. See p. 95.

Selenography. A description of the surface of the moon, as geography is a description of the earth's surface. We might call it lunar geography but for the etymological absurdity.

Sexigesimal. Counting by sixties. Applied to those denominate systems in which one unit is sixty times the next inferior one, as the usual subdivision of time and arc.

GLOSSARY OF TECHNICAL TERMS.

Sextant. The sixth part of a circumference. Also an instrument much used in practical astronomy and navigation, for the ready measurement of the angular distance of two points, or of the altitude of a heavenly body.

Sidereal. Relating to the stars. *Sidereal time* is time measured by the diurnal revolution of the stars. Each unit of sidereal time is about $\frac{1}{365}$th part shorter than the usual one. See p. 150.

Signs of the Zodiac. The twelve equal parts into which the ecliptic or zodiac was divided by the ancient astronomers. These signs, beginning at the vernal equinox, are:

Aries, the Ram.
Taurus, the Bull.
Gemini, the Twins.
Cancer, the Crab.
Leo, the Lion.
Virgo, the Virgin.

Libra, the Balance.
Scorpius, the Scorpion.
Sagittarius, the Archer.
Capricornus, the Goat.
Aquarius, the Water-bearer.
Pisces, the Fishes.

Solstices (*standing-points of the sun*). Those points of the ecliptic which are most distant from the equator, and through which the sun passes about June 21st and December 21st. So called because the sun, having then attained its greatest declination, stops its motion in declination, and begins to return towards the equator. The two solstices are designated as those of summer and winter respectively, the first being in 6 hours and the second in 18 hours of right ascension.

Sothic Period. That in which the Egyptian year of 365 days corresponded in succession to all the seasons. The equinoctial year being supposed to be $365\frac{1}{4}$ days, this period would be 1461 years, but it is really longer. See p. 47.

Speculum (*a mirror*). The concave mirror of a reflecting telescope.

Stationary. Applied to those aspects of the planets occurring between the periods of direct and retrograde motion when they appear for a short time not to move relatively to the stars.

Synodic. Applied to movements or periods relative to the sun. The synodic movement of a planet is the amount by which its motion exceeds or falls short of that of the earth round the sun, while its synodic period is the time which elapses between two consecutive returns to inferior or superior conjunction, or to opposition.

Syzygy. The points of the moon's orbit in which it is either new moon or full moon. The line of the syzygies is that which passes through these points, crossing the orbit of the moon.

Terminator. The bounding line between light and darkness on the moon or a planet.

Transit (*a passing across*). The passage of an object across some fixed line, as the meridian, for example, or between the eye of an observer and an apparently larger object beyond, so that the nearer object appears on the face of the more distant one. Applied especially to passages of Mercury and Venus over the disk of the sun, and of the satellites of Jupiter over the disk of the planet.

Trepidation. A slow oscillation of the ecliptic, having a period of 7000 years, imagined by the Arabian astronomers to account for the discordance in the determinations of the precession of the equinoxes. In consequence of this motion the equinox was supposed to oscillate backward and forward through a space of about twenty degrees. The trepidation continued to figure in astronomical tables until the end of the sixteenth century, but it is now known to have no foundation in fact.

Umbra (*a shadow*). That darkest part of the shadow of an object where no part of the luminous object can be seen. Also, the interior and darkest part of a sun-spot.

Vertical, Angle of. The small angle by which the real direction of the earth's centre from any point on its surface differs from that which is directly downward, as indicated by the plumb-line. It arises from the elipticity of the earth, vanishes at the equator and poles, and attains its greatest value of about 12' at the latitude of 45°

Vortex (*a whirlpool*); pl. *Vortices.* The theory of vortices is that which assumed the heavenly bodies to be carried round in a whirling fluid. See p. 72.

Zenith. The point of the celestial sphere which is directly overhead, and from which a plumb-line falls. The *geocentric zenith* is the point in which a straight line rising from the centre of the earth intersects the celestial sphere. It is a little nearer the celestial equator than the apparent or astronomical zenith, owing to the ellipticity of the earth. See *Vertical, Angle of.*

Zodiac. A belt encircling the heavens on each side of the ecliptic, within which the larger planets always remain. Its breadth is generally considered to be about sixteen degrees—eight degrees on each side the ecliptic. In the older astronomy it was divided up into twelve parts, called *signs of the zodiac.*

INDEX.

Abbe, distribution of the nebulæ 452
    parallax of Sirius 536
Aberration of light described 211
Acceleration of moon's motion 96
Adams determines moon's acceleration. 96
    investigates motions of Uranus..... 360
Aërolites, description of 386, 388
Airy, his water telescope 214
    density of the earth.................. 46
Algol a variable star 426
Apparition, circle of perpetual 11
Argelander catalogues the stars 414
Argus, η, a variable star 428
Aristarchus attempts to measure the distance of the sun 22
Asten, motion of Encke's comet......... 382
Asteroids (see also *Planets*, small).... 323, 530
Astrolabe described 105, 549
Astronomer Royal, duties of 160
Attraction of a mountain 85
    of small masses...................... 81
Aurora, description of 301
    height, nature, etc. 304
    periodicity of........................ 249
    spectrum of.......................... 305
Auwers, motion of Sirius and Procyon.. 439

Baily determines density of earth....... 84
Baily's beads explained................. 314
Barker, spectrum of Aurora............. 305
Bayer system of naming stars........... 415
Bernoulli (J.) sustains theory of vortices. 80
Bessel, parallax of 61 Cygni............. 206
Black drop in transits of Venus 179
    its cause............................ 181
Blanchini, his great telescope........... 112
    rotation of Venus.................... 291
Bode's law of planetary distances 233
Bond discovers satellite of Saturn....... 350
    intensity of moonlight............... 317
    investigates rings of Saturn......... 350

Books, list of, for reference............. 542
Bradley attacks stellar parallax......... 204
    detects aberration of light.......... 211
Brahe (Tycho), his obs. and system...... 66
Brünnow, researches in stellar parallax.. 208

Calendar, history, etc. 44
    Julian and Gregorian................ 49
Cassegrainian telescope................. 124
Cassini discovers satellites of Saturn.... 352
    theory of Saturn's rings..............350
Cavendish, density of the earth.......... 82
Cayley determines moon's acceleration.. 98
Challis searches for Neptune............ 360
Chromosphere of the sun................ 256
    its violent movements, etc. 262
Chronograph described 155
Circles of the celestial sphere 146
Clark (Alvan), his telescopes............ 137
    discovers companion of Sirius 138
Clusters of stars........................ 441
Comet, great, of 1680 374
    of 1682 (Halley's) 375
    of 1843 379
    its near approach to sun... 259
    of 1858 (Donati's).......... 319
    views of 368, 380
    of Biela............. 378, 396
    of Encke................... 381
Comets, aspects of, etc. 365
    development......................... 366
    relations to meteors 391
    motions............................. 369
    number.............................. 373
    orbits of, their form................. 369
    physical constitution of.............. 398
    remarkable, description of.......... 374
    tails of, repelled by the sun......... 400
Constellations, antiquity of names....... 414
    description of........................ 417
Copernicus founds modern astronomy... 51

INDEX.

	PAGE
Copernicus publishes his system	53
his system explained	54
represents eccentricity of orbits	60
his distances of the planets	60
estimate of his work	61
work condemned by Inquisition	72
Cornu measures velocity of light	218, 220
Corona of the sun described	252
its probable nature	253
its spectrum	257
Cosmogony, the system of	491
Cycle, the Metonic	48
Dean determ. transatlantic longitude	159
Delaunay, secular acceleration of moon	97
Density of the earth	84
Descartes' theory of vortices	72
Donati's comet, description of	379
views of	368, 380
Draper, his great telescope	135
photograph of the moon	313
theory of the solar spectrum	562
Earth, density of	84
figure of, view of Ptolemy	82
on Newton's theory	86
the French investigations	87
theory of its fluidity	299
difficulties of this theory	300
temperature of interior	298, 511
secular cooling of	511
Easter, how determined	48
Eastman, view of total eclipse in 1869	253
Eccentric in ancient astronomy	41
Eclipses, geometrical explanation	24
classification	25
duration of	28
seasons and periodic recurrence	29
total, phenomena of	252
of 1869, general view of	253
observations of	257
Ecliptic, description of	15
obliquity explained	61
Elements of the planetary orbits	528, 552
Encke determines solar parallax	181
investigates resisting medium	379
Epicycles, ancient system of	37
explained by Copernicus	54
Equator, celestial	12, 147
Evection discovered by Ptolemy	43
Eye-piece of telescope	118
Faculæ of the sun	553
Faye, constitution of the sun	273
his comet, motions of	383
Fizeau measures velocity of light	217
Foucault measures velocity of light	218

	PAGE
Galaxy, or Milky Way, its aspect	416
Galileo reinvents the telescope	106
discovers phases of Venus	290
satellites of Jupiter	336
resolves the Milky Way	408
Galle, parallax of asteroids	200
optical discoverer of Neptune	361
Gentil, his unfortunate voyage	180
Gilliss, expedition to Chili	174
Glacial epoch, its possible cause	241
Glasenapp, velocity of light	214
Gnomon, its use by the ancients	104
Golden number	48
Gould determ. transatlantic longitude	159
Gravitation not newly discovered	43
how generalized by Newton	76
universal law of	81
exerted by small masses	81
explains motion of the planets	98, 100
Grubb constructs Melbourne telescope	132
Hall observes spot on Saturn	341
Halley discovers secular accel. of moon	96
total eclipse in 1715	252
periodicity of his comet	375
proposes obs. of transit of Venus	176
Hansen, moon's secular acceleration	97
solar parallax	182
Harkness, spectrum of the corona	257
observes meteoric shower	389
Herschel, his telescopes	126
discovery of Uranus	354
of two satellites of Uranus	355
his star gauges	466
structure of the universe	468
nebular hypothesis	495
Song of the Telescope	127
Hilgard determ. transatlantic longitude	159
Hipparchus observes motions of planets	40
catalogues the stars	413
Holden investigates satellites of Uranus	357
Hooke, problem of stellar parallax	203
Horrox first observes transit of Venus	175
Huggins, appearance of sun's surface	239
motion of stars in line of sight	456
spectrum of nebulæ	447
of new star	435
Huyghens prep. the way for gravitation	73
discovers rings of Saturn	342
Inquisition condemns work of Copernicus	72
Intra-Mercurial planets, supposed	100, 281
pretended observations of	287
Jansen supposed inventor of telescope	107
Janssen analyzes solar protuberances	254

INDEX. 561

Jupiter, the planet................ 331
   appearance of surface.......... 332
   light and activity of........... 334
   rotation of, on axis............ 335
   satellites of................... 336

Kant, structure of the universe...... 462
   founds nebular hypothesis..... 493
Kepler investigates motions of planets.. 68
   first two laws of planetary motion.. 69
   third law...................... 70
   structure of the universe...... 461

Lambert, structure of the universe..... 465
Langley, appearance of the sun......... 238
   heat of the sun................ 239
   on the sun's constitution...... 280
Laplace, cause of moon's acceleration... 96
   nebular hypothesis............. 495
Lassell, his great telescopes........... 131
   discovery of satellites...... 356, 364
Latitude, how determined astronomically 148
Leverrier investigates motion of Mercury 100
   discovery of Neptune........... 359
Libration of the moon.................. 307
Light, motion of....................... 210
   time of coming from sun........ 213
   velocity of, measured.......... 215
Lipperhey an inventor of the telescope.. 107
Lockyer analyzes sun's protuberances... 255
Longitude, terrestrial, how found...... 150
   the transatlantic.............. 159
Loomis, periodicity of the aurora, etc... 249
Lovering, periodicity of the aurora..... 249
Lyman investigates atmosphere of Venus 294

Mars, the planet....................... 320
   aspect of...................... 321
   maps of.................... 322, 323
   rotation of.................... 322
Maskelyne, attraction of mountain...... 85
Maxwell, theory of Saturn's rings...... 350
Mercury, the planet.................... 283
   ancient theory of.............. 40
   aspect and rotation............ 284
   motion of perihelion....... 100, 286
   transits of.................... 285
Meridian circle described.............. 152
Meteoric showers....................... 385
           radiant point of........ 390
           produced by comet...... 396
Meteors and shooting-stars............. 384
   how caused..................... 387
   combustion of, by motion....... 388
   orbits of...................... 394
   relations to comets............ 391
Metonic cycle.......................... 48

Milky Way described.................... 416
Möller, motion of Faye's comet......... 383
Month, origin of.................... 45–47
Moon, revolution and phases............ 21
   acceleration of its motion..... 96
   unexplained changes of motion.. 98
   path among the stars........... 23
   nodes, motion of............... 23
   eclipses of, how caused........ 24
   gravitation of, found by Newton.... 76
   investigations of the ancients. 42
   atmosphere..................... 314
   surface described.............. 311
   distance and magnitude......... 306
   figure, rotation, and libration. 307
   changes of surface............. 316
   light and heat of.............. 317
   effect on the earth............ 319
Music of the spheres................... 4

Nasmyth, appearance of the sun......... 238
Nebulæ, appearance of.................. 444
   views of.................... 448, 450
   distribution................... 450
   great, of Orion................ 445
   gaseous nature of.............. 447
Nebular hypothesis..................... 493
      reached by reasoning backward from the present... 499
   conclusions respecting......... 514
Neptune, history of its discovery...... 353
   physical aspect of............. 364
   satellite of................... 364
Newall, his great telescope............ 138
Newton (H. A.), meteoric showers....... 391
Newton (Sir I.), his work.............. 74
   laws of motion................. 75
   theory of comets............... 402

Olbers, hypothesis of the explosion of a planet......................... 324, 326
Orbits of the planets.............. 528, 530

Parallax, definition of................ 165
   annual......................... 170
   solar, measures of............. 171
         from transit of Venus......... 175
         most probable value........... 200
         list of papers on............. 537
   stellar, efforts to find....... 202
         list of measures.............. 535
Peirce, rings of Saturn................ 350
   perturbations of Neptune....... 363
   theory of comets............... 403
Photosphere, its appearance............ 238
   light and probable nature...... 263
Pickering, intensity of sunlight....... 239

INDEX.

Planets, the seven, of the ancients 14
 order of distance, ancient............ 40
  modern....... 231, 235
 laws of their motion............... 69, 93
 secular variations of orbits.......... 95
 aspects of............................ 235
 distances and masses................ 233
 of other suns........................ 516
 supposed intra-Mercurial........... 100
 small, fill gap between Mars and Jupiter...................... 323
  earlier discoveries............ 324
  number and mass.............. 328
  elements of orbits............. 530
Pleiades, map of.......................... 442
Plurality of worlds....................... 516
Pole of the heavens...................... 10
Precession of the equinoxes.............. 19
 explained by Copernicus............ 62
 cause of.............................. 88
Proctor, arrangement of the stars....... 476
Prominences. See *Protuberances*.
Protuberances of the sun................. 252
 spectroscopic observation of........ 254
Ptolemy, his system of the world........ 32
 his answers to objectors............. 35
 his relations to Copernicus.......... 58
 his catalogue of stars............... 413
Pythagoras, crystalline spheres of....... 3
 his supposed system............... 4, 52

Radiant point of meteors................ 390
Refraction, astronomical................. 300
Reich, density of earth................... 84
Resisting medium, indications of........ 381
 researches relating to............... 382
Rings of Saturn.......................... 341
Rittenhouse observes transit of Venus... 294
Roemer searches for stellar parallax..... 203
Rosse, his great telescope................ 131
 heat of the moon.................... 318

Saros, or period of eclipses.............. 30
Satellites of Jupiter..................... 336
 of Saturn............................ 351
 of Uranus............................ 355
 of Neptune........................... 364
Saturn, the planet....................... 338
 aspect of............................. 339
 rotation on axis..................... 340
 remarkable spot on.................. 341
 rings........................... 341, 350
 old views of......................... 343
 phases of............................ 344
 satellites of.......................... 351
Schiaparelli theory of meteors........... 393
Schönfeld catalogue of variable stars.... 429

Schwabe, periodicity of sun-spots........ 248
Seasons, explanation of................... 63
Secchi, temperature of the sun........... 241
 on the sun's constitution............ 265
 view of lunar crater................. 315
 spectrum of nebula.................. 447
Secondary spectrum in telescope........ 236
Seidel, photometric researches.......... 413
Sirius, brilliancy of....................... 413
 companion of........................ 439
Solar system, relation to the stars....... 101
 structure of......................... 231
 plan of.............................. 236
Spectroscope described................... 223
Spectrum analysis explained............. 227
Sphere, celestial, described............... 7
 circles of............................ 146
Spheres, crystalline, of Pythagoras...... 3
Stars (see also *Universe*)................. 407
 arrangement of, in space....... 460, 478
 binary systems of.................... 438
 catalogues of........................ 413
 changes among them................ 459
 clusters of........................... 441
 constellations, formation of......... 414
  description of........ 417
 double.............................. 436
 light of, how graded................. 411
 magnitudes of, apparent............. 410
  intrinsic............. 483
 number of, visible................... 410
 motions of, apparent........... 452, 484
  in line of sight......... 456
 names of, how given................. 415
 nearest.............................. 207
 new, explanation of................. 430
 nature of............................ 433
 observations of some............... 432
 parallaxes of.................... 202, 535
 probable orbits of some............. 476
 systems of........................... 476
 shooting. See *Meteors*.
 variable.............................. 426
Stone corrects solar parallax............. 199
Struve (O.), changes in rings of Saturn.. 347
 inner satellites of Uranus........... 356
Struve (W.) investigates stellar parallax............................ 205
 parallax of α Lyræ................... 207
 structure of the universe............ 474
Sun, age of................................ 509
 appearance of....................... 237
 atmosphere of....................... 240
 constitution of...................... 258
 brightness of, as a star.............. 483
 contraction of, probable............. 507
 distance of, most probable value of.. 200

INDEX. 563

	PAGE
Sun, distance of, methods of finding	196
gaseous theory of	264
heat of, quantity radiated	241
how maintained	247, 505
law of radiation	501
motion, apparent annual	14
probable real	454
parallax, how measured	171-201
most probable value	201
list of papers on	538
rotation on axis, law of	249
spots, their appearance	242
their periodicity	248
appear as cavities	245
surroundings of	251
temperature	241
Telescope, origin of	106
Galilean form of	108
principle of construction of	108
magnifying power of	110, 139
aberration, defect of	110
achromatic	114
how mounted for use	118
reflecting, how made	121
great ones of modern times	125
list of the principal	521
Thomson, rigidity of the earth	298
Tides, how produced	90
friction of, retards earth	98
Time, mean and apparent	162
sidereal	150
See also *Calendar*.	
Titus, law of planetary distances	263
Transits of stars, how observed	154
of Venus, law of recurrence	175
old observations	178
in 1874	183, 190
in 1882, where visible	194
of Mercury	285
Tycho Brahe, his work	66

	PAGE
Ulugh Beigh, catalogue of stars	413
Universe, structure of	460
stability of, not necessary	490
Uranus, the planet	353
old observations of	355
satellite of	355
deviations of its motion	358
Venus, ancient theory of motion	39
general description	289
phases	290
supposed axial rotation	291
atmosphere	293
spectrum	295
visibility of dark side	296
satellite of, suspected	296
Vogel, photographic measures of sun's rays	239
rotation of Venus	293
spectrum of aurora	305
views of Encke's comet	367
Vortices, theory of	72
Walker, motions of Neptune	362
Week, days of the	46
Wheatstone revolving mirror	218
Winnecke, parallax of a star	209
Wolf, periodicity of sun-spots	248
Wright, spectrum of zodiacal light	406
Year, sidereal and tropical	20
Young, constitution of the sun	276
researches in spectrum analysis	257
Zodiac, definition of	15
signs of	16
Zodiacal light	289, 405
Zollner, law of sun's rotation	250
nature of photosphere	264
intensity of moonlight	317
theory of comets	402

EXPLANATION OF THE STAR MAPS.

THESE maps show all the stars to the fifth magnitude inclusive between the north pole and 40° south declination, the middle of each map extending to 50° declination. They therefore include all the stars which can be readily seen with the naked eye in our latitudes, except the very smallest. They are, for the most part, founded on Heis's *Atlas Cœlestis* and the catalogue accompanying it.

To recognize the constellations on the maps, reference may be had to the descriptions on pp. 418–426. To find what constellations are on the meridian at any hour of any day in the year, it will be necessary to calculate the sidereal time by the precepts on p. 151: the corresponding hour of right ascension is then to be sought around the margin of Map I., and at the top and bottom of the other maps. Then, if Map I. be held with this hour upwards, it will show the exact position of the northern constellations, while on Maps II.-V. it will show the position of the meridian. Each of these four last maps extends about from the zenith to the south horizon.

The several dates on the ecliptic show the positions of the sun during its apparent annual course as described in part i., chap. i., § 3, and explained on pp. 54, 55. The apparent path of the moon in 1877 is marked out, in order to illustrate § 6, p. 21.

To illustrate precession, the position of the equator 2000 years ago is shown on Map II., where it can be compared with the present position, marked 0° on the sides of Maps II.-V. For the same object the circle which the celestial pole seems to describe around the pole of the ecliptic in 25,000 years is shown on Map I.

The small circles marked here and there on the maps show the positions of the more remarkable nebulæ and star clusters, a list of which is given in No. III. of the Appendix.

ADDENDUM II.—THE SATELLITES OF MARS.

WHILE this work is passing through the press, one of the most remarkable telescopic discoveries of the century has been made by Professor Asaph Hall, of the Naval Observatory, Washington. On the night of August 11th, 1877, he was searching in the neighborhood of Mars for a possible satellite, and found a small object about 80 seconds east of the planet. Cloudy weather prevented further observation at that time, but on the night of the 16th it was again seen, and two hours' observation showed that it followed the planet in its apparent orbital motion. This showed conclusively that it was not a fixed star, and must therefore be a satellite of Mars, unless, by chance, one of the group of small planets between Mars and Jupiter happened to occupy this position. Examining an ephemeris, it was found that the small planet Europa was calculated to be only 2 or 3 degrees distant from Mars; and if the ephemeris were erroneous by this amount, the object observed might be this very body. This seemed extremely improbable, but the possibility of it was sufficient to deter Professor Hall from announcing his discovery until the question could be settled by another observation. A rough calculation from the observed positions of the satellite, and the known mass of Mars, showed that the period of revolution would probably not be far from 29 hours; and that if the object were a satellite, it would be hidden during most of the following night, but would reappear near its original position towards morning. On the other hand, if the object were the small planet Europa, it would, on the next evening, be a little south-east of the planet.

The following night was beautifully clear, and when Mars rose high enough to be well seen, the telescope was pointed at it. A small star was soon seen quite near the computed position of the hypothetical small planet, while no satellite was visible. But a few minutes of observation with the micrometer showed that Mars was passing by this object, and that the latter was therefore a fixed star, and not the moving object seen on the preceding night. The appearance of the satellite was therefore looked for with much confidence, and at four o'clock on the following morning it emerged from the rays of the planet as predicted, so that no reasonable doubt of its character could remain.

But this was not all. The reappearance of the satellite was followed by the appearance of another object, much closer to the planet, which proved to be a second and inner satellite. The reality of both objects was abundantly confirmed by observations on the following nights, not only at Washington, but at the Cambridge Observatory, by Professor Pickering and his assistants, and at Cambridgeport, by Messrs. Alvan Clark & Sons.

ADDENDUM II.—THE SATELLITES OF MARS.

The most extraordinary feature of the two satellites is the proximity of the inner one to the planet, and the rapidity of its revolution. The shortest period hitherto known is that of the inner satellite of Saturn—22 hours 37 minutes. But the inner satellite of Mars goes round in 7 hours 38 minutes. Its distance from the centre of the planet is about 6000 miles, and from the surface less than 4000. If there are any astronomers on Mars with telescopes and eyes like ours, they can readily find out whether this satellite is inhabited, the distance being less than one-sixtieth that of the moon from us.

That kind of near approach to simple relationships between the times of revolution is found here which we see in the satellites of Jupiter and Saturn. The inner satellite of Mars revolves in very nearly one-fourth the period of the outer one, these times being,

Outer satellite	30h. 14m.
One-fourth this period	7h. 33½m.
Period of inner satellite	7h. 38m.

These satellites may also be put down as by far the smallest heavenly bodies yet known. It is hardly possible to make anything like a numerical estimate of their diameters, because they are seen in the telescope only as faint points of light; and, having no sensible surface, no such thing as a measure of the diameters is possible. The only datum on which an estimate can be founded is the amount of light which they give. The writer judged the magnitude of the outer one to be between the eleventh and twelfth. According to the estimate of Zöllner, Mars itself, at this opposition, is three magnitudes brighter than a first-magnitude star. The difference of brilliancy between Mars and the outer satellite is, therefore, represented by thirteen or fourteen orders of magnitude. From this, it would follow that Mars gives from 200,000 to 500,000 times as much light as the satellite; and if both are of the same light-reflecting power, the diameter of the satellite would be from 6 to 10 miles. It may be as small as 5 miles, or as great as 20, but is not likely to lie far without these limits. The inner satellite is much brighter than the outer one, and its diameter probably lies between 10 and 40 miles.

The following are the first rough elements of the apparent orbit of the outer satellite, supposing it circular. The distance of the inner one is 33″.

THE OUTER SATELLITE.

Major semi-axis of apparent orbit seen at distance [9.5930]	82″.5 ± 0″.5
Minor semi-axis of apparent orbit seen at distance [9.5930]	27″.7 ± 2″.
Major semi-axis of orbit seen at distance unity	32″.3
Position angles of apsides of apparent orbit	70°, 250° ± 2°
Passage through the west apsis ($p=250°$), Aug. 19, 16h.6, W. m. t.	
Period of revolution	30h 14m ± 2m
Hourly motion in areocentric longitude	11°.907
Inclination of true orbit to the ecliptic	25°.4 ± 2°
Longitude of ascending node	82°.8 ± 3°
Position of pole of orb t in celestial sphere	Long. 352°.8 R.A. 316.1
	Lat. +64.6 Decl. +53.8

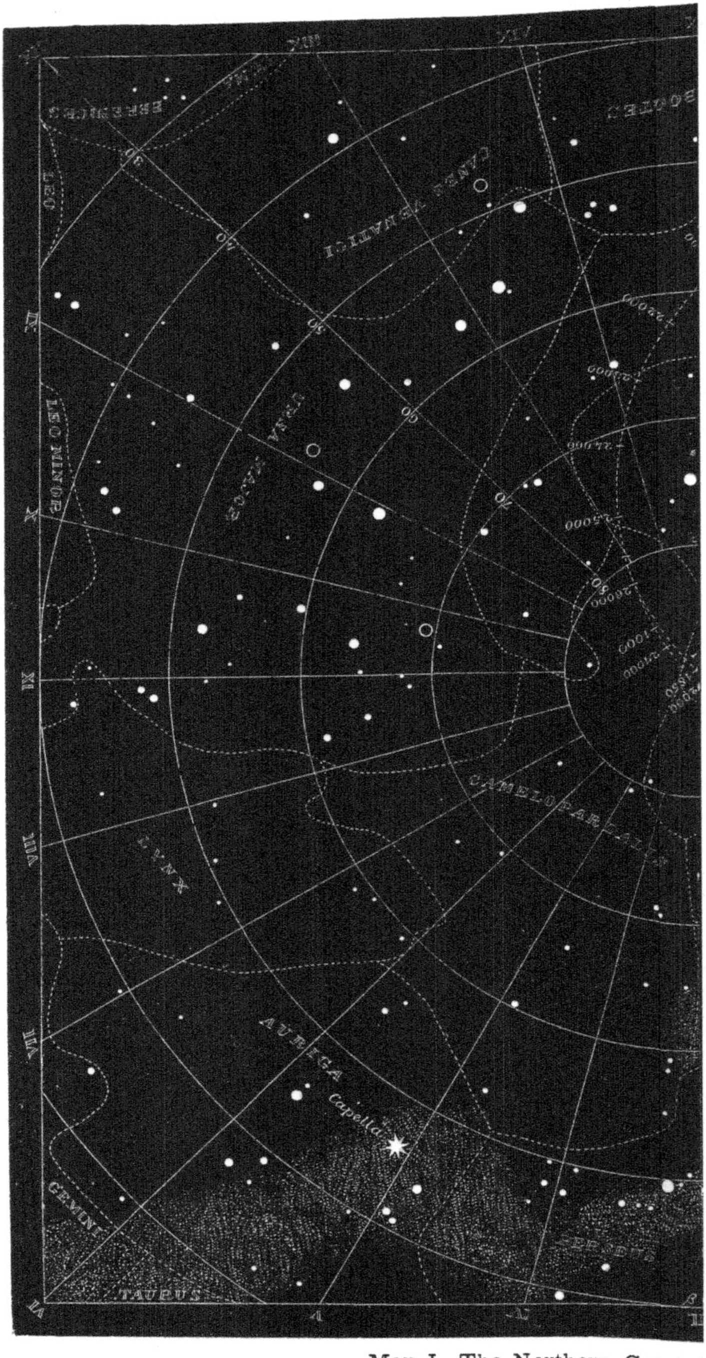

Map I.—The Northern Constel

lations within 50° of the Pole.

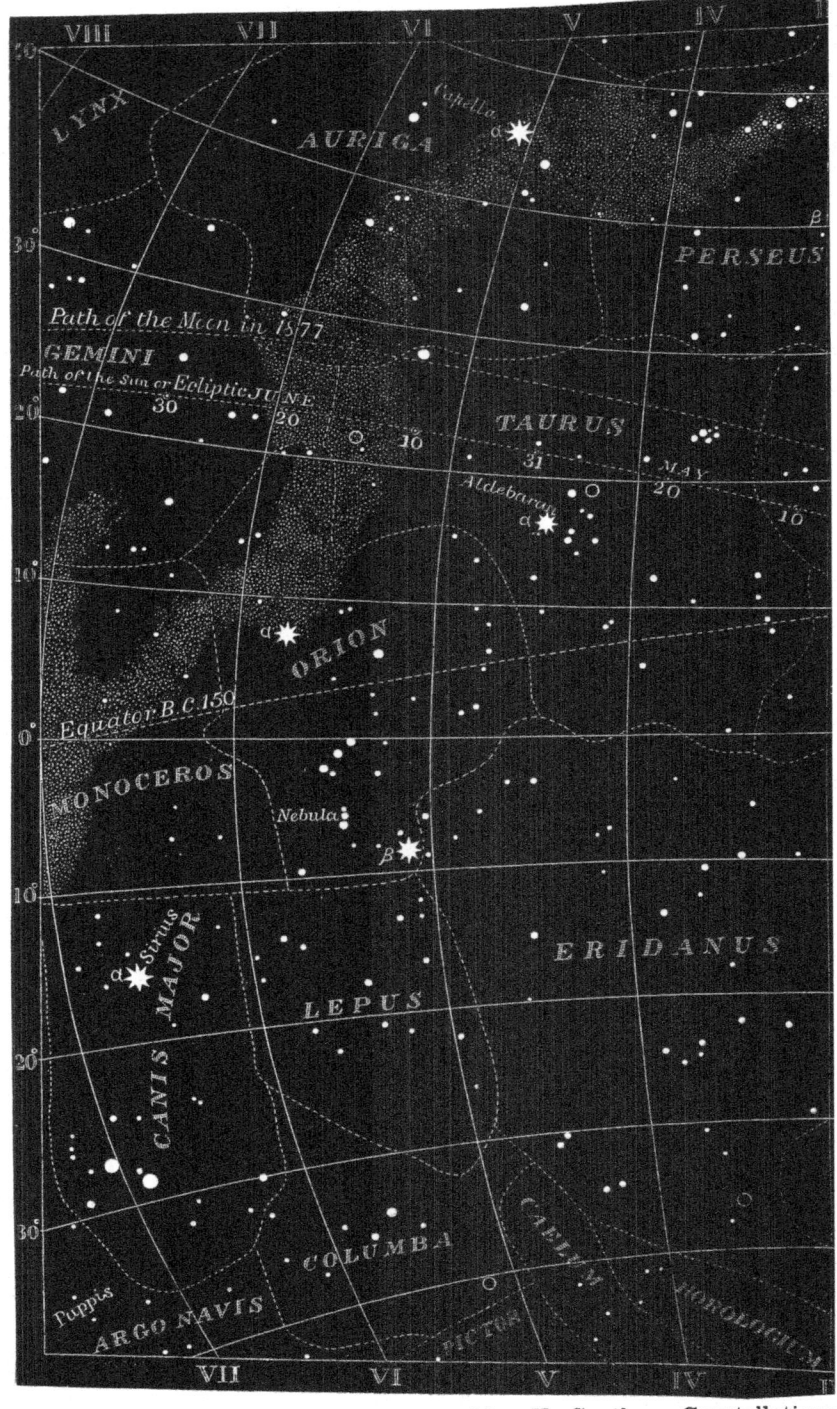

Map II.—Southern Constellations

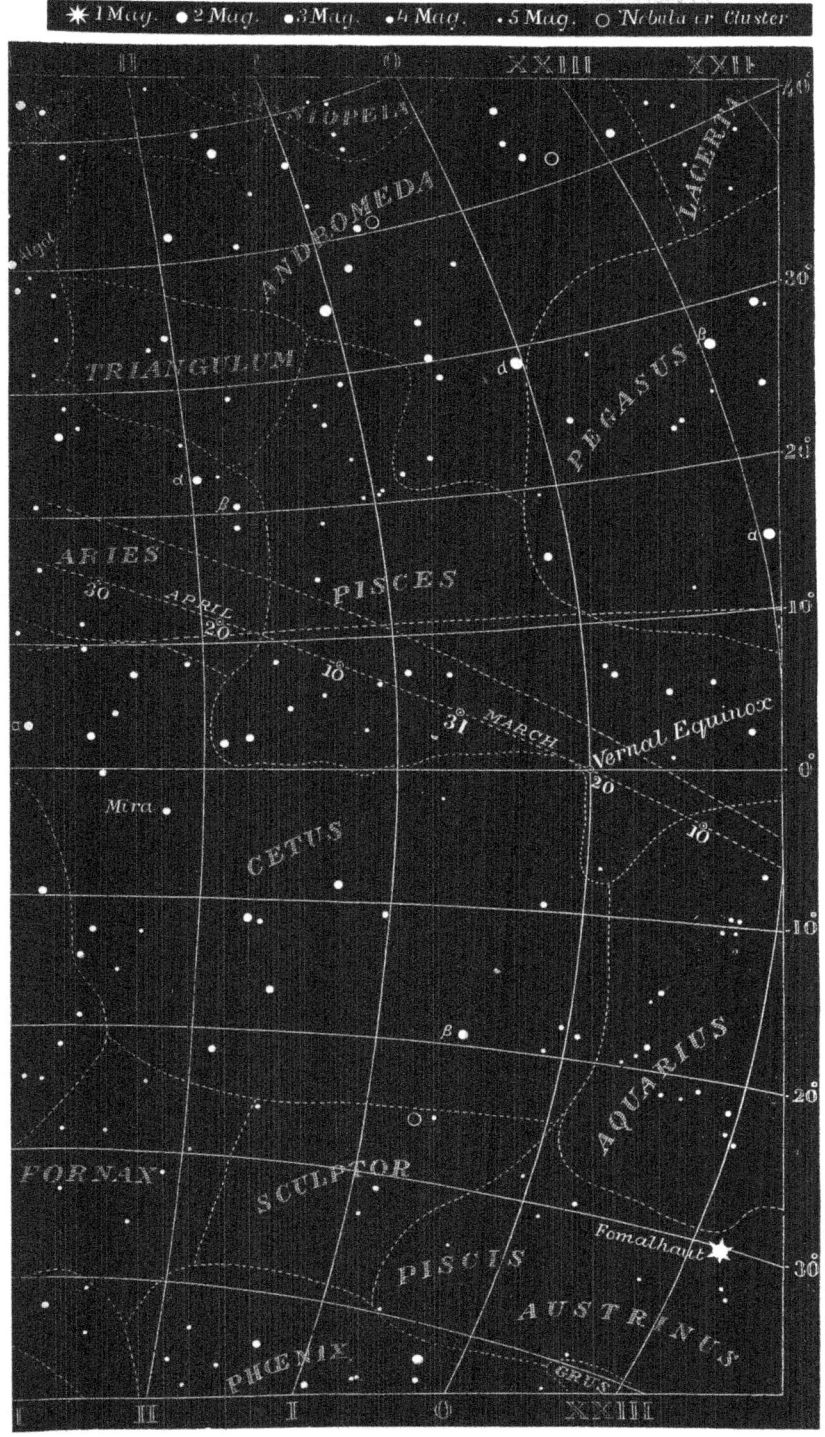

visible in Autumn and Winter.

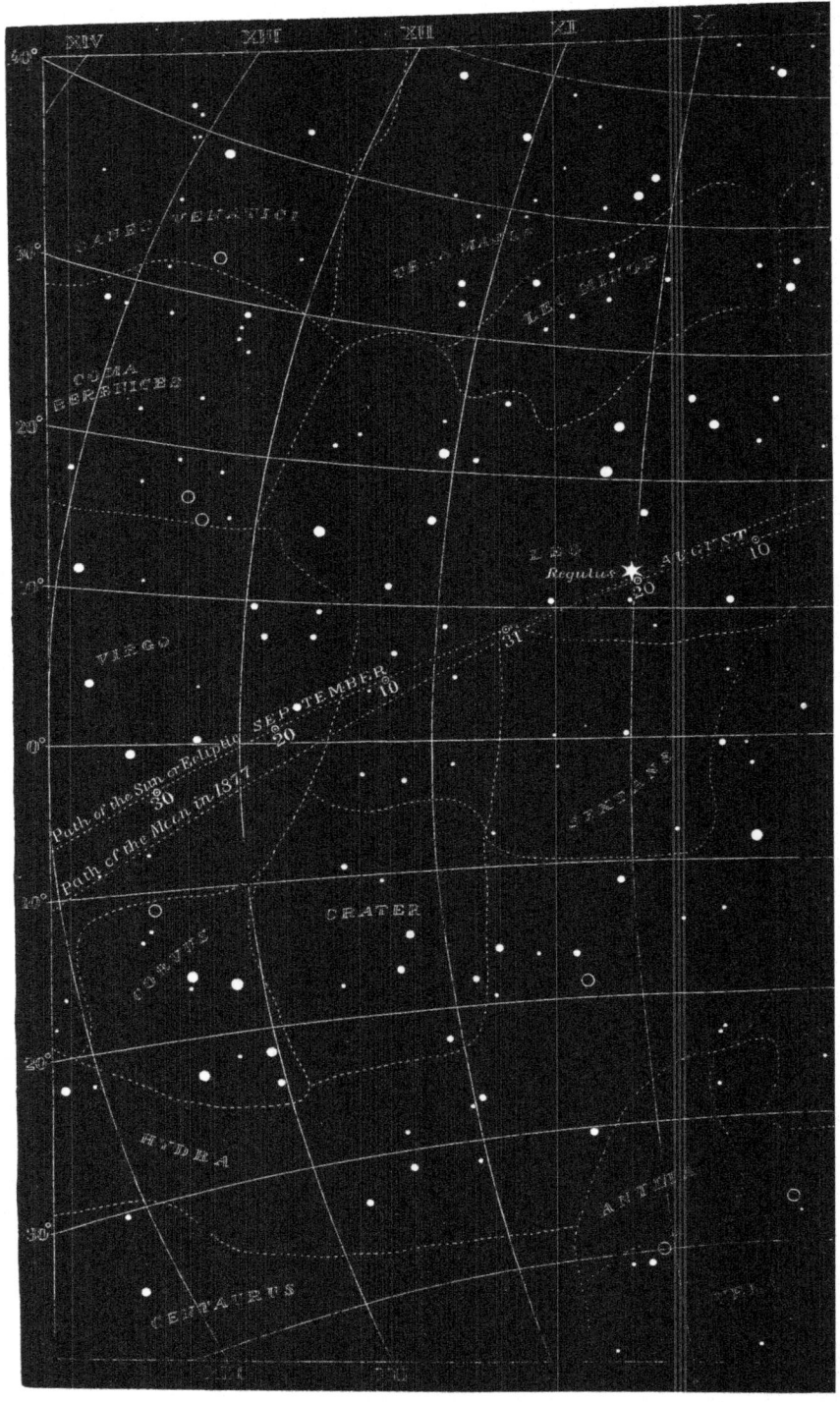

Map III.—Southern Constellation

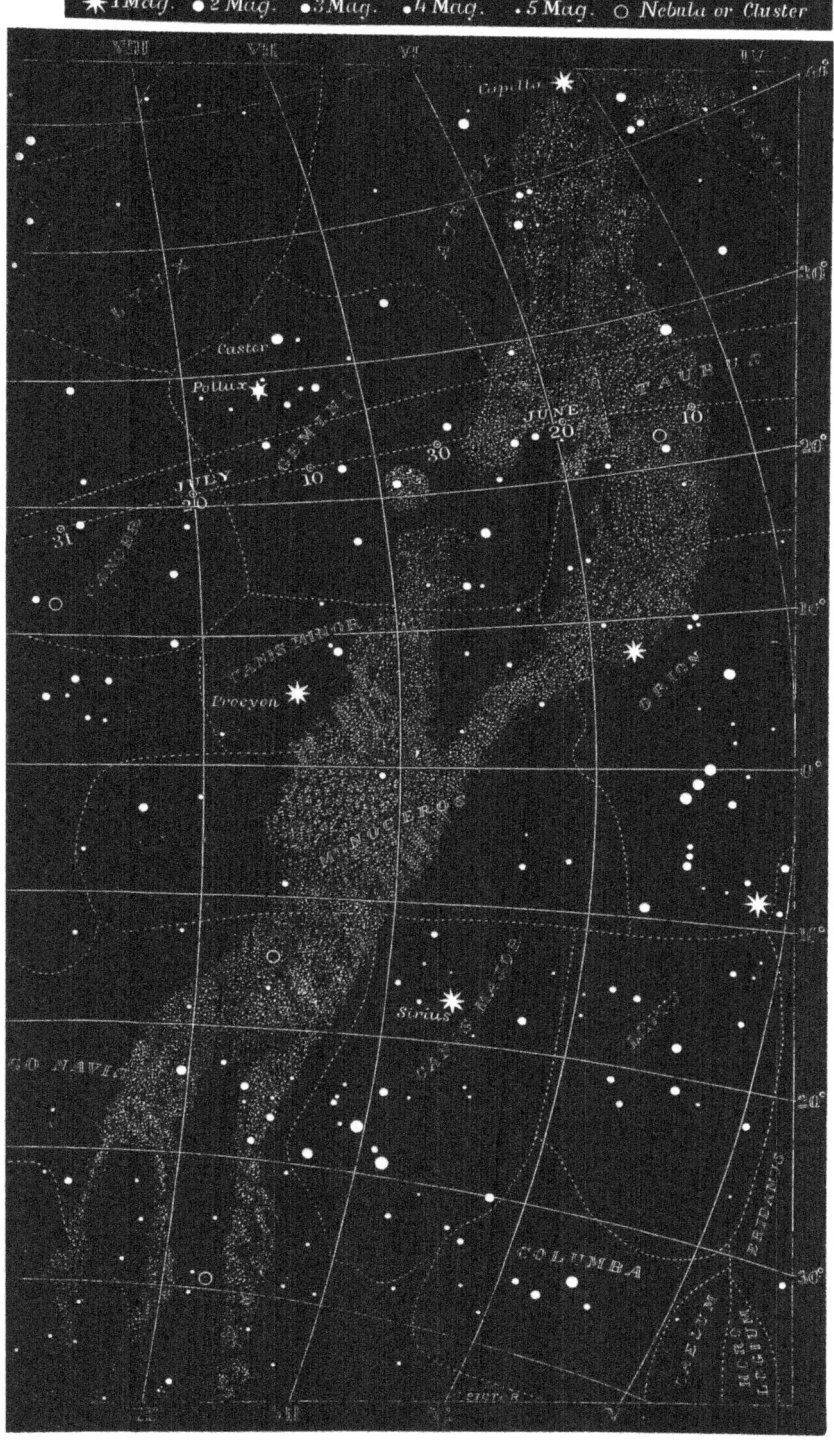

s visible in Winter and Spring.

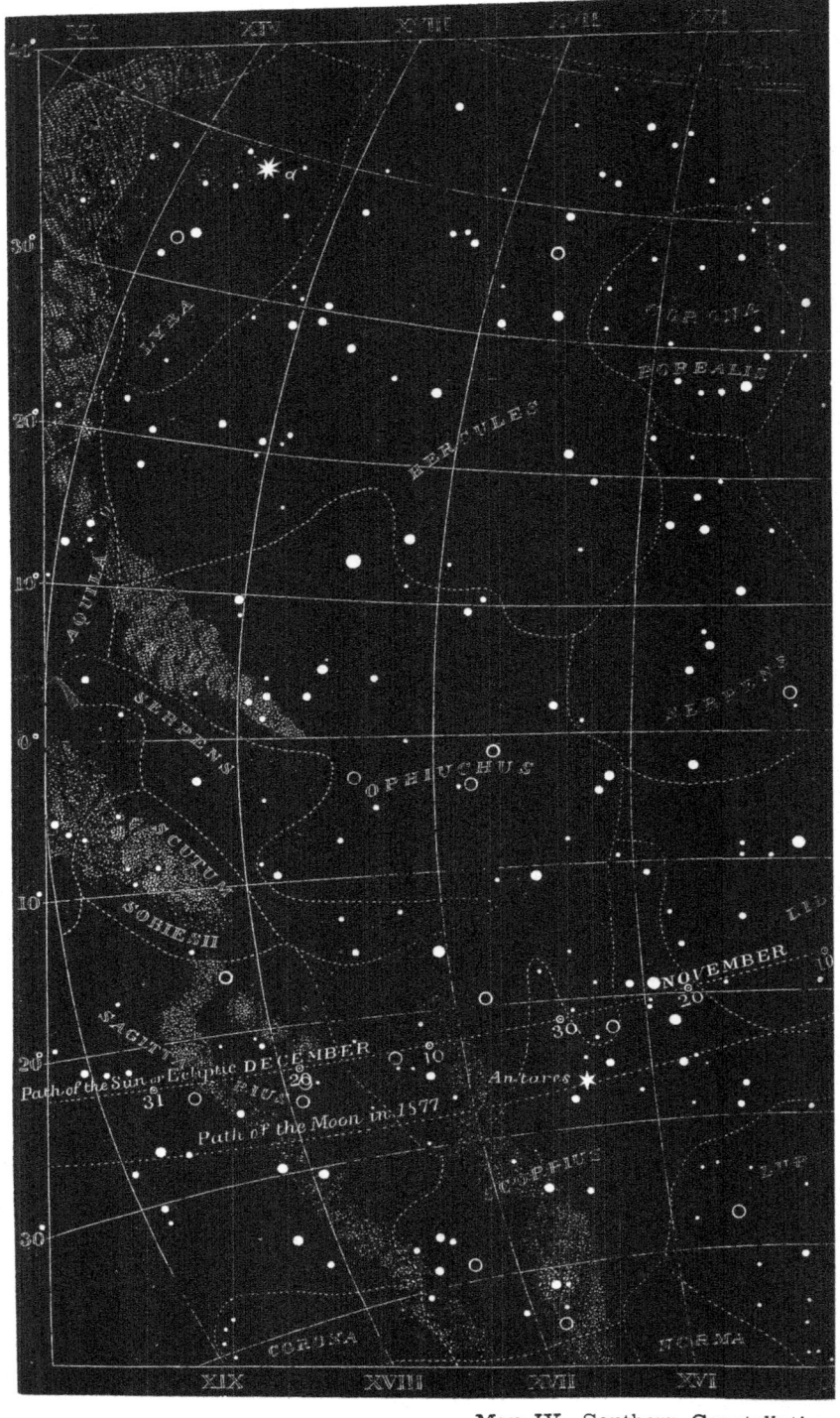

Map IV.—Southern Constellations

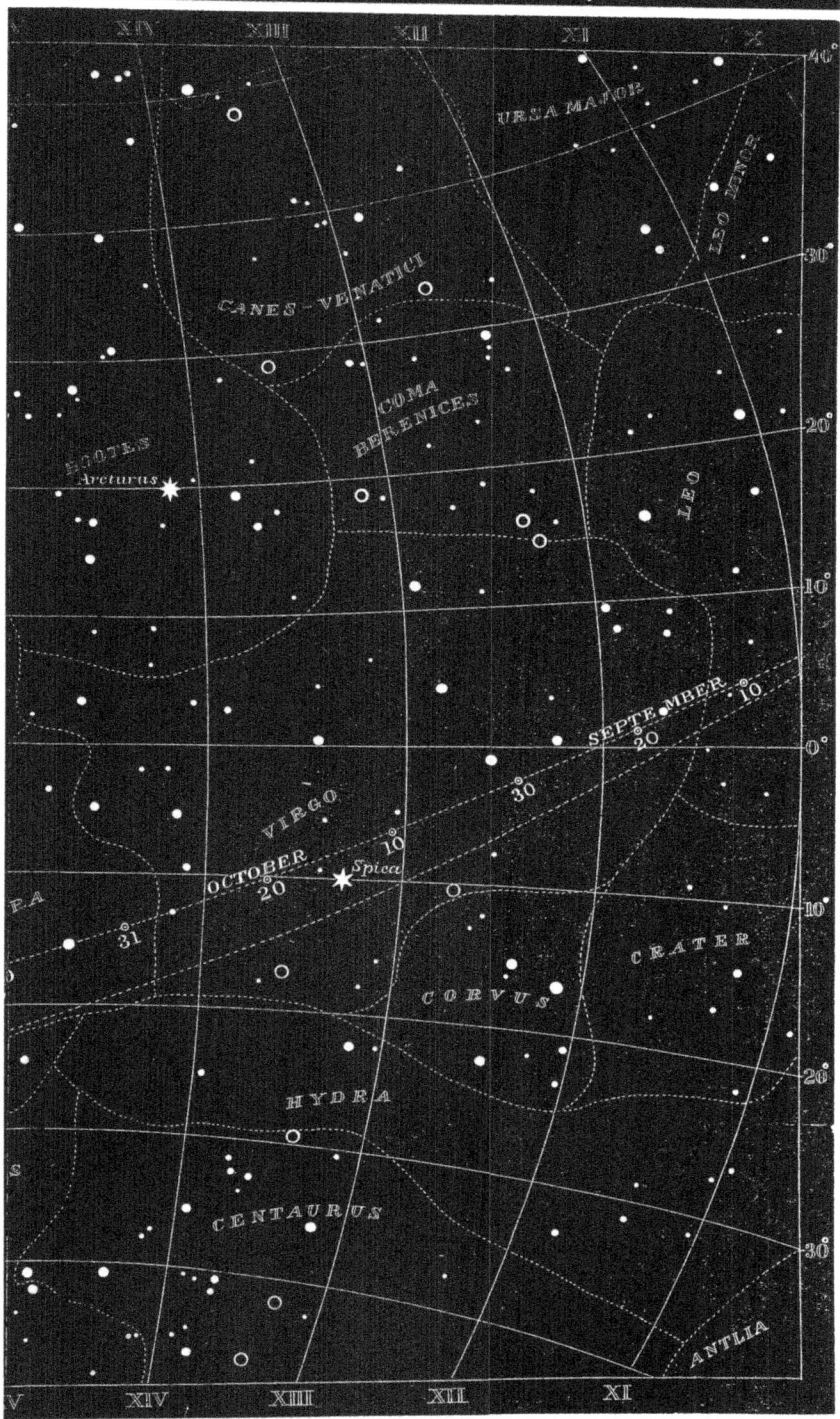

visible in Spring and Summer.

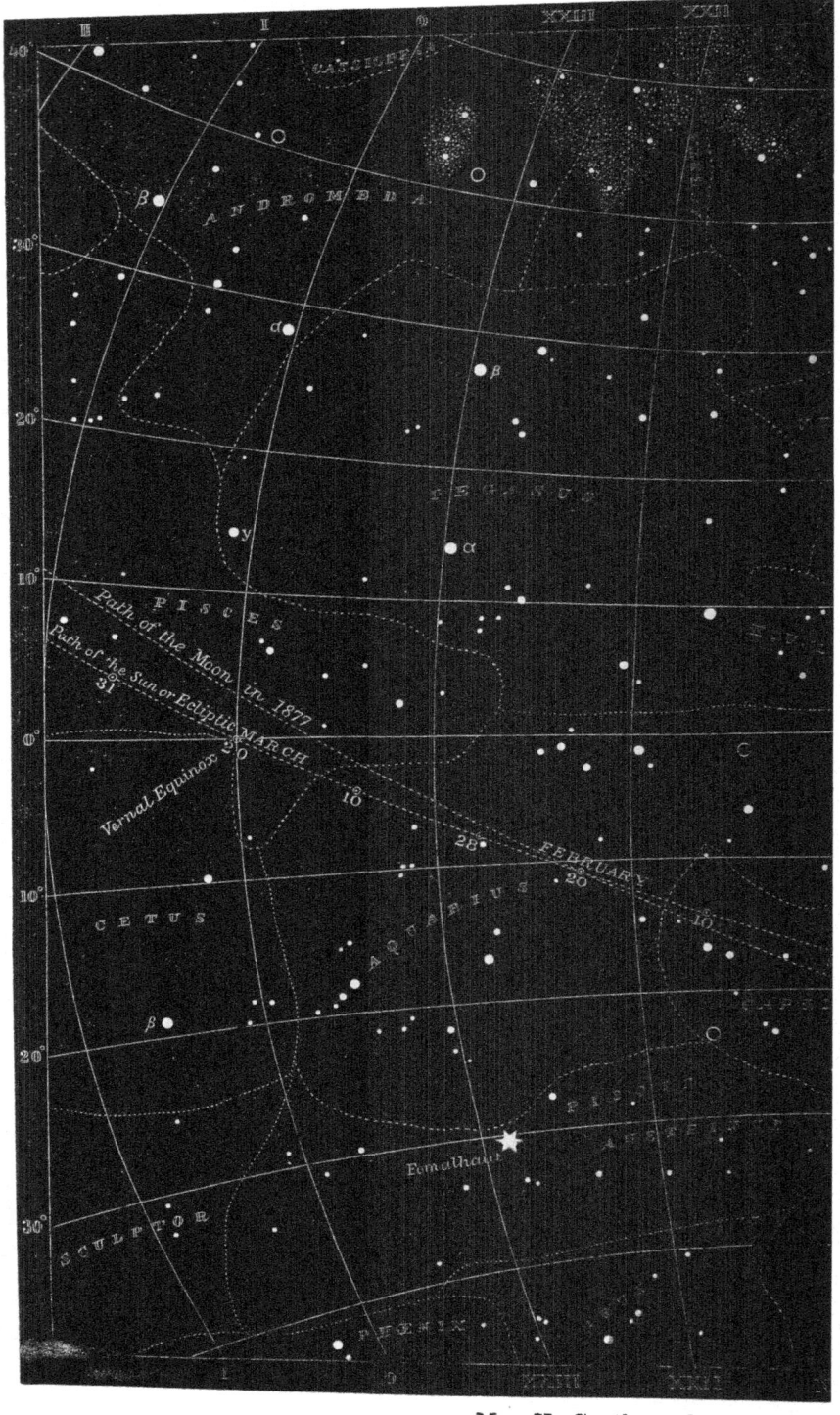

Map V.—Southern Constellations

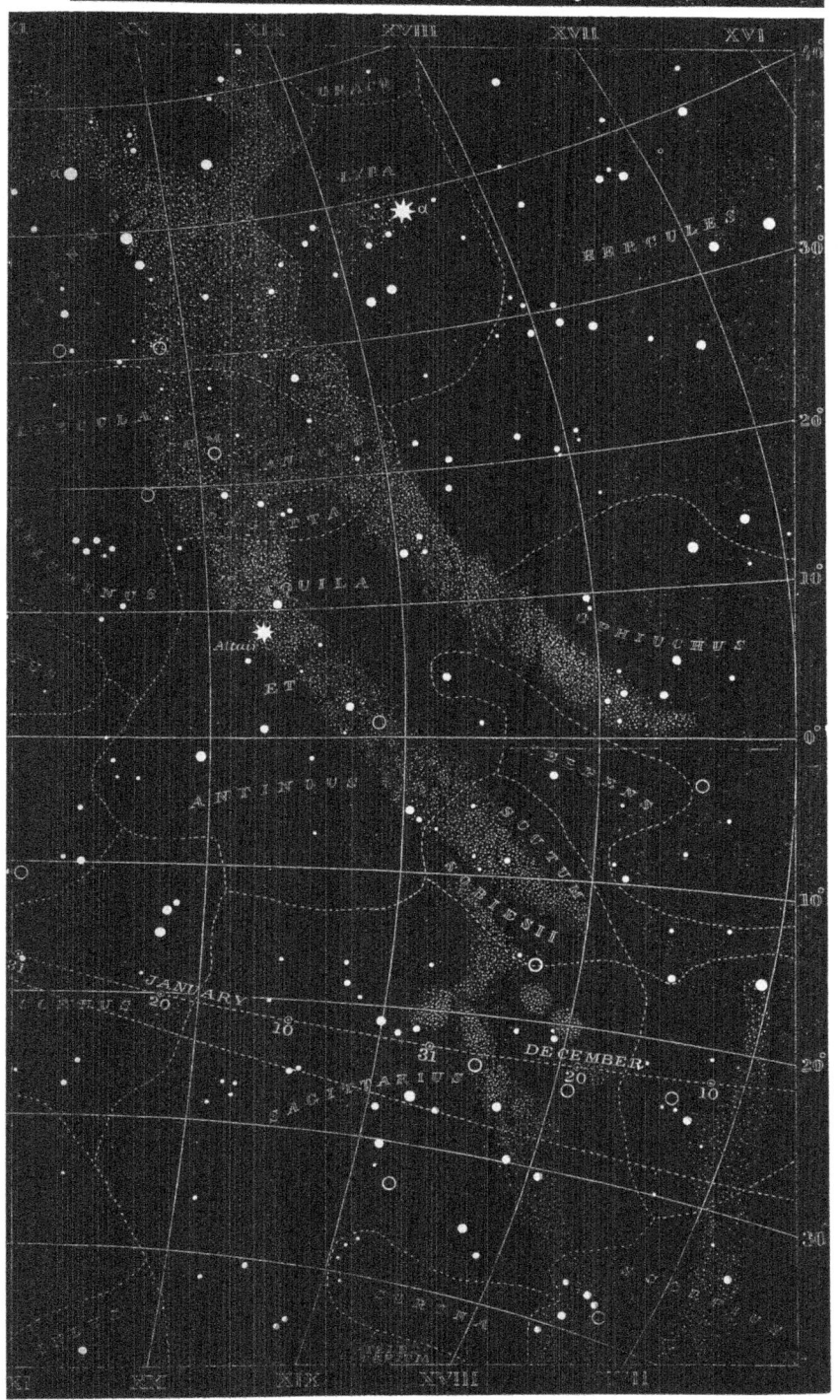

visible in Summer and Autumn.

Printed by Printforce, United Kingdom